INTERFACIAL PHENOMENA IN BIOLOGICAL SYSTEMS

SURFACTANT SCIENCE SERIES

CONSULTING EDITORS

Volume 1: NONIONIC SURFACTANTS, edited by Martin J. Schick (*see also Volumes 19 and 23*)

Volume 2: SOLVENT PROPERTIES OF SURFACTANT SOLUTIONS, edited by Kozo Shinoda (*out of print*)

Volume 3: SURFACTANT BIODEGRADATION, by R. D. Swisher (*See Volume 18*)

Volume 4: CATIONIC SURFACTANTS, edited by Eric Jungermann (*see also Volumes 34 and 37*)

Volume 5: DETERGENCY: THEORY AND TEST METHODS (*in three parts*), edited by W. G. Cutler and R. C. Davis (*see also Volume 20*)

Volume 6: EMULSIONS AND EMULSION TECHNOLOGY (*in three parts*), edited by Kenneth J. Lissant

Volume 7: ANIONIC SURFACTANTS (*in two parts*), edited by Warner M. Linfield (*out of print*)

Volume 8: ANIONIC SURFACTANTS–CHEMICAL ANALYSIS, edited by John Cross

Volume 9: STABILIZATION OF COLLOIDAL DISPERSIONS BY POLYMER ADSORPTION, by Tatsuo Sato and Richard Ruch

Volume 10: ANIONIC SURFACTANTS–BIOCHEMISTRY, TOXICOLOGY, DERMATOLOGY, edited by Christian Gloxhuber (*out of print*)

Volume 11: ANIONIC SURFACTANTS–PHYSICAL CHEMISTRY OF SURFACTANT ACTION, edited by E. H. Lucassen-Reynders (*out of print*

Volume 12: AMPHOTERIC SURFACTANTS, edited by B. R. Bluestein and Clifford L. Hilton

Volume 13: DEMULSIFICATION: INDUSTRIAL APPLICATIONS, by Kenneth J. Lissant

Volume 14: SURFACTANTS IN TEXTILE PROCESSING, by Arved Datyner

Volume 15: ELECTRICAL PHENOMENA AT INTERFACES: FUNDAMENTALS, MEASUREMENTS, AND APPLICATIONS, edited by Ayao Kitahara and Akira Watanabe

Volume 16: SURFACTANTS IN COSMETICS, edited by Martin M. Rieger

Volume 17: INTERFACIAL PHENOMENA: EQUILIBRIUM AND DYNAMIC EFFECTS, by Clarence A. Miller and P. Neogi

Volume 18: SURFACTANT BIODEGRADATION, Second Edition, Revised and Expanded, by R. D. Swisher

Volume 19: NONIONIC SURFACTANTS: CHEMICAL ANALYSIS, edited by John Cross

Volume 20: DETERGENCY: THEORY AND TECHNOLOGY, edited by W. Gale Cutler and Erik Kissa

Volume 21: INTERFACIAL PHENOMENA IN APOLAR MEDIA, edited by Hans-Friedrich Eicke and Geoffrey D. Parfitt

Volume 22: SURFACTANT SOLUTIONS: NEW METHODS OF INVESTIGATION, edited by Raoul Zana

Volume 23: NONIONIC SURFACTANTS: PHYSICAL CHEMISTRY, edited by Martin J. Schick

Volume 24: MICROEMULSION SYSTEMS, edited by Henri L. Rosano and Marc Clausse

Volume 25: BIOSURFACTANTS AND BIOTECHNOLOGY, edited by Naim Kosaric, W. L. Cairns, and Neil C. C. Gray

Volume 26: SURFACTANTS IN EMERGING TECHNOLOGIES, edited by Milton J. Rosen

Volume 27: REAGENTS IN MINERAL TECHNOLOGY, edited by P. Somasundaran and Brij M. Moudgil

Volume 28: SURFACTANTS IN CHEMICAL/PROCESS ENGINEERING, edited by Darsh T. Wasan, Martin E. Ginn, and Dinesh O. Shah

Volume 29: THIN LIQUID FILMS, edited by I. B. Ivanov

Volume 30: MICROEMULSIONS AND RELATED SYSTEMS: FORMULATION, SOLVENCY, AND PHYSICAL PROPERTIES, edited by Robert S. Schecter and Maurice Bourrel

Volume 31: CRYSTALLIZATION AND POLYMORPHISM OF FATS AND FATTY ACIDS, edited by Nissim Garti and Kiyotaka Sato

Volume 32: INTERFACIAL PHENOMENA IN COAL TECHNOLOGY, edited by Gregory D. Botsaris and Yuli M. Glazman

Volume 33: SURFACTANT-BASED SEPARATION PROCESSES, edited by John F. Scamehorn and Jeffrey H. Harwell

Volume 34: CATIONIC SURFACTANTS: ORGANIC CHEMISTRY, edited by James M. Richmond

Volume 35: ALKYLENE OXIDES AND THEIR POLYMERS, by F. E. Bailey, Jr. and Joseph V. Koleske

Volume 36: INTERFACIAL PHENOMENA IN PETROLEUM RECOVERY, edited by Norman R. Morrow

Volume 37: CATIONIC SURFACTANTS: PHYSICAL CHEMISTRY, edited by Donn N. Rubingh and Paul M. Holland

Volume 38: KINETICS AND CATALYSIS IN MICROHETEROGENEOUS SYSTEMS, edited by M. Grätzel and K. Kalyanasundaram

Volume 39: INTERFACIAL PHENOMENA IN BIOLOGICAL SYSTEMS, edited by Max Bender

OTHER VOLUMES IN PREPARATION

INTERFACIAL PHENOMENA IN BIOLOGICAL SYSTEMS

edited by
Max Bender
Professor of Chemistry (Emeritus)
Fairleigh Dickinson University
Teaneck, New Jersey

Marcel Dekker, Inc. New York • Basel • Hong Kong

Library of Congress Cataloging-in-Publication Data

Interfacial phenomena in biological systems/edited by Max Bender.
p. cm. -- (Surfactant science series; v. 39)
Includes bibliographical references and index.
ISBN 0-8247-8436-7
1. Surface chemistry. 2. Membranes (Biology) 3. Biological transport. I. Bender, Max. II. Series.
QP517.S87B46 1991
574.19'283--dc20 91-12675
CIP

This book is printed on acid-free paper.

Marcel Dekker, Inc.
270 Madison Avenue, New York, New York 10016

Current printing (last digit):
10 9 8 7 6 5 4 3 2 1

PRINTED IN THE UNITED STATES OF AMERICA

Many thanks are given to Fairleigh Dickinson University and its founder, Dr. Peter Sammartino, where the atmosphere for me was one of intellectual freedom. And, to my wife, May, for her understanding and assistance.

Preface

The biological system in its very nature is an extremely intricate and complex network of different kinds of interfaces. Historically, the recognition of colloid and surface science as a particular branch of physics and chemistry originated with Thomas Graham's papers in 1861 and 1864 stemming from his investigations of biological products. Properties seen to be unique to colloids—namely, particle size, shape, and structure; surface area, shape, and structure; physical and chemical interfacial interactions, including sorption and aggregation; electrical charge and electrokinetics; transport; and interaction with electromagnetic radiation—are at the very heart of the workings of biological systems, an intimate array of interfacial phenomena.

Applying concepts from the physical sciences to understand the complex living system is no easy matter. One must be careful to use appropriate principles of physical chemistry and not be misled into introducing theory that has evolved from relatively simple models, no matter the brilliance of the derivation of this theory. The relevance of biological parameters may be more weighty than expected. Not to be ignored are the difficulties which can occur due to investigatory probes and methodology altering the state of the medium under observation from normalcy including the bringing on of inanimateness.

Thus there is a continuous need for painstaking study, searching out the variables for understanding the "miraculous" workings of a living thing. How far can one go in applying physics concepts based on inanimate models, including those of random chaotic processes, to physiologically energized translocation; appropriate alteration of molecular conformation; electrogenics; and maintaining levels of intelligence whereby there is a low entropic profile of intricate structure and highly organized processes? What should be considered appropriate live-model physicochemical concepts?

In this book, experts present findings from different avenues of research demonstrating interfacial phenomena in biology. Integration of interfacial knowledge with that from genetic, biochemical, kinetic, thermodynamic, and electrical research should give deeper insight into the mechanisms of living processes.

The text is arranged into parts on the basis of the following subdivisions: "Biological Surfaces," "Membranes," "Corpuscular," "Interactions," and "Transport." Particular physical and chemical principles are shared by more than one of these areas. While the grouping of chapters in the biosurfaces category is reasonably clear-cut, arrangement of the other chapters within specific groups does not readily follow. This is because membranes, corpuscles (generalized from cells to any particulates), interaction, and transport all have important relevance in various manifestations throughout the book. However, the categorization furnishes particular emphasis to each subject.

"Biological surfaces" include the skin, oral areas, bone, manmade materials (metals, alloys, and polymers), autografts and allografts, tissue substitutes, the eye, endothelia of blood vessel, valve and heart, lungs, intestines, etc. This volume has chapters pertinent to skin, the oral cavity, manmade materials, and endothelium.

The two chapters under "Corpuscular" have to do with cells: One pertains to the cell membrane and the other to the structure of cell contents. At the same time, in one of the transport chapters, physiologically energized translocating electrically charged particulates are ascribed as a basis of electrogenics.

The proteins are basic to all the categories, playing complex and important roles in biosurfaces, in membranes, as particulates, in interactions and in transport. Thus they are featured throughout the book. As the original colloid, proteins are in every sense representative of colloidal and interfacial chemistry. Considering this and all the other descriptions of biological interfacial effects, this volume complements the Surfactant Science Series, which so far has covered only nonbiological systems.

The audience for which the book is intended of course includes the researchers in industry and academe who are active in the life sciences. People in the physical chemical sciences should also be interested, as they will gain an introduction to and knowledge of interfacial phenomena in the life sciences. The book will be of use to chemists, biologists, and physiologists in research, development, and production. The unusual need the book fulfills is to make clear to participating scientists, and even those who are just interested, the importance of interfacial phenomena in living processes, furnishing ideas for further study and appealing to the imagination. It will also be useful as a text in graduate and undergraduate courses in chemistry and the life sciences.

Max Bender

Contributors

Rolf W. Attström Faculty of Odontology, Lund University, Malmö, Sweden

Robert E. Baier Faculty of Health Sciences, State University of New York, Buffalo, New York

Max Bender Department of Chemistry, Fairleigh Dickinson University, Teaneck, New Jersey

Martin Blank Department of Physiology and Cellular Biophysics, Columbia University, New York, New York

Donald E. Brooks Department of Pathology and Chemistry, University of British Columbia, Vancouver, British Columbia, Canada

Steven Chin IBM Corporation, Kingston, New York

Enrico Clementi IBM Corporation, Kingston, New York

Giorgina Corongiu IBM Corporation, Kingston, New York

S. Doglia Dipartimento di Fisica, Università degli Studi di Milano, Milano, Italy

Stig E. Friberg Department of Chemistry, Clarkson University, Potsdam, New York

Nissim Garti Casali Institute of Applied Chemistry, The Hebrew University of Jerusalem, Jerusalem, Israel

E. Del Giudice Istituto Nazionale Fisica Nucleare (INFN), Milano, Italy

James Given* IBM Corporation, Kingston, New York

Per-Olof J. Glantz Faculty of Odontology, Lund University, Malmö, Sweden

L. B. Goldsmith** Department of Chemistry, Clarkson University, Potsdam, New York

Anthony G. Gristina† Department of Orthopedic Surgery, Wake Forest University Medical Center, Winston-Salem, North Carolina

Johan Janzen Department of Pathology, University of British Columbia, Vancouver, British Columbia, Canada

I. Kayali‡ Department of Chemistry, Clarkson University, Potsdam, New York

George C. Lie§ IBM Corporation, Kingston, New York

Shlomo Magdassi Casali Institute of Applied Chemistry, The Hebrew University of Jerusalem, Jerusalem, Israel

Marko S. Markov Department of Biophysics, Sofia University, Sofia, Bulgaria

Toni L. Meador¶ Department of Orthopedic Surgery, Wake Forest University Medical Center, Winston-Salem, North Carolina

Anne E. Meyer Center for Biosurfaces, State University of New York, Buffalo, New York

---

Current affiliations:
*Department of Chemistry, State University of New York, Stony Brook, New York.
**Dow Chemical Company, Midland, Michigan.
†Musculoskeletal Sciences Research Institute, Herndon, Virginia.
‡College of Science and Technology, Jerusalem, Israel.
§Dow Corning Corporation, Midland, Michigan.
¶Bowman Gray School of Medicine, Winston-Salem, North Carolina.

Michele Migliore* IBM Corporation, Kingston, New York.

M. Milani Dipartimento di Fisica, Università degli Studi di Milano, Milano, Italy

Quentin N. Myrvik** Department of Microbiology and Immunology, Wake Forest University Medical Center, Winston-Salem, North Carolina

Paul T. Naylor† Department of Orthopedic Surgery, Wake Forest University Medical Center, Winston-Salem, North Carolina

Piero Procacci†† IBM Corporation, Kingston, New York

Linda D. Rhein Colgate—Palmolive Company, Piscataway, New Jersey

F. Anthony Simion Colgate-Palmolive Company, Piscataway, New Jersey

H. Suhaimi Department of Chemistry, Clarkson University, Potsdam, New York

G. Vitiello Dipartimento di Fisica, dell'Università di Salerno, Salerno, Italy

Leo Vroman Department of Chemical Engineering, Columbia University, New York, New York

James C. Weaver Harvard-MIT Division of Health, Sciences, and Technology, Massachusetts Institute of Technology, Cambridge, Massachusetts

Current affiliations
*Institute for Interdisciplinary Applications of Physics, National Research Council, Palermo, Italy.
**Musculoskeletal Sciences Research Institute, Herndon, Virginia.
†Orthopedic Surgery, Inc., Knoxville, Tennessee.
††Dipartimento di Chimica, Università di Firenze, Firenze, Italy.

Contents

Preface v
Contributors ix

Part I: Biological Surfaces

1 Lipid Structure of Stratum Corneum 3
Stig E. Friberg, L. B. Goldsmith, I. Kayali, and *H. Suhaimi*

2 Surfactant Interactions with Skin 33
Linda D. Rhein and *F. Anthony Simion*

3 The Influence of Electrolyte at Frog Skin Surfaces: Deductions on the Origin of Physiological Electric Circuits 51
Max Bender

4 Interfacial Phenomena with Reference to the Oral Cavity 77
Per-Olof J. Glantz, Rolf W. Attström, Anne E. Meyer, and *Robert E. Baier*

5 Biomaterial Surface Domains: Bacterial Versus Tissue Cells 105
Anthony G. Gristina, Quentin N. Myrvik, Paul T. Naylor, and *Toni L. Meador*

6 Proteins in Blood Plasma at Interfaces 137
Leo Vroman

Part II: Membranes

7 Contribution of Water in the Stabilization of Biological Membranes 153
Marko S. Markov

8 Electromagnetic Field Influence on Membranes 171
Marko S. Markov

9 A Critical Reevaluation of the "Nonspecific" Adsorption of Plasma Proteins and Dextrans to Erythrocytes and the Role of These in Rouleaux Formation 193
Johan Janzen and *Donald E. Brooks*

Part III: Corpuscular

10 Electroporation 253
James C. Weaver

11 Dynamic Mechanism for Cytoskeleton Structures 279
E. Del Giudice, S. Doglia, M. Milani, and *G. Vitiello*

Part IV: Interactions

12 Surface Activity of Proteins 289
Shlomo Magdassi and *Nissim Garti*

13 Supercomputer Simulations of the Interaction of Biomolecules 301
Enrico Clementi, Steven Chin, Giorgina Corongiu, James Given, George C. Lie, Michele Migliore and *Piero Procacci*

Part V: Transport

14 Membrane Transport: Insights from Surface Science 337
Martin Blank

15 Transport Mechanisms, Physiological Communication, and Intelligence 367
Max Bender

Index 401

I

Biological Surfaces

1
Lipid Structure of Stratum Corneum

STIG E. FRIBERG, L. B. GOLDSMITH*, I. KAYALI†, and H. SUHAIMI Department of Chemistry, Clarkson University, Potsdam, New York

I.	Introduction	3
II.	Structure of skin	4
	A. Keratinization	4
	B. Stratum corneum lipids	7
III.	Model for the Lipid Structure	11
	A. Experimental support of the model	11
	B. Location of lipid components in the structure	16
IV.	Crystalline/Liquid Crystalline Transition and the Appearance of Skin	26
V.	Water Transport Through Stratum Corneum	26
	References	30

I. INTRODUCTION

Interfacial phenomena play a decisive role in all biological structures, transport, and reactions; this volume provides a multitude of examples of their importance. Perhaps the most easily perceived interface is the body surface against the surrounding atmosphere, the skin. The skin is not only the largest organ of the body, but its presence is decisively essential for retaining the life functions. It

---

Current affiliation:
*Dow Chemical Company, Midland, Michigan.
†College of Science and Technology, Jerusalem, Israel.

acts not only as a defense against the assault of a variety of chemical substances, but must also serve as a defense against cold and hot temperatures by its isolation and temperature-monitoring characteristics. The skin serves as a protector against ultraviolet radiation and prevents desiccation in a dry atmosphere as well as preventing water uptake and accompanying swelling during a swim or bath.

II. STRUCTURE OF SKIN

The skin is formed from material arriving from the inner part of the body, resulting in a structure consisting of two major layers, the dermis and the epidermis. These two layers on top of the subcutaneous tissue (hypodermis) host a whole series of regulatory units for the active functions of the skin. Figure 1 shows the hair follicles, the sweat glands, the sebaceous glands, and the nerve fibers.

The epidermal layer (Fig. 2) is the site of gradual changes in the shape of the cells—from standing columns (20 μm high, 10 μm in diameter) in the basale layer, to more and more flattened cells in the spinosum and granulosum layers, to the extremely elongated forms in the corneum layer [3].

In parallel with these changes in shape, essential changes also occur in the structure of the cell; they may be characterized as a gradual transition from the complete cells of stratum basale to the essentially lifeless protein aggregates embedded in a lipid mortar in the stratum corneum. The specific changes are of four types: keratinization, keratohyelin synthesis, formation of a peripheral envelope, and generation of the lipid intercellular region [4,5].

A. Keratinization

Keratinization refers to the synthesis of the principal fibrous proteins of the differentiating cell or keratinocyte. Epidermal keratinocytes contain three types of protein: (1) actin containing microfilaments with a diameter of about 7 nm, (2) intermediate filaments with a diameter of about 10 nm, and (3) tubulin containing microtubules with diameters ranging from 20 to 25 nm [6].

The structural proteins are synthesized in the stratum basale and eventually comprise two-thirds of the total mass of the terminally differentiated stratum corneum [6]. Enlargement and elongation of the keratin filaments occur as cell differentiation progresses. The fibrous keratin protein assumes an α-helix conformation and is believed to contribute substantially to the mechanical behavior of the epidermis.

A dense amorphous material called keratohyalin is the distinguishing characteristic of the stratum granulosum. This heterogeneous substance, discovered by Auffhammer in 1869 [7], gradually encloses

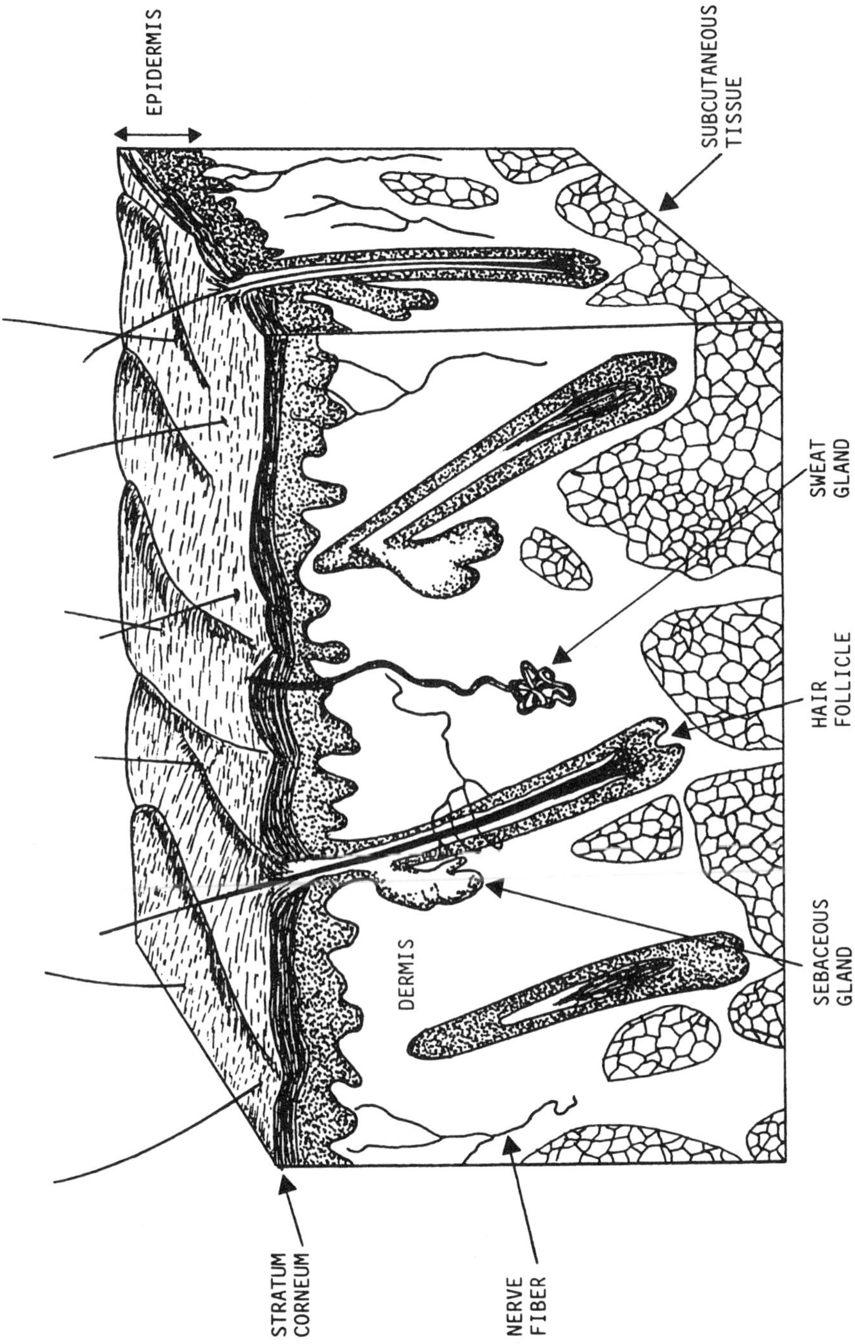

FIG. 1 Cross section of human skin. (From Ref. 1.)

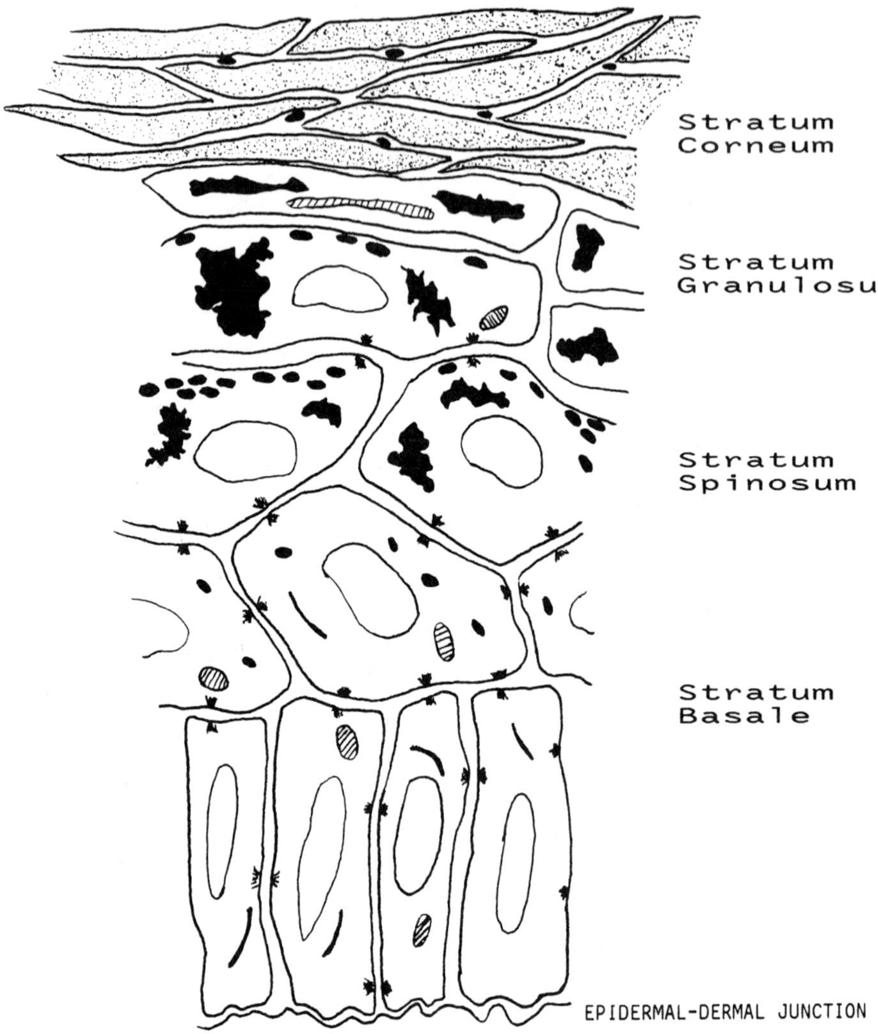

FIG. 2 The epidermis may be subdivided into four major layers: stratum basale, stratum spinosum, stratum granulosum, and stratum corneum. (From Ref. 2.)

the cytoplasmic filament of the cells. The keratohyalin mass enlarges as the cell undergoes differentiation, gradually filling the interstices of the stratum corneum cell. Due to the heterogeneity of keratohyalin, isolation of keratohyalin proteins yields different relative values for the amount of various amino acids present [8].

The keratinized cell is protected from chemical and mechanical destruction by a highly cross-linked peripheral envelope. Cells found without the stratum basale and stratum spinosum have a plasma membrane with a trilaminar structure as found in most mammalian cell membranes. The terminally differentiated stratum corneum cell is unable to maintain the continual synthesis of such a membrane. As the cell moves up to the stratum granulosum there is an increase in size and mass [9,10] along with a thickening of the inner leaflet of the plasma membrane [11]. By the time the cell reaches the stratum corneum an envelope is formed that consists of amorphous particles 40 to 60 Å in size [12]. At least three protein precursors and two models of cross-linking are found in the envelope, which has a thickness of approximately 20 nm [13,14].

B. Stratum Corneum Lipids

The stratum corneum lipids are found in the regions between the terminally differentiated keratinized cells. These lipid-rich intercellular domains originate with the secretion of small lamellar granules (100 × 300 nm) which first appear in the stratum spinosum [5]. Matoltsy and Parakkal [15] suggested in 1965 that these granules appeared in the regions of the Golgi apparatus and then migrated to the perimeter of the cell. The granules are then pushed out into the intercellular space between flattened stratum granulosum cells. Investigators suggest [16] that the function of lamellar granules is to transfer lipids from within the stratum granulosum cells to the intercellular regions. Two cellular events are associated with the exocytosis of lamellar granules: (1) fusion of the lamellar granule with the plasma membrane of the cell, and (2) actual secretion of the lamellar granule into the intercellular region [4]. The cellular processes described above along with the disappearance of such cellular components as Golgi complexes and mitochondria can be seen in Fig. 3.

The stratum corneum lipids may be grouped into two main categories: (1) simple lipids that hydrolyze to yield one or two different types of product, or (2) complex lipids that yield three or more hydrolysis products per mole [18]. Simple lipids are composed of fatty acids and alcohols. Glycerol is the alcohol that is usually found in simple lipids but long-chain alcohols or glycerols may be found as well. Examples of simple lipids are triglycerides, steroids (i.e., cholesterol), wax esters, and waxes (i.e., squalene) (Fig. 4). Complex lipids may be further categorized into three main divisions: (1) glycerophospholipids, (2) glycerolipids, and (3) sphingolipids. Examples of each of the three divisions are shown in Fig. 5.

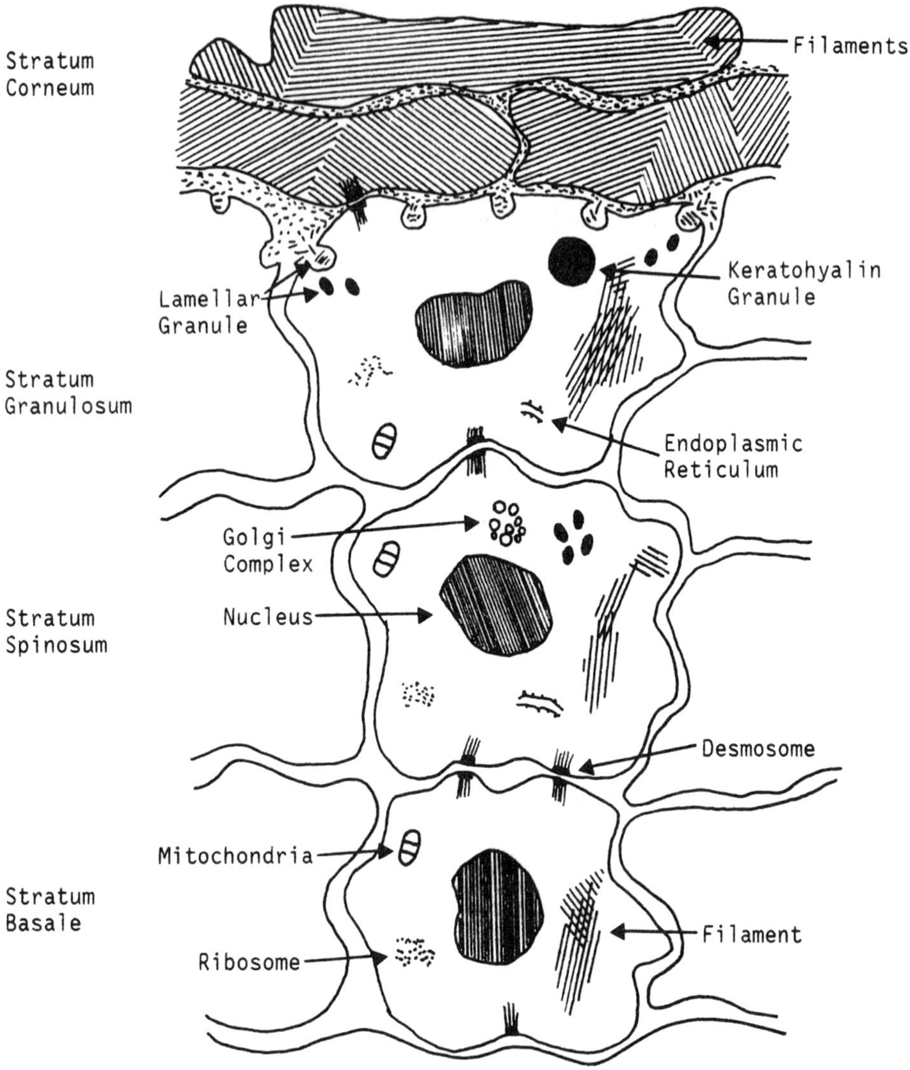

FIG. 3 As the epidermal cell progresses toward the skin surface, intercellular changes occur. Cellular organelles are replaced with protein filaments. (From Ref. 17.)

Lipids found in cells serve two major functions: (1) they serve as structural entities, and (2) they provide a means of storing energy. The hydrocarbon chains of the triglycerides provide energy storage

Steroids

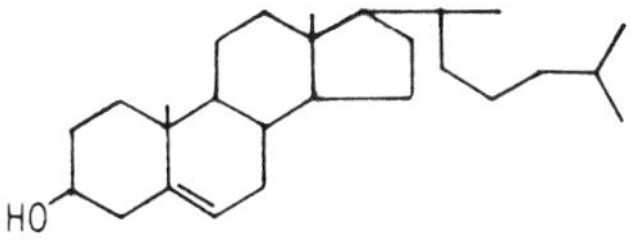

Cholesterol

Waxes

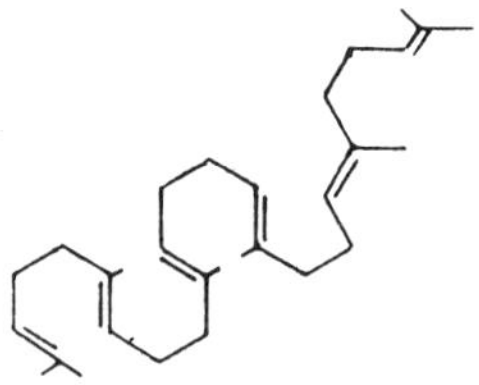

Squalene

Wax Esters

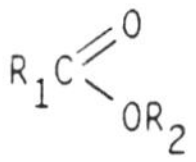

R_1 = long chain fatty acid
R_2 = long chain alcohol

Triglycerides

CH_2OCOR_1
|
$CHOCOR_2$
|
CH_2OCOR_3

FIG. 4 Compounds such as steroids, waxes, wax esters, and triglycerides are classified as simple lipids.

for the cell [19]. The hydrophobic parts of certain lipids form and maintain cell membranes and separate the organelles within the cell. In the stratum corneum the organelles, such as the mitochondria, Golgi bodies, and endoplasmic reticulum, are no longer contained within the now-dead cell and thus their maintenance is no longer required. As epidermal cells mature, proceed upward through the epidermis, and eventually die, the function and therefore the composition of the lipids changes drastically. For example,

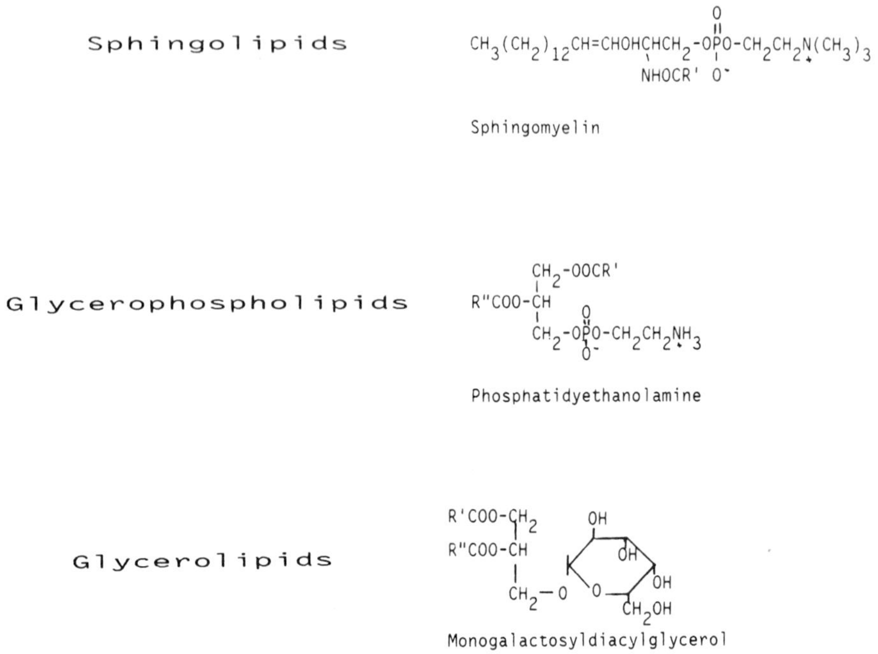

FIG. 5 Complex lipids may be classified as (1) sphingolipids, (2) glycerophospholipids, and (3) glycerolipids.

the content of the phospholipids found in pig epidermis decreased from 45% of the total lipid in the stratum basale and stratum spinosum to 25% of the total lipid in the stratum granulosum to less than 5% of the total lipid in the stratum corneum [19]. Kooyman first observed this decrease in phospholipids in 1932 [20].

Lampe [21] notes three events that occur in the transformation of the stratum granulosum to the stratum corneum: (1) depletion of phospholipids with enrichment of neutral lipids, (2) generation and maintenance of sphingolipids, and (3) increased content of sphingolipids with longer chain lengths. Table 1 together with Fig. 6 show the variation in lipid composition of the epidermis as determined by Lampe et al. [21]. Lampe and Elias [22] have also conducted a very thorough characterization of human stratum corneum lipids. Table 2 lists the composition of lipid found to be present in abdominal human stratum corneum [22].

III. MODEL FOR THE LIPID STRUCTURE

It was first suggested by Elias in 1979 [23] that the lipids present in the intercellular regions of the stratum corneum would form a bilayer structure as depicted in Fig. 7. Initial attempts by Friberg and Osborne to obtain a layered structure by combining the lipids characterized by Elias in Table 2 along with water resulted in a heterogeneous mixture of liquid fats, undissolved solids, and water [24]. Bilayers were not obtained in the mixture until the pH of the skin was taken into consideration.

The pH value has the most drastic influence on the organization of the free fatty acids. At low pH they exist as acids; the interaction with water is minimal and the acids appear as a separate phase. At high pH, on the other hand, the acids are transformed to ionic carboxylates (soaps), which are very water soluble and form small spherical aggregates, micelles, in an aqueous solution. None of these cases apply to the conditions in skin, where the pH is in the range 4.5 to 6.0. For these pH values the fatty acids exist as a mixture of the acids and their ionized counterparts and the combination spontaneously forms a lamellar structure with water according to Fig. 8. This soap—acid—water lamellar structure, a liquid crystal, was used [24] as a basis to form the layered structure earlier envisioned by Elias [23] and discussed extensively by Downing and collaborators [25,26].

A. Experimental Support of the Model

The model was tested against stratum corneum separated from healthy skin using the trypsin method. The small-angle x-ray diffraction pattern of this intact stratum corneum is given with that of the model in Fig. 9. The two diffractograms agree well, both showing an interlayer distance of approximately 60 Å. The intact stratum corneum contains other substances in addition to the lipids, thus, the possibility that the x-ray pattern originates from other parts of the layer must be evaluated. Extraction by methanol/chloroform removes the lipids from the stratum corneum [27] and the extraction of the lipids in this manner removed the pronounced reflection at 60 Å. Hence it appears reasonable to assume that this reflection is due to the lipids in stratum corneum. Furthermore, it appears highly probable that an x-ray spacing of 60 Å describes a layered structure considering the preponderance of layered structures among lipids [28].

The result of the x-ray analysis of model lipids in combination with corneocytes further indicates the presence of a lipid layered

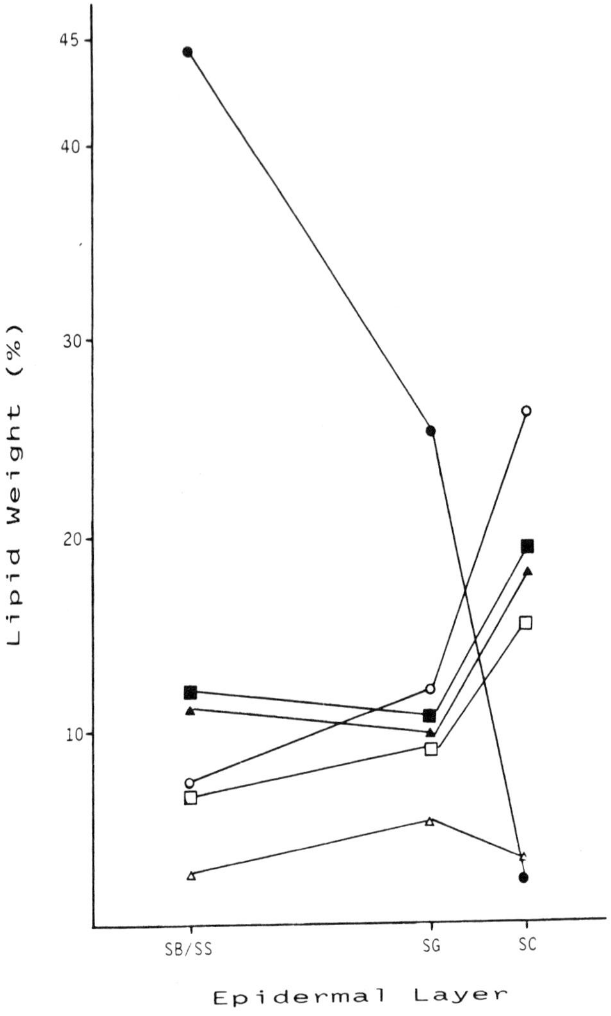

FIG. 6 Variation of lipid composition within the epidermis as investigated by Lampe [21]. ● Phospholipids; △ cholesterol sulfate; □ free fatty acids; o sphingolipids; ▲ free sterols; ■ non-polar.

structure in the stratum corneum. The x-ray diffraction patterns of the lipids alone and when combined with delipidized corneocytes gave similar small-angle x-ray diffraction patterns. This result

TABLE 1 Lipid Variations Within the Epidermis (wt % of Total Lipids)[a]

Lipid component	SB/SS	SG	Whole SC	Outer SC
Polar lipids	44.5	25.3	4.9	2.3
Cholesterol sulfate	2.4	5.5	1.5	3.4
Neutral lipids	51.0	56.5	77.7	68.4
Free sterols	11.2	11.5	14.0	18.8
Free fatty acids	7.0	9.2	19.3	15.6
Triglycerides	12.4	24.7	25.2	11.2
Sterol/wax esters	5.3	4.7	5.4	12.4
Squalene	4.9	4.6	4.8	5.6
n-Alkanes	3.9	3.8	6.1	5.4
Sphingolipids	7.3	11.7	18.1	26.6

Source: Ref. 21.
[a]SB, stratum basale; SS, stratum spinosum; SG, stratum granulosum; SC, stratum corneum.

TABLE 2 Abdominal Human Stratum Corneum Neutral Lipid Composition (wt % of Total Lipids)

Free sterols	14.0
Free fatty acids	19.3
C14:0	3.8
C16:0	36.8
C16:1	3.6
C18:0	9.9
C18:1	33.1
C18:2	12.5
C20:0	0.3
Triglycerides	25.2
Nonpolar lipids	16.3
Sterol/wax esters	6.1
Squalene	6.5
n-Alkanes	3.7
Sphingolipids	18.1

Source: Ref. 22.

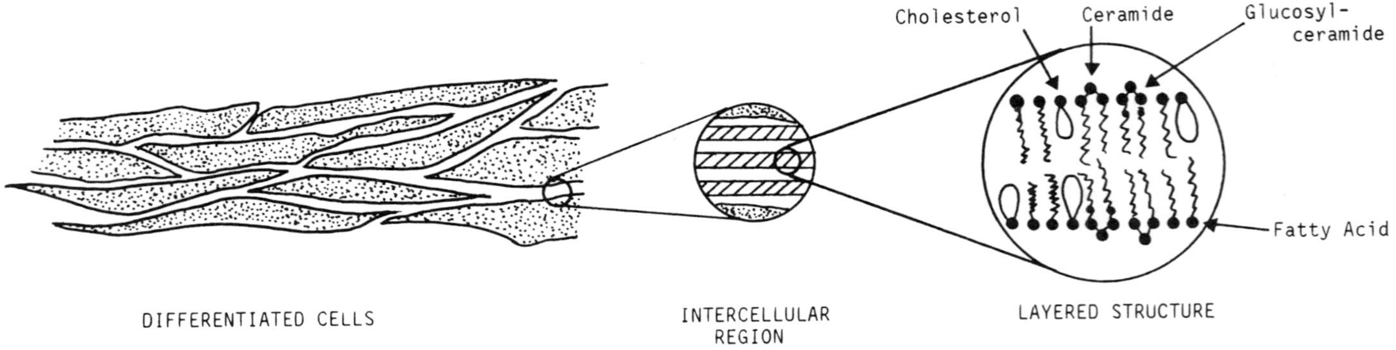

FIG. 7 Layered structure of epidermal lipids as proposed by Elias [23].

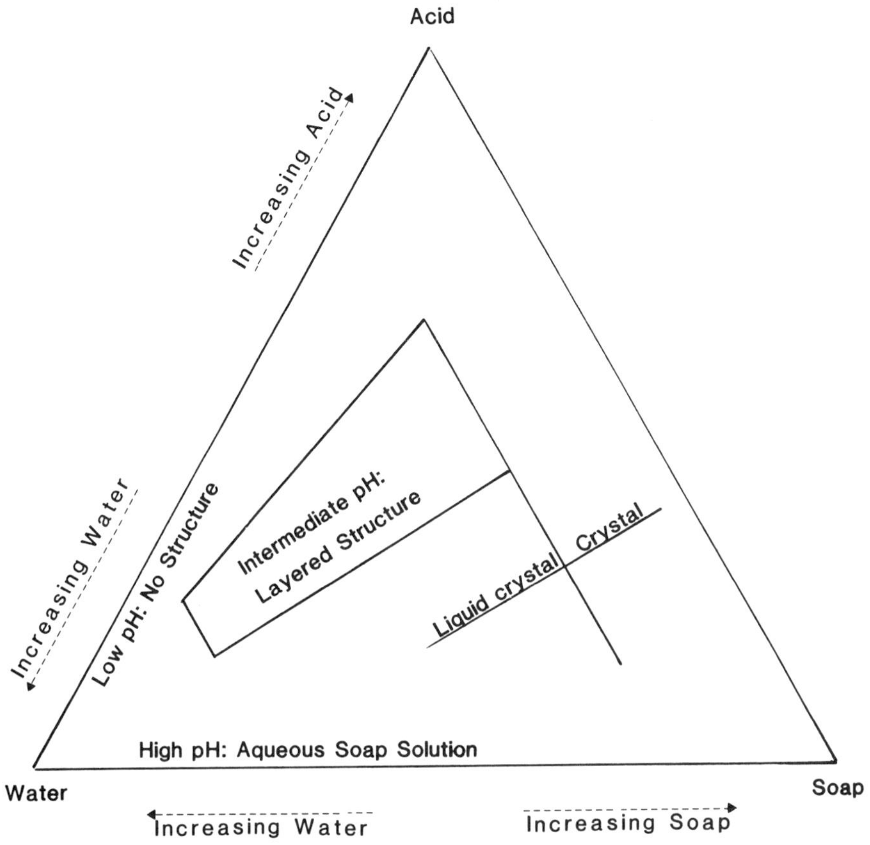

FIG. 8 Fatty acids form a layered structure with water only when they are neutralized to the correct pH.

shows that the lipid layered structure is not significantly perturbed by the presence of the corneocytes. Hence model investigations on lipids, independent of the corneocytes, may be expected to have relevance for the lipid organization in stratum corneum. With this conclusion supported by the experimental results it is justified to examine the specific information that the model can provide and the relevance of the findings for different phenomena experienced under in vivo conditions.

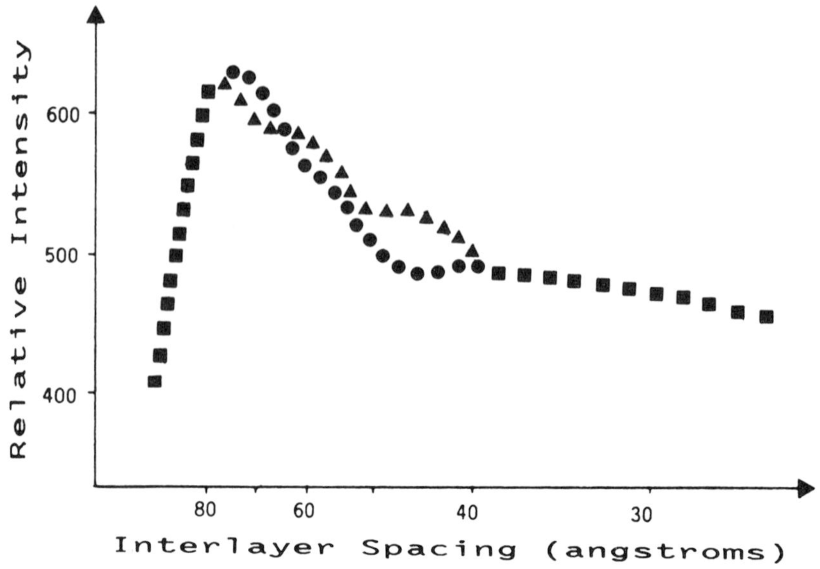

FIG. 9 Comparison of the small-angle x-ray diffraction pattern of the model epidermal lipid containing 32% water to that of human stratum corneum. ▲ Model epdermal lipid containing 32% water; ● Human stratum corneum; ■ regions of spectral overlap. (From Ref. 24.)

B. Location of Lipid Components in the Structure

Useful information about the location of individual lipids is obtained from small-angle x-ray diffractometry. The interlayer spacing of a layered packing is calculated directly from x-ray reflections. The packing of individual atoms is not revealed by the x-ray diffraction pattern, because these layered structures are not crystalline and short-range order is absent. Instead, the reflections in the small-angle region give the distance between the electron-rich polar layers in the structure.

Small-angle diffractograms also give an indication of the location of added compounds, despite the paucity of information. The evaluation is obtained from a comparison of the original layer spacing with that obtained after addition of a component. If the added component is dissolved in the aqueous layer, the interlayer spacing increases as shown in Fig. 10. Similarly, a location between the methyl group layers of the original hydrocarbon chains increases the d-spacing (Fig. 11). Inclusion of the compound in the space between the original lipid molecules results in an unchanged interlayer spacing provided that the ratio of water to lipid is kept constant by adding a corresponding amount of water (Fig. 12). With no addition of water, addition of an amphiphile to mix the lipids will cause a reduction of interlayer spacing.

The degree of penetration of a lipid into the bilayer may be calculated by using the relationship

$$\frac{V_i}{V_a} = \frac{\Delta d^i_{calc}}{d_a}$$

where

V_i = molar volume of lipid added
V_a = molar volume of amphiphiles already present in bilayer prior to addition of lipid

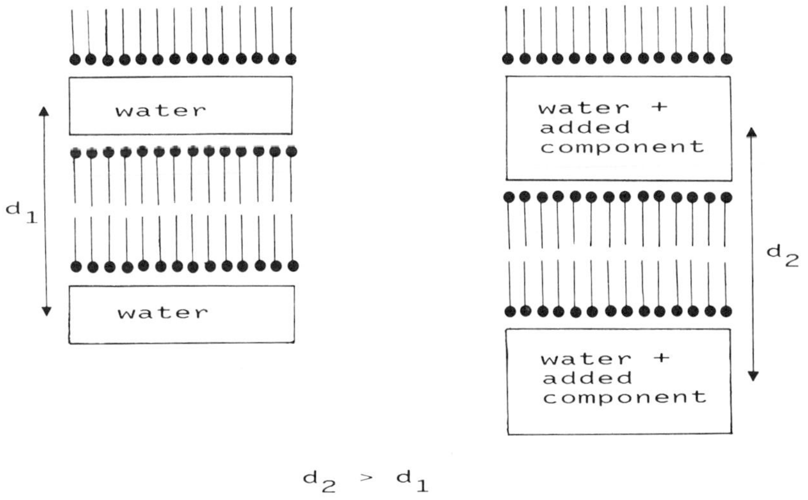

FIG. 10 An increase in interlayer spacing is obtained when the added component is dissolved within the aqueous layer.

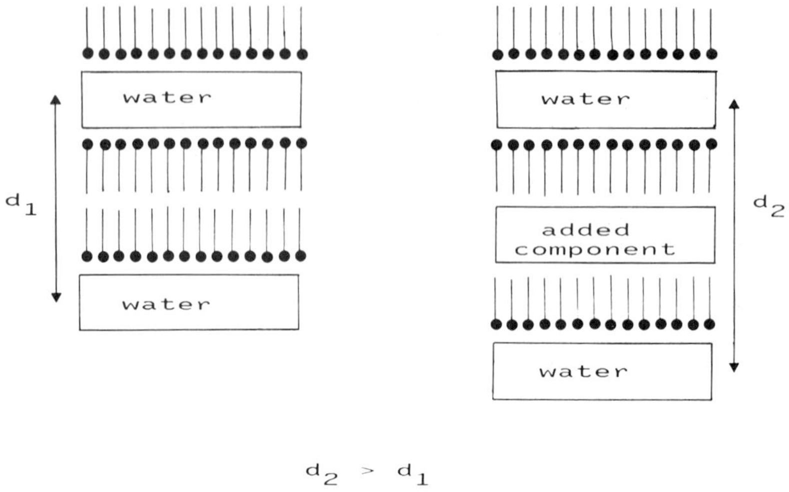

FIG. 11 When the added component preferentially locates between the methyl groups of the hydrocarbon chains, an increase in interlayer spacing results.

d_a = d-spacing of the bilayer without the added lipid (Fig. 13a)

Δd^i_{calc} = distance that the d-spacing would increase by if all of the added lipid were to become oriented between the methyl groups of the hydrocarbon chains (Fig. 13b)

The calculated increase in d-spacing is compared to the experimental increase in d-spacing d^i_{exp} (Fig. 13c), before and after the lipid is added. The percentage of penetration of the lipid into the bilayer is given by

$$\frac{\Delta d^i_{calc} - \Delta d^i_{exp}}{\Delta d^i_{calc}} \times 100\% = \text{percentage of penetration}$$

The percentage of penetration represents the amount of added lipid that is incorporated within the layered structure. If the percentage of penetration is less than 100%, a portion of the added lipid remains between the methyl groups of the hydrocarbon chains and the remaining portion is incorporated within the bilayer. If the percentage of penetration exceeds 100%, added lipid not only locates within

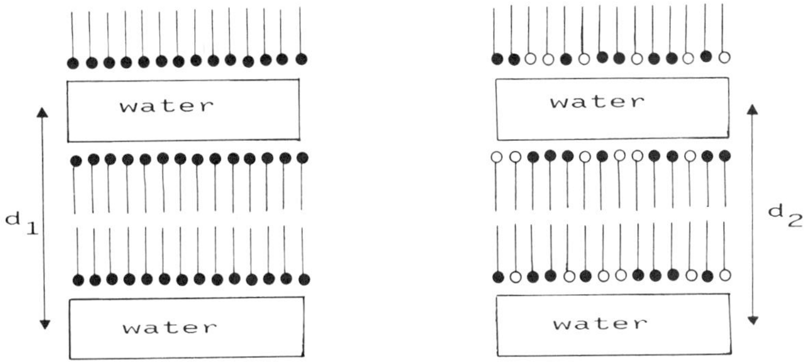

FIG. 12 If the added component orients within the bilayer among the amphiphiles, equivalent interlayer spacings are obtained with constant water content. $d_2 = d_1$ providing water content remains constant; $d_2 < d_1$, if water is not added to the system.

the bilayer, but draws with it another component, which was formerly located between the methyl groups. Only when the percentage of penetration is equal to zero is the added lipid found completely between the methyl groups, as shown in Fig. 13b.

This percentage of penetration is useful as a qualitative measure; a quantitative use is not possible due to the far-reaching assumptions made of no change in tilt as well as no change in individual order after addition of the extra component. Change of order has less influence on interlayer spacing than may be felt intuitively. According to Seelig [29], the effective length of a hydrocarbon chain in an amphiphilic bilayer may be estimated as

$$\langle L \rangle = 0.625\left(n + \sum_{i=1}^{n} S^i\right) \tag{1}$$

in which $\langle L \rangle$ is the effective chain length, n the number of carbon atoms, and S^i the order parameter of the CD_2 group. S^i is directly proportional to the quadrupole splitting, arising from a deuteriated proton site on the lipid molecule:

$$\Delta\nu = \frac{3}{4}\chi\ S^i \tag{2}$$

where χ is the quadruple coupling constant; S^i is typically on the order of 0.2 to 0.3. Equation 1 clearly demonstrates that even an

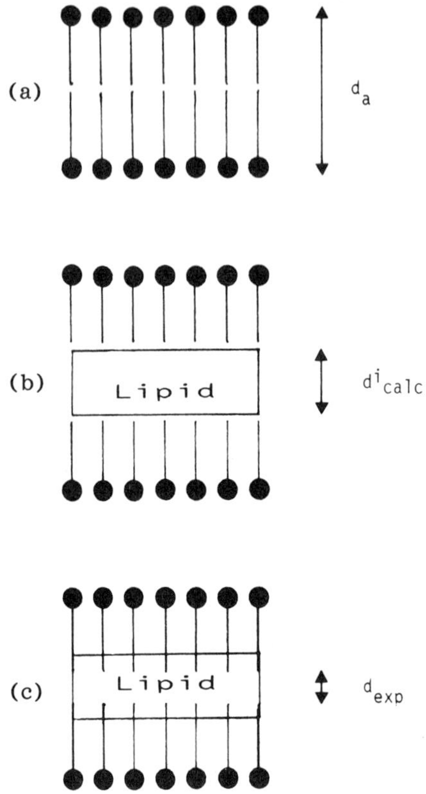

FIG. 13 (a) Original d-spacing of the layer with no lipid added; (b) d^i_{calc} is the distance by which the d-spacing would increase if all of the added lipid were to be oriented between the methyl groups of the hydrocarbon chains; (c) d^i_{exp} is the actual d-spacing due to the added lipid.

average reduction of order parameters by 0.1, which would bring the system close to instability [30], would reduce the interlayer spacing by only about 2 to 3 Å.

With these qualifications taken into account, a reasonable estimation may be made of the location of an added component to the stratum corneum lipid structure. However, the x-ray results are also able to give information concerning the interplay between water and the added components. The results on stratum corneum lipids shown in Table 3 and Figs. 14 and 15 are a good illustration [31]. The model contains lipids according to Table 3. The fatty

TABLE 3 Composition of Model Epidermal Lipid

Component	Source	Purity (%)	Wt % in mixture
Free fatty acids[a]			17
Myristic	Aldrich	99.5	10
Linoleic	Sigma	99	15
Oleic	Sigma	99	55
Palmitic	Sigma	99	5
Palmitoleic	Sigma	99	10
Stearic	Sigma	99	5
Phosphatidylethanolamine	Avanti Polar Lipids	99	5
Cholesteryl sulfate	Research Plus	98	4
Cholesterol	Aldrich	98	17
Triolein	Sigma	98	22
Oleic acid palmityl ester	Sigma	98	5
Squalene	Aldrich	98	5
Pristane	Aldrich	96	4
Ceramides	Sigma	99	21

[a]To produce a lamellar structure, free fatty acid mixtures were neutralized to 41% using NaOH with a water content of 32%. (These partially neutralized FFA are the main structure-forming compounds; hence the liquid crystal formed by this mixture is called the host.) After the addition of the remainder of the lipids, the water content was adjusted to remain at 32 wt %.

acids were saponified to sodium soaps to 41 mol % and water was added. With only unsaturated fatty acids present, a lamellar liquid crystal is formed in a wide water content range. The optical pattern (Fig. 14) is typical of a lamellar liquid crystal.

The small-angle x-ray results from the stratum corneum lipid model are shown in Fig. 15, and a superficial examination of the values seems to leave what appears to be fairly trivial information. The interlayer spacing versus the water content for the unsaturated free fatty acids increases from 46.50 Å to 48.25 Å when the water content is enhanced from 30% to 40%. Addition of the saturated fatty acids and of the phosphatidyl ethanolamine leads to an increase of the interlayer spacing, and addition of the cholesterol

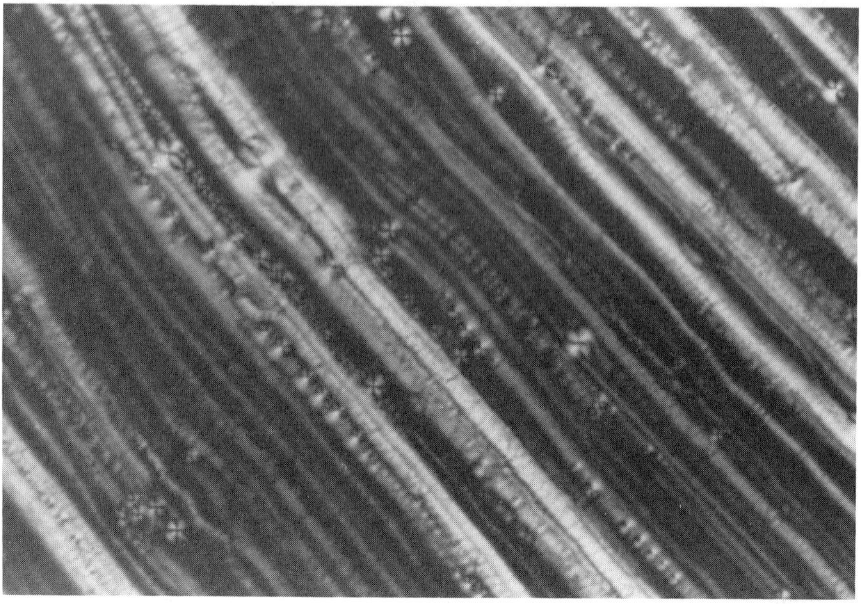

FIG. 14 Microphotograph of a typical lamellar liquid crystal.

gives a further addition of the spacing and an enhanced dependence on the water content.

The real significance of these x-ray data is obvious first when the extrapolation is made to zero water content. After this extrapolation the influence of the water is eradicated and the interlayer spacing reflects the conditions in the lipid layer only. Figure 15 makes obvious the difference in action of the cholesterol to the other components. The components added prior to cholesterol are located in the lipid part of the structure and their addition causes an increase in the interlayer spacing of the lipids (Fig. 13b). Hence part of them reside in the space between the methyl group layers of the original fatty acid molecules. Furthermore, because the slopes of the curves against water content are similar, a conclusion may be drawn that the presence of these lipids has no influence on the penetration of water into the lipid part.

The cholesterol, on the other hand, has an entirely different role. The increase that is observed in the interlayer spacing in the water content range investigated is *not* due to cholesterol congregating into the space between the methyl groups. On the contrary, the extrapolation to zero water content shows the interlayer spacing of the lipids to be reduced to the original value of

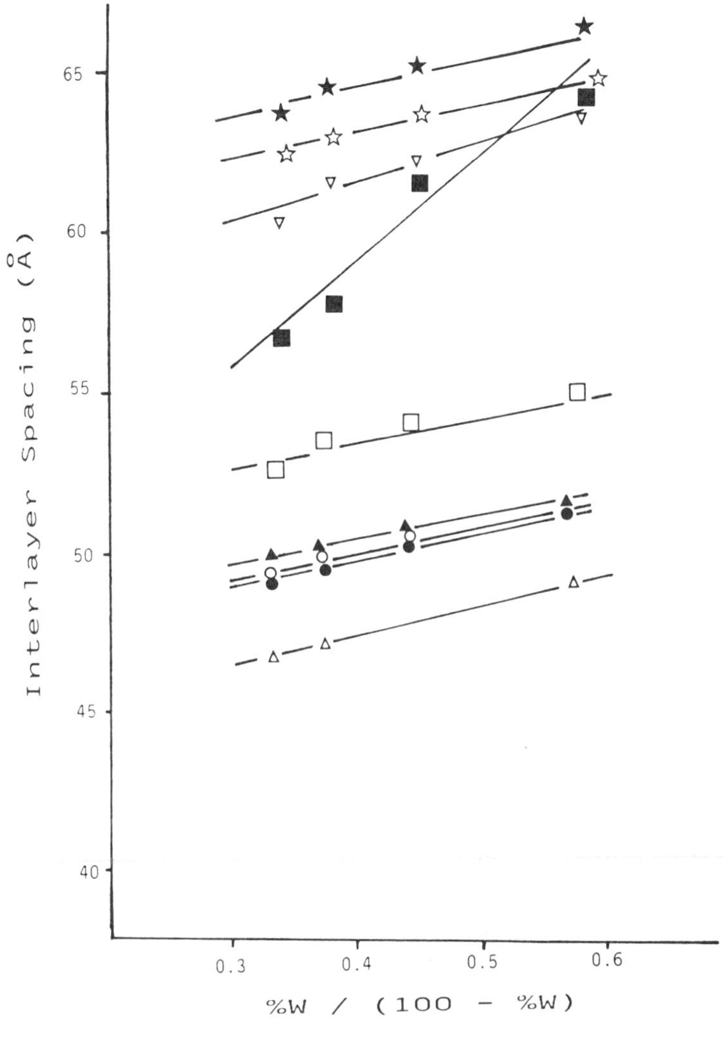

FIG. 15 Interlayer spacing as a function of water content for the systems. △, Unsaturated FFA; o, myristic + △; ●, palmitic + o; ▲, stearic + ●; □, phosphatidyl ethanolamine (PE) + ▲; ■, cholesterol + □; ▽, ceramide + ■; ☆, oleic acid palmityl ester + ▽; ★, squalene + ☆.

that of the lamellar liquid crystal of unsaturated fatty acids. The implication is obvious. The cholesterol resides entirely in the space between the fatty acid chains, and its presence also causes all the molecules located earlier between the methyl groups to relocate to penetrate the hydrocarbon chains of the original compounds.

Against this background the strong increase in interlayer spacing with water content after addition of the cholesterol may be properly intrepreted. The increase is not related to the location of compounds in the hydrocarbon layer at all; the reason for the strong increase is that the presence of cholesterol prevents the water from penetrating the lipid part of the structure.

These conclusions are supported by the optical patterns. The original lamellar liquid crystal of the unsaturated fatty acids shows typical features; the oily streaks are a poignant part (Fig. 16a). After addition of the saturated fatty acids and the phosphatidylethanolamine the patterns are less well defined (Fig. 16b) and of

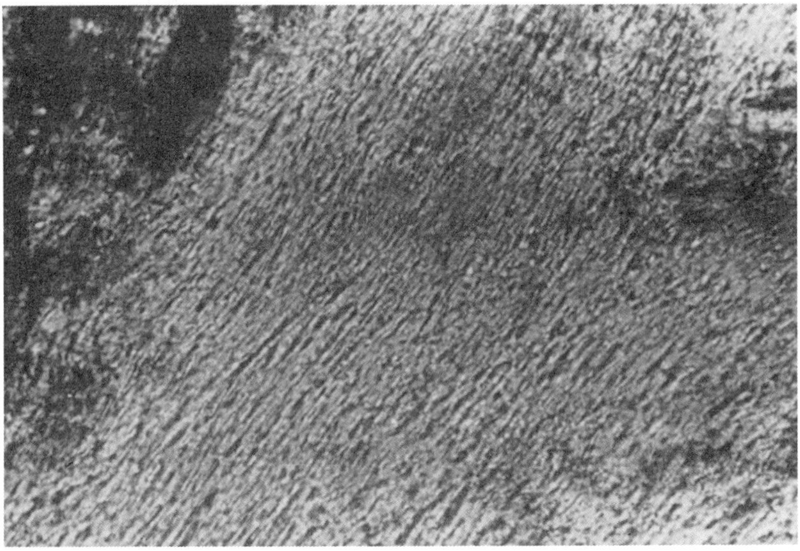

(a)

FIG. 16 (a) The unsaturated fatty acid/soaps give a lamellar liquid crystal (32% water); (b) with addition of the saturated fatty acid/soaps, the optical pattern indicates microcrystals embedded in a liquid crystalline matrix (32% water); (c) addition of cholesterol gives an optical pattern typical of a lamellar liquid crystal with well-defined oily streaks (32% water).

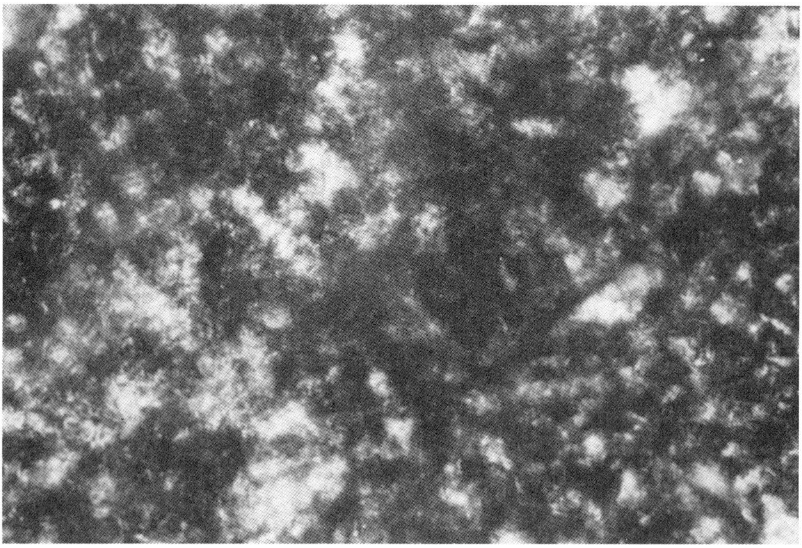

(b)

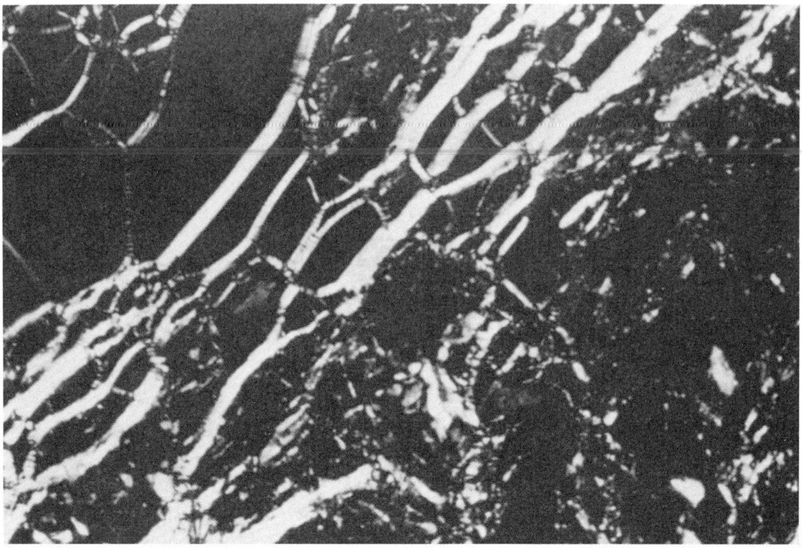

(c)

a character found when a crystalline long-chain compound is dispersed in a liquid crystal, causing a perturbation of the dislocation pattern. Addition of cholesterol restores the liquid crystalline pattern (Fig. 16c), which is well in accordance with the interpretation of the x-ray diffraction pattern.

IV. CRYSTALLINE/LIQUID CRYSTALLINE TRANSITION AND THE APPEARANCE OF SKIN

The liquid crystalline transition strongly depends on the water content of the structure. For a simple system such as the carboxylic acid/sodium carboxylate/water combination, the rule is straightforward. There is a lower limit of water content below which the liquid crystals do not exist (Fig. 8). In this region crystals are formed. It is essential to realize that the change of properties is profound in the crystal/liquid crystal transition. Lipid crystals are a dry white powder; a lipid liquid crystal is a creamy vehicle with properties resembling those of soft butter.

Comparison of the cracked and "powdery" aspect of dry skin (Fig. 17a) with the glossy and smooth appearance of humid skin (Fig. 17b) leads to the conclusion that a new mechanism of skin moisturizing has developed. The presence of water in the skin influences its appearance in an indirect manner because it changes the *organization* of the lipids. The white, dry powder of crystals is replaced by a soft liquid crystal, changing the appearance drastically (Fig. 17). The lipids of the stratum corneum are interesting not only because they influence the appearance of the skin, but they are also decisive for transport throughout the layer [27,32]. The transport of water is of particular interest.

V. WATER TRANSPORT THROUGH STRATUM CORNEUM

The stratum corneum is the layer that protects the body against dehydration; removal of a few micrometers of thickness results in a 25- to 50-fold increase in dehydration rate. The evaporation rate from skin has been studied extensively [33] and there is compelling evidence that lipids are the decisive element, but investigations that attempted to find the specific lipids responsible for the resistance failed [34]. The results of Smith et al. [27] are interesting because they show the lipids from plantar stratum corneum cells to be considerably less efficient in reducing water transport through a reaggregated stratum corneum than are the lipids from stratum corneum in the calf. Unfortunately, no structural analysis of the

(a)

(b)

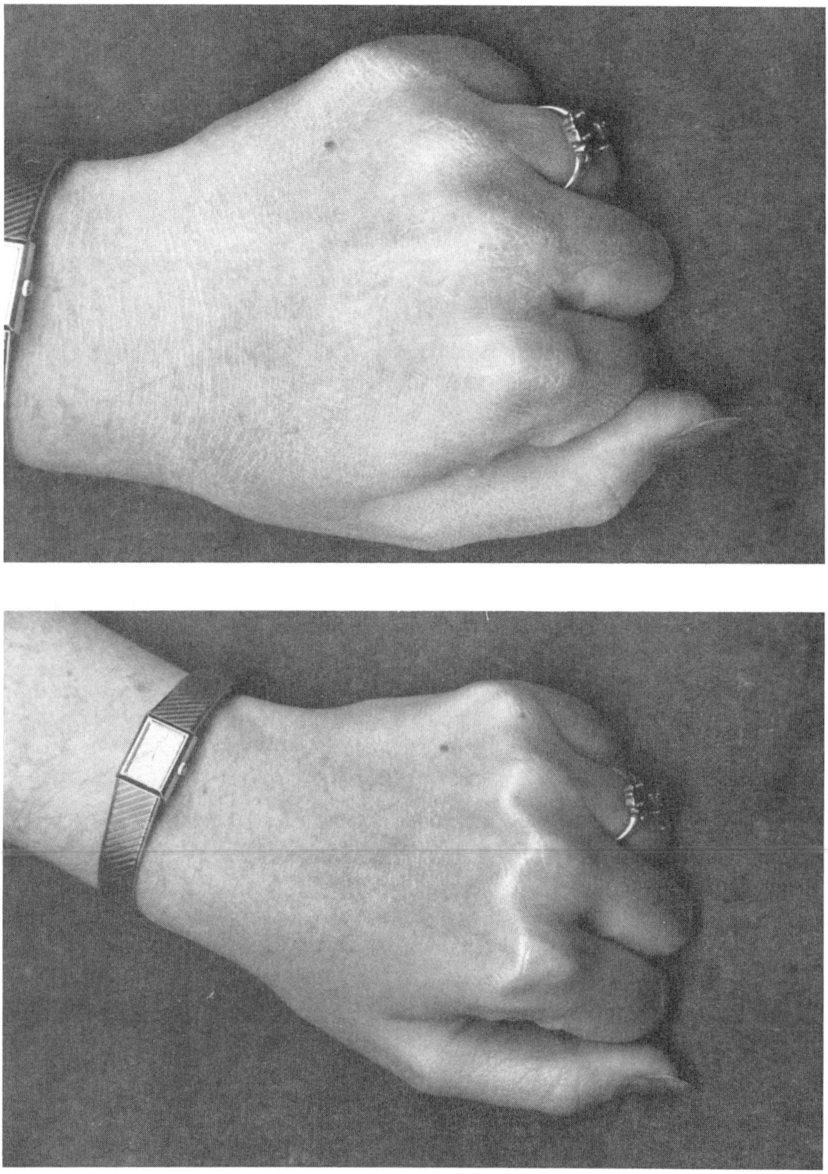

FIG. 17 (a) At low water contents the lipids of the stratum corneum exist as a mixture of crystalline materials, giving skin a rough, dry appearance; (b) at intermediate water contents the lipids of the stratum corneum associate into a lamellar structure, resulting in a smooth glossy appearance.

lipids was made. Blank has shown a reduction in water transport at reduced hydration levels [33]. Only small variations in transport were shown in atmospheres between 0 and 80% relative humidity.

Elias [35] has demonstrated that the problems of dry scaly appearance and water barrier breakdown that accompany essential fatty acid deficiency can be corrected by topical application of linoleic acid. The model for stratum corneum [24] has given some pertinent information that relates directly to the useful results by Elias [35]. Model layers 1/2 mm thick were exposed to atmospheres with controlled humidity and the evaporation rate of water was measured.

The diffusion coefficient may be calculated by using the approximation [36]

$$\frac{M_t}{M_\infty} = 4\left(\frac{Dt}{\pi L^2}\right)^{1/2}$$

where M_t is the amount of material diffused out of the membrane at time t and M_∞ is the amount of material originally present in the membrane having thickness (L). Plotting for M_t/M_∞ versus $t^{1/2}$ gave the diffusion coefficient values in Table 4 from the slope of initial rates.

The diffusion coefficients are one to two orders of magnitude greater than those found in intact stratum corneum; an expected result in view of the fact that the lipids in stratum corneum are only 10% of the total and are embedded in a matrix of keratinous aggregates.

The relative values of the coefficients in Table 4 show two features of interest. One characteristic is the fact that the lamellar structure with only free fatty acids present showed a lower diffusion than the one from the structure with all the lipids present. An inevitable conclusion is that the hydrophobic material gathered between the methyl groups of the fatty acid hydrocarbon chains does not play a decisive role for water transport. It appears probable that the resistance to water transport lies with the free fatty acids or, rather, with the lamellar structure. The evaporation rates for a combination with only saturated fatty acids coincided exactly with those from a free water surface; that is, the saturated free acids alone form no barrier to water transport.

An important feature of the values in Table 4 is the enhanced diffusion of water for samples in higher humidity. This has been observed earlier [33] and is an expected consequence of the geometry of the lipid—water structure in the stratum corneum. With enhanced humidity and hence increasing water content the thickness of

TABLE 4 Diffusion Coefficients for Water in Layered Models for Stratum Corneum Lipids ($cm^2/s \times 10^{-8}$)

Relative humidity	FFA/soap[a]	Model lipid	Oleic acid/ oleaote
6.2	5.9	8.3	8.7
32.4	7.1	8.9	11.0
80.3	8.1	15.7	12.2
92.5	11.5	24.7	20.8
97.0	9.7	28.0	15.7

[a]Fatty acid/soap combination (see Table 3 and its footnote.)

the water layers in the lamellar structure (Fig. 10) will increase. An increase in the water layer thickness means that the fraction of water more intimately bound to the polar groups is reduced versus the fraction of more "free" water at a distance from the polar groups. This free water has less resistance to diffusion than that for the "bound" water close to the polar groups, and the diffusion coefficient will increase.

For extremely high humidity a reduction is found in the diffusion coefficient (Table 4), a phenomenon also found for the stratum corneum [33]. This decrease in the transport is understood if the stability of the layered structure (Fig. 10) is considered. When the relative humidity approaches 100%, the chemical potential of water is close to that of water in its native state. The consequence of such a high chemical potential is that the layered conformation (Fig. 10) is moved close to its limit of stability. The interface between polar groups and water layers will now be less defined, more dislocations will occur in the overall pattern, and hence the transport of water will be impeded.

As pointed out earlier, the essential fatty acid deficiency symptoms [37,38] of dry scaly skin and enhanced water transport rates are related to the linoleic acid deficiency. Realizing the fact that unsaturated fatty acids are necessary for the formation of a liquid crystalline structure, it appears reasonable to relate the syndrome to the degree of crystallinity of the hydrocarbon chain. Too low an amount of unsaturated fatty acids leads to crystallinity and thus to enhanced water transport, as noted above.

It is useful to compare the transport through these layered structures with the one through monomolecular layers. Investigations of these phenomena 20 years ago [39] showed that long-chain saturated alcohol monolayers were more efficient in preventing water evaporation than were monomolecular layers of unsaturated compounds. However, this rule was valid only if the monomolecular layer was intact. A cracked layer diminished the resistance to evaporation to a few percent of the value for the intact layer.

Combining the present results with these earlier ones [39] appears to give a useful and reasonable pattern. A mix of unsaturated and saturated free fatty acids provides a better barrier to water transport than do the unsaturated ones alone because of less mobility in the structure. A layered structure of saturated fatty acids only would, if perfect, give excellent barrier function, but the presence of cracks is highly derogatory to barrier efficiency.

ACKNOWLEDGMENT

This research was funded by a grant from Dow Chemicals, Midland, Michigan.

REFERENCES

1. V. Vannini and G. Pogliani, *The Color Atlas of Human Anatomy*, Beekman House, New York, 1980, pp. 13–14.
2. W. Montagna and P. F. Parakkal, *The Structure and Function of Skin*, 3rd ed., Academic Press, New York, 1974, p. 46.
3. G. F. Odland and T. M. Reed, in *Ultrastructure of Normal and Abnormal Skin* (Alvin S. Zelickson, ed.), Lee & Febiger, Philadelphia, 1967, p. 54.
4. P. M. Elias, *J. Invest. Dermatol.* *80*:44 (1983).
5. G. F. Odland, in *Biochemistry and Physiology of the Skin*, Vol. 1 (Lowell A. Goldsmith, ed.), Oxford University Press, New York, 1983 p. 8.
6. P. M. Steinert and J. S. Cantieri, in *Biochemistry and Physiology of the Skin*, Vol. 1 (Lowell A. Goldsmith, ed.), Oxford University Press, New York, 1983, p. 135.
7. H. Auffhammer, *Verh. Phys. Med. Ges. Wurzburg, New Jersey* *1*:192 (1868) (quoted by Waldeyer, 1882).
8. I. A. Bernstein, in *Biochemistry and Physiology of the Skin*, Vol. 1 (Lowell A. Goldsmith, ed.), Oxford University Press, New York, 1983, p. 170.
9. I. Stern, *J. Invest. Dermatol.* *59*:251 (1972).
10. H. J. Yardley, *Br. J. Dermatol.* *95*:621 (1976).

11. G. M. Gray, *Front Matrix Biol.* *9*:83 (1981).
12. A. G. Matoltsy, in *Biochemistry of Cutaneous Epidermal Differentiation* (M Seiji and I. A. Bernstein, eds.), University of Tokyo Press, Tokyo, 1977, p. 93.
13. A. G. Matoltsy, *J. Cell Biol.* *47*:593 (1970).
14. R. H. Rice and H. Green, *Cell* *11*:427 (1977).
15. A. G. Matoltsy and P. F. Parakkal, *J. Cell Biol.* *24*:297 (1965).
16. P. W. Loert, D. T. Downing, et al., *J. Invest. Dermatol.* *83*: 193–195 (1984).
17. A. G. Matoltsy and P. F. Parakkal, in *Ultrastructure of Normal and Abnormal Skin* (Alvin A. Zelickson, ed.), Lee & Febiger, Philadelphia, 1967, p. 78.
18. W. W. Christie, *Lipid Analysis*, 2nd ed., Pergamon Press, Oxford, 1982.
19. N. Nicolaides, *Science* *186* 19 (1974).
20. D. J. Kooyman, *Arch. Dermatol. Syph.* *25*:444 (1932).
21. M. A. Lampe, M. L. Williams, and P. M. Elias, *J. Lipid Res.* *24*:131 (1983).
22. M. A. Lampe, P. M. Elias, et al., *J. Lipid Res.* *24*:120 (1983).
23. P. M. Elias, B. E. Brown, et al., *J. Invest. Dermatol.* *73*:339 (1979).
24. S. E. Friberg and D. W. Osborne, *J. Dispersion Sci. Technol.* *6*(4):485–495 (1985).
25. P. W. Wertz, W. Abraham, L. Lauderman, and D. T. Downing, *J. Invest. Dermatol.* *87*:582 (1986).
26. K. C. Madison, D. C. Swaitendwben, P. W. Wertz, and D. T. Downing, *J. Invest. Dermatol.* *88*:714 (1987).
27. W. P. Smith, M. D. Christenson, S. Necht, and E. H. Gans, *J. Invest. Dermatol.* *78*:7 (1982).
28. L. Larsson, in *Food Emulsions* (S. E. Friberg, ed.), Marcel Dekker, New York, 1976, p. 39.
29. H. Schindler and J. Seelig, *Biochemistry* *14*:2283 (1975).
30. A. J. I. Ward, *Mol. Cryst. Liq. Cryst.* (In press).
31. S. E. Friberg, H. Suhaimi, and L. Goldsmith (in preparation).
32. P. M. Elias, N. S. McNutt, and D. S. Friend, *Anat. Rec.* *189*:577 (1977).
33. I. H. Blank, J. Maloney III, A. G. Emslie, I. Simon, and C. Apt, *J. Invest. Dermatol.* *82*:188 (1984).
34. T. M. Sweeney and D. T. Dowing, *J. Invest. Dermatol.* *55*: 135 (1970).
35. P. M. Elias, B. E. Brown, and V. A. Fiboh, *J. Invest. Dermatol.* *74*:230 (1980).
36. E. R. Cooper and B. Berner, in *Methods in Skin Research* (D. Skerrow and C. J. Skerrow, eds.), Wiley, New York, 1985, Chap. 15.
37. P. M. Elias and B. E. Brown, *Lab. Invest.* *39*:574 (1978).

38. C. Prottey, *Br. J. Dermatol.* *94*:579 (1976).
39. V. K. LaMer, ed., *Retardation of Evaporation by Monolayers: Transport Processes*, Academic Press, New York, 1962.

2

Surfactant Interactions with Skin

LINDA D. RHEIN and F. ANTHONY SIMION Colgate—Palmolive Company, Piscataway, New Jersey

I.	Introduction	33
II.	Surfactants Elicit Multiple In Vivo Skin Effects	34
III.	Molecular Basis of the In Vivo Cutaneous Tightness Sensation	35
IV.	Molecular Basis of Surfactant-Induced Skin Roughness	37
V.	Surfactant Structure Effects on Stratum Corneum Swelling and the Relationship to Irritation	40
VI.	Surfactant Interactions that Counter Skin Irritation	44
VII.	Percutaneous Absorption of Surfactants	46
	Summary	48
	References	48

I. INTRODUCTION

Surfactants are commonly used in household and personal care products. These products come in contact with the skin for variable lengths of time. When this happens, certain undesirable effects may result, such as skin dryness, tightness, and roughness. Sometimes more severe reactions may occur: redness (erythema), swelling (edema), and pain (i.e., a characteristic inflammatory response).

It is important for researchers to understand the fundamental nature of surfactant interactions with this biological surface, which form the molecular basis of these undesirable skin effects. Products can then be formulated that will not elicit such negative skin effects, but rather, will produce more desired effects, such as softness, smoothness, and skin moisturization.

To achieve the above, researchers must separate and explore the individual molecular events that control surfactant–skin interactions. These are (1) the physicochemical behavior of surfactants in solution, (2) interactions with stratum corneum components, (3) diffusion rates through the stratum corneum, (4) interactions with living epidermal cells affecting growth and differentiation processes, (5) action on dermal components initiating the inflammatory response, and (6) initiation of the immune response. For the purpose of this book, we focus on the first three aspects of surfactant action which control their effects on skin. We review what is currently known about the initial interactions of surfactants with the skin surface and relate them to their biological consequences (i.e., irritation and/or skin feel effects in vivo). Cooper and Berner [1] have reviewed many aspects of physicochemical interactions of surfactants with skin. We focus here on the fundamental relationship of these interactions to in vivo cutaneous effects.

II. SURFACTANTS ELICIT MULTIPLE IN VIVO SKIN EFFECTS

When a consumer's skin is exposed to surfactant-containing products such as soaps, liquid cleansers, dishwashing detergents, or cosmetics for a brief amount of time (on the order of minutes), certain immediately perceived skin effects emerge. These include the sensations of tightness, dryness, and roughness. Consumers frequently associate these immediate negative skin effects with skin damage, although these effects usually occur in the absence of clinically defined skin inflammation (characterized by erythema or redness, swelling, and pain). Increasing exposure time to hours (or days) may induce inflammation.

The difference in the time required to elicit these different classes of skin effects suggests that they may be induced by separate individual molecular mechanisms. Because the immediate negative skin effects (tightness, dryness, and roughness) occur after brief exposure to the surfactant, they probably result from the direct action of surfactants on the stratum corneum. Inflammation which occurs after exaggerated treatment or longer exposure results from penetration of the surfactant (or secondary mediators) into the living tissue, causing tissue destruction and initiation of the inflammatory response. For instance, there is a positive correlation between the ability of surfactants to elicit erythema and their ability to release such mediators as arachidonic acid and its metabolites from keratinocytes in culture [2]. Arachidonic acid metabolites (e.g., prostaglandins and leukotrienes) have been implicated in mediating clinically defined skin inflammation [3,4].

III. MOLECULAR BASIS OF THE IN VIVO CUTANEOUS TIGHTNESS SENSATION

Kawai and Imokawa [5] studied how adsorption of surfactants to skin relates to the onset of cutaneous tightness (self-assessed by human subjects). They also explored how the ability of surfactants to remove skin lipids controls skin tightness. They reported that most anionic surfactants produce skin tightness within 1 to 5 min after a 10-s wash. The authors identified several surfactants that differ in their inherent ability to elicit tightness. They line up as follows:

SLS > AEOS-2EO > TEA cocoylglutamate > K myristate > TEA monolaurylphosphate

Most Tightness — Least Tightness

(See Table 1 for definitions of surfactant abbreviations.)

In an attempt to define a mechanism for eliciting tightness, the authors explored the ability of these surfactants to extract lipids and to adsorb to the skin. They quantified removal of cholesterol and squalene as representative skin lipids. Lipid extraction was assessed by attaching cups containing the surfactant solution (1%) to the ventral arm for 10 min and quantifying these lipids in the solution using gas chromatography. They found that the ability of the surfactants to remove skin lipids correlates with the order in which they induced skin tightness (the order was that shown above, with a Pearson correlation coefficient $r = 0.94$ for cholesterol removal and 0.89 for squalene; both correlations were highly significant, $p < 0.01$). Additionally, the authors showed that if the skin was first washed with diethyl ether followed by 5% SLS, the tightness sensation was considerably enhanced compared to treatment with either chemical alone. We suggest that removal of the lipid with diethyl ether potentiates the interaction of the SLS with the nonlipid components (presumably proteinaceous) of the stratum corneum by making more binding sites available. The resulting effect is increased tightness. The lipids in the skin are discussed further in Chapter 1.

TABLE 1 Surfactants

Type	Structure
Sodium dodecyl sulphate (SLS)	$C_{12}H_{25}OSO_3Na$
Sodium polyoxyethylene-dodecyl sulphate (AEOS-nEO, n = 2, 3, 6, 9, or 12)	$C_{12}H_{25}(OCH_2CH_2)_nOSO_3Na$
Sodium lauroyl isethionate (Igepon AC)	$C_{11}H_{23}-CO\overset{CH_3}{N}CH_2CH_2SO_3Na$
Sodium dodecylbenzene-sulphonate (LAS)	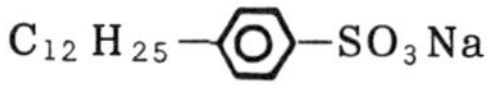
Polyoxyethylenedodecyl ether ($C_{12}(EO)_n$, n = 3, 6, 17)	$C_{12}H_{25}(OCH_2CH_2)_nOH$
Dodecyltrimethylammonium bromide	$C_{12}H_{25}-\overset{CH_3}{\underset{CH_3}{N^+}}-CH_3Br^-$
N,N-Dimethyldodecylamine oxide (LDAO)	$C_{12}H_{25}-\overset{CH_3}{\underset{CH_3}{N^+}}-O^-$
Cocoamidopropyl betaine	$C_{11}H_{23}CONH(CH_2)_3-\overset{CH_3}{\underset{CH_3}{N^+}}-CH_2COO^-$
Alfa Olefin Sulfonate (AOS)	$C_9H_{19}CH_2CH=CHSO_3Na$ and $C_9H_{19}CH(OH)CH_2CH_3SO\ Na$
TEA Cocoylgultamate	$^-OOCCH_2\underset{HNCO(CH_2)_{10}CH_3}{CH}COO^-\quad (TEA^+)_2$

(continued)

TABLE 1 (Continued.)

Type	Structure
TEA Monolauryl Phosphate	$C_{12}H_{25}OPO_3^{-}TEA^{+}$
Potassium Myristate (K Myristate)	$C_{13}H_{27}COO^{-}K^{+}$
Lauryl Alcohol	$C_{12}H_{25}OH$
Triethanolamine (TEA)	$(HO-C_2H_4)_3-N$

Kawai and Imokawa [5] also measured, for the same surfactants, the binding reaction to skin. This was done by quantifying bound indigo carmine dye. This dye cannot bind to the skin if anionic surfactant already occupies the positively charged sites on the keratin. After treatment with surfactant using the cup method described above, the skin was exposed to the dye in a second cup for 1 minute. After removal and rinsing, the residual dye bound to the skin was quantified using a color-difference reflectance meter. Dye binding is inversely proportional to the binding of surfactants and the resulting order of surfactant adsorption was as follows:

SLS > AEOS-2EO > TEA cocoylglutamate = K myristate ≥ TEA monolaurylphosphate

Most Adsorption ... Least Adsorption

Thus surfactant adsorption to skin correlates with the surfactants' ability to elicit skin tightness (correlation coefficient $r = 0.99$, $p < 0.01$). From the foregoing studies, it can be hypothesized that the cutaneous tightness sensation induced by surfactants is related fundamentally to two events: removal of skin lipids, and adsorption of surfactants to sites on the keratin.

IV. MOLECULAR BASIS OF SURFACTANT-INDUCED SKIN ROUGHNESS

Imokawa and co-workers [6–9] probed the molecular mechanisms by which surfactants induce skin roughness. They showed a relationship between the propensity of surfactants to elicit skin roughness and two molecular events: (1) extent of surfactant binding to skin and (2) ability to denature proteins. Skin roughness was induced using the cup circulation method. This involves circulating a surfactant solution through a cup attached to the forearm skin for 10 minutes once a day for 4 consecutive days. This method was designed to simulate an actual "use" situation. The extent of roughness was assessed by a trained clinician.

The authors [6] found a significant correlation ($r = 0.82$, $p < 0.01$, calculated from Fig. 8 in ref. 6) between extent of surfactant-induced skin roughness and the protein denaturing potency of the surfactants (measured as a change in specific rotation of bovine serum albumin). In an alkyl sulfate homologous series, both skin roughness *and* protein denaturation were maximal for the C_{10} or C_{12} alkyl chain lengths [6]. Similar patterns were found for LAS [6], AOS [6], and soap [9] homologous series (abbreviations described in Table 1). The comparative extent of skin roughness for different types of commercially available surfactants also parallels their protein denaturing potency. The order obtained for both intensity of roughness and denaturing potency can be summarized as follows:

SLS > LAS > AOS ≥ AEOS-3EO > $C_{12}(EO)_7$

Most Roughness and Denaturation	Least Roughness and Denaturation

It is known that the first step in the denaturing interaction of surfactants with bovine serum albumin is ionic adsorption onto the protein [10]. Indeed, there is a significant correlation between surfactant adsorption (binding) to skin callus and ability to denature bovine serum albumin ($r = 0.93$, $p < 0.01$; Fig. 9, ref. 7). Fifteen different surfactants were tested in that study. This suggests that surfactant adsorption onto stratum corneum keratin (and their resulting propensity to denature proteins) may be a factor in the fundamental mechanism of skin roughening, as it appears to be for induction of tightness.

The relationship between surfactant binding to skin and the ability of surfactants to induce skin roughness was probed further [8, 9,11]. Adsorption of surfactants to skin in vivo (on humans) or in vitro (using callus) was quantified using indigo carmine dye, while the cup circulation treatment method was used to induce skin roughness. There is a correlation between in vitro binding and the ability to induce skin roughness ($r = 0.92$) [11]. In vitro binding to callus showed the following rank order of adsorption [8].

SLS > LAS > AOS > AEOS-3EO

Binds Most	Binds Least

Similar results were found for the order in which these surfactants induce skin roughnes [8,11]. Additionally, in vivo adsorption of these surfactants to skin paralleled the results above [11].

Figure 1 shows the adsorption isotherms for the homologous series of alkyl sulfates to callus measured in vitro. It is apparent that they follow Langmuir's model, suggesting a limited number of saturable binding sites. However, it was further noted that the apparent saturation of the isotherms occurs at approximately the critical micelle concentration (CMC) for each surfactant. As the concentration of surfactant monomers levels off around the CMC, this would limit the number of monomers (or submicellar species) available to

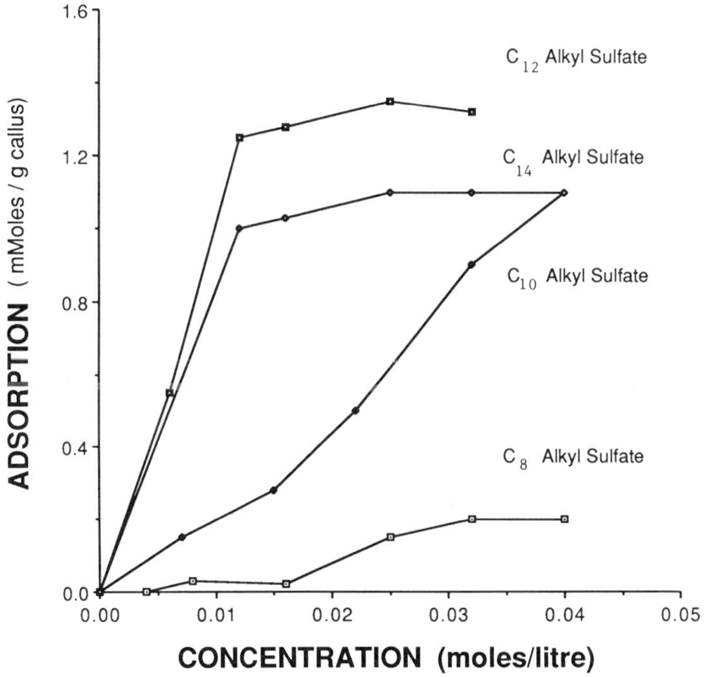

FIG. 1 Adsorption isotherms for alkyl sulfates with isolated callus powder in vitro. Experimental conditions were: callus, 48 to 80 mesh, 30 mg surfactant, 10 ml, pH 7.0, 6-h incubation at 40°C. (From ref. 11; by permission of Munksgaard Publishers Ltd., Copenhagen, Denmark.)

bind to the callus. These data are also consistent with the hypothesis that the surfactant monomer rather than the micelle is the species that interacts with the skin. Thus the saturation phenomenon observed in the isotherm may be related to this particular solution property of surfactants.

We wanted to compare the findings noted above for skin roughness caused by surfactants with their ability to elicit clinical irritation. The comparative clinical irritation potential (manifested as erythema) of some of these surfactants has also been determined (Table 2 and ref. 12). This was done using a 24-h patch test. Clinical irritation requires a longer exposure time to elicit an inflammatory response in the dermis resulting in the erythema. The surfactants line up as follows:

SLS > LAS > AEOS-3EO > polyoxyethylene alkyl ether

Most Erythema	Least Erythema

Thus for surfactants where comparisons can be made, their comparative propensity to elicit erythema appears similar to their ability to induce cutaneous roughness (and tightness, although data are sparse for this attribute). This suggests that these effects all have at least some common molecular basis. Binding to keratin and denaturation of protein are part of this common molecular basis, and results to date suggest that lipid removal by surfactants is also a factor because it potentiates binding. We can then speculate that the greater the surfactant binding to keratin, the greater the extent of denaturation of keratin protein; once these events occur extensively, the surfactant will penetrate through the stratum corneum into the living tissue. In the living tissue, their denaturing activity continues, but this time an imflammatory reaction sets in to repair the damage.

V. SURFACTANT STRUCTURE EFFECTS ON STRATUM CORNEUM SWELLING AND THE RELATIONSHIP TO IRRITATION

Another in vitro response of stratum corneum to surfactants has been probed in depth. Putterman and co-workers [13] and researchers in our laboratory [14] have shown that exposure of isolated stratum corneum to anionic surfactants induces continuous swelling of the membrane. The swelling occurs in all dimensions, but is substantially greater in the thickness dimension. The extent of swelling

TABLE 2 Relationship Between Surfactants' Clinical Irritation Potential and Their Ability to Cause Stratum Corneum to Swell

Surfactant	Skin irritation in vivo[a]	Stratum corneum swelling[b] (% change in length)
LAS	Severe within 1 day	10.0
SLS	Severe within 1 day	11.8
AEOS-3EO	Mild to moderate within 5 days	4.0
AEOS-6EO	Mild to moderate within 5 days	1.5
AEO-12EO	None after 5 days	1.5
Tween-20	None after 5 days	0.2

[a]From Blake-Haskins et al. [16]; human subjects, 10% concentration, pH 7, Duhring chambers.
[b]Methodology is given in ref. 14. Results shown are summarized from this work. Concentration was approximately 75 mM.
Source: Refs. 14 and 16; reproduced by permission of the *Journal of the Society of Cosmetic Chemists.*

induced by various surfactants parallels their propensity to elicit clinical erythema (Table 2); that is, more irritating surfactants caused more swelling. Associated with the swelling reaction was curling and twisting of a strip of stratum corneum, which dramatically decreased the width of the treated strip [15]. This curling phenomenon and stratum corneum swelling respond similarly to surfactants and they both directly parallel the clinical irritation potential of various anionic surfactants; that is, the greater the twisting, the more the swelling, the greater the irritation potential [15]. More recently we have used an alternative membrane, cross-linked collagen (Helitrex Corp.), to track swelling. This membrane behaved surprisingly similarly to stratum corneum [16].

Rhein et al. [14] found that surfactant-induced stratum corneum swelling was dependent on surfactant concentration, increased with incubation time, and was reversible under conditions of lower concentration (1%) and shorter incubation times (1 to 6 h). Because swelling is reversible, it probably results from interactions of surfactants with stratum corneum proteins rather than lipids (the latter would probably be removed irreversibly during treatment). Indeed, Putterman et al. [13] showed that pretreatment with lipid solvents would potentiate surfactant-induced stratum corneum swelling probably by exposing more binding sites on the protein. The reversible twisting and curling of the stratum corneum may also be caused by conformational changes in proteins resulting from their denaturation.

When stratum corneum (and collagen) swelling was measured as a function of anionic surfactant concentration, the swelling induced appeared to saturate at the CMC [14,16]. Again, this suggests that the surfactant monomer rather than the micelle is the species that interacts with the proteins and that the surfactant solution properties probably control the reaction. Systematic structural studies of the surfactant-induced swelling reaction show that stratum corneum swelling parallels the known irritation potential of structurally related surfactants. Maximum swelling was induced by surfactants with C_{12} or C_{14} alkyl chains in homologous series of several anionic surfactants; for example, alkyl benzene sulfonates, α-olefin sulfonates, alkyl sulfates, and paraffin sulfonates (see Fig. 2). This parallels their known in vivo clinical irritation, which demonstrates C_{12} maximal activity [6,8,9,17]. Ethoxylation of alkyl sulfates reduces the swelling response (see Table 2 and ref. 14), which also parallels the known reduced irritation of these ethoxylated variants [15,16]. The presence of divalent cations reduces or inhibits the swelling response to anionic surfactants. Schrader's in vivo clinical results [18] suggest that magnesium lauryl sulfate is also clinically less irritating than sodium salt. Cationic surfactants have an even more interesting effect. Alkyl trimethylammonium bromides, which are clinically irritating [9], inhibit stratum corneum and collagen swelling. Positive charges obviously change the nature of the interaction of the surfactants with the skin surface.

The reason for the maximum activity (whether quantified as swelling, protein denaturing potency, induction of skin roughness, or clinical irritation) that usually occurs for the C_{12} or C_{14} homolog of a series for various surfactants has been the subject of much discussion. Robbins and Fernee [19] explain the phenomenon (based on Breuer's model [20]), as indicating the importance of hydrophobic interactions to the swelling and/or adsorption mechanism. In essence, the shorter alkyl chains do not interact as efficiently as C_{12} with the hydrophobic regions on the proteins. As chain length increases above C_{12}, penetration into the keratin matrix would be more difficult.

However, Cooper and Berner [1] have not fully agreed with this model. They observed that Imokawa et al. [6] reported a C_{12} maximum activity for surfactant effects on optical rotation of bovine

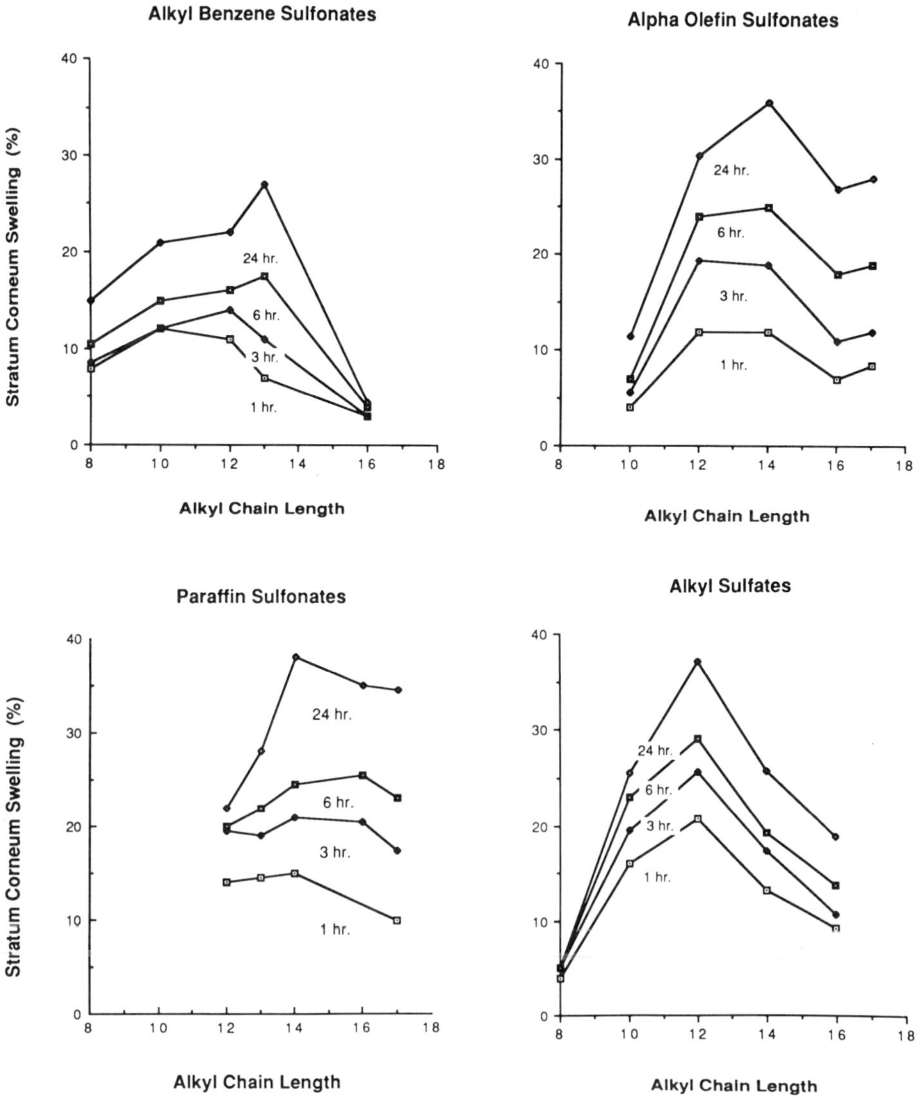

FIG. 2 Dependence of swelling of human stratum corneum on alkyl chain length for various surfactants. Studies were all run at 1% (w/w) surfactant. Time periods shown are incubation time of the treatments, after which readings were taken. Swelling values are percent increases from hydrated lengths. Means differing by greater than 2% are significantly different ($p \leq 0.05$). Note that swelling results for one homologous surfactant series in this figure cannot be compared with results for a different homologous series because they were run in separate experiments. (Data from ref. 14; reproduced by permission of the *Journal of the Society of Cosmetic Chemistry*.)

serum albumin (BSA), yet penetration into a cross-linked matrix is not a requisite for surfactant interaction with bovine serum albumin. Thus these authors searched for a mechanism other than inhibition of matrix penetration to explain the reduced swelling and adsorption for the higher alkyl chain lengths. They proposed a mathematical model to explain this effect based on binding, critical micelle concentration, solubility, and partitioning into the stratum corneum. Hydrophobic sites on the keratin might optimally bind a specific hydrophobic alkyl chain length of C_{12} (in the case of alkyl sulfates). Beyond this chain length there may be no thermodynamic advantage to binding of longer alkyl chains. As the alkyl chain is extended, the CMC is reduced, resulting in less monomer available to interact with the keratin protein. Thus the reduced activity of the longer chains probably results from the lower overall CMC (reducing the available monomer) and reduced solubility of the surfactant in the aqueous phase.

VI. SURFACTANT INTERACTIONS THAT COUNTER SKIN IRRITATION

Recent information suggests that surfactants do not react with skin in an additive fashion to produce their various effects. In fact, addition of a mild surfactant to an irritating one often results in *lower* reactivity than is elicited by the irritating surfactant alone. This occurs despite the higher total concentration of the system after the second surfactant is added. As an example, our laboratory [14] explored the effects of mixing ethoxylated alcohol sulfates and amphoterics with SLS on the swelling of isolated stratum corneum (Table 3). Clearly, the swelling induced by the mixed systems was considerably lower than that caused by 1% SLS alone, despite the higher level of actives in the mixed system. The same effect was validated in an in vivo patch test for SLS and AEOS-6EO [21]. The reduction in surfactant activity can possibly be caused by either of two mechanisms. The surfactants may interact in solution, forming mixed micelles that sequester the irritating surfactant, preventing its interaction with the skin [22]. Alternatively, the mild surfactant may compete with the irritating surfactant for binding sites on the skin's surface. If tested at saturating levels, this would result in a reduced response.

Dominguez et al. [23] showed that mixtures of betaine and SLS elicited less ocular irritation than did either surfactant alone. They ascribed this synergism to interactions between the surfactants in solution, reducing the amount of either species available to elicit irritation. Similarly, Faucher and Goddard [24] demonstrated a reduction in SLS adsorption to human hair keratin when a nonionic surfactant (Tergitol 15-S-9) was added to the system, even though

TABLE 3 Effects of Ethoxylated Alcohol Sulfates and Amphoterics on SLS-Induced Stratum Corneum Swelling

Experiment	Treatment (6-h incubation)	Stratum corneum swelling (% increase from hydrated length)
1	1% SLS	46.1 ± 3.62
	1% SLS + 1% AEOS-3EO[a]	21.6 ± 1.29
	1% SLS + 1% AEOS-6EO[a]	19.1 ± 3.60
2	1% SLS	46.8 ± 4.08
	1% SLS + 1% LDAO[a]	7.7 ± 1.72
	1% SLS + 0.5% LDAO[a]	25.8 ± 8.44
	1% SLS + 1% betaine[a]	21.0 ± 4.45
	1% SLS + 0.5% betaine[a]	34.5 ± 4.25

[a]Values are means ± SD. AEOS-3EO, AEOS-6EO, lauryl dimethyl amine oxide (LDAO), and cocoamidopropyl betaine when tested *alone* at 1% (w/w) all produced small or negligible increases in swelling above the hydrated lengths (i.e., equal to or less than 5% increase in 6 h). All treatments caused a significant reduction in swelling compared to 1% SLS treatment alone (≤ 0.01).
Source: Ref. 14; reproduced by permission of the *Journal of the Society of Cosmetic Chemicals*.

the total concentration of the system increased with the added nonionic. As very little nonionic surfactant was adsorbed to the hair, they suggest that the Tergitol and SLS are interacting in solution, preventing SLS binding. Inclusion of the nonionic in the system will lower the CMC, thus reducing the amount of free SLS monomer available to interact with skin.

Surfactants in mixtures can also compete for binding sites on the skin. In an earlier study, Dominguez et al. [25] showed competition between betaine and SLS in their binding to skin callus. This competition was affected by the pH and alkyl chain length of the betaine. More research is needed to identify more precisely the major factors controlling interactions to reduce the skin response.

VII. PERCUTANEOUS ABSORPTION OF SURFACTANTS

The role of percutaneous absorption of surfactants in eliciting irritation is unclear. Clinically, mild surfactants such as nonionics penetrate the stratum corneum rapidly, yet do not cause irritation. Thus there are obviously other criteria for inducing irritation. Not only must the surfactant penetrate the stratum corneum, but it must be able to elicit irritation once it has penetrated. Few studies have been reported comparing the adsorption and penetration of surfactants through the skin. In a comprehensive study, Howes [26] compared the adsorption of various anionic surfactants into the skin. Overall he found, as have others [27], that while anionic surfactants bind well to the skin, only low levels of surfactants penetrate through the stratum corneum. The penetration rate of surfactants generally increases with time; this probably reflects a progressive increase in damage to the barrier by surfactants. For instance, 10% SLS did not penetrate initially (i.e., it showed a lag time); however, after about 24 h, penetration began to occur and increased rapidly as the stratum corneum barrier broke down [26].

Adsorption (entrapment) of anionic surfactants in the epidermis was also explored. Results [26] show that although small amounts of surfactants penetrate the skin, considerable amounts are deposited in the stratum corneum. After applying the surfactants as aqueous samples (25 mM) for 15 minutes, autoradiography revealed *heavy* deposition of SLS, isethionate, and dodecylbenzene sulfonate in the stratum corneum. The autoradiography also showed that deposition occurs both through the interfollicular membrane and via hair follicles. For a homologous series of soaps, the amounts penetrating after washing rat skin with 60 mM soap solution was highest for the C_{12} homolog. This lineup was similar to results for adsorption of these soaps to epidermis.

Permeability constants for penetration of anionic surfactant through isolated epidermis were calculated for the homologous soap series, isethionate, SLS, LAS, and nonionic (Table 4). The constant (at 6 hours for human epidermis) was highest for C_{10} and C_{12} soaps and lowest for longer-chain soaps, SLS and LAS. Again Howes [26] showed that the permeability constant for the anionic surfactants increases with time of exposure. This increase may be caused by the surfactants degrading the permeability barrier, thereby increasing the penetration rate.

For guinea pig, Prottey and Ferguson [27] confirmed that laurate soap had a substantially higher permeability constant than that for SLS or isethionate but was similar to that of the mild nonionic $C_{12}(EO)_3$ (Table 4). The permeability constants through in vivo

TABLE 4 Rates of Percutaneous Penetration of Pure Surfactants

Surfactant	Permeability constants ($\times 10^{-6}$ cm/min): Isolated human epidermis in vitro[a], 6 h	Isolated human epidermis in vitro[a], 24 h	Guinea pig skin in vivo, 24 h[b]
Soaps			
C_{10}	5.4	18.6	—
C_{12}	18.2	25.0	4.6
C_{14}	1.6	9.4	—
C_{16}	0.1	0.2	—
C_{18}	0.1	0.1	—
Igepon AC	0.7	0.9	0.3
SLS	0.1	1.8	0.065
AEOS-3EO	—	—	0.44
$C_{12}(EO)_3$	—	—	4.52
$C_{12}(EO)_6$	—	—	0.75
Lauryl alcohol	—	—	0.51
LAS	0.1	0.1	—

[a]From ref. 24.
[b]From ref. 25.
Source: Refs. 24 and 25; reproduced by permission of the *Journal of the Society of Cosmetic Chemists.*

guinea pig skin were reasonably similar to those of Howes [26], which were determined using isolated human epidermis (see Table 4). Prottey and Ferguson also showed that nonionic surfactants such as $C_{12}(EO)_3$ penetrate the stratum corneum significantly more rapidly than does SLS (Table 4). However, this nonionic ethoxylate is significantly milder than SLS. Similarly, mild AEOS-3EO and sodium isethionate have higher initial permeability constants than those of the more irritating SLS. This suggests that penetration of a surfactant through the stratum corneum is not the sole criterion for eliciting irritation.

Faucher et al. [28] also studied permeation of SLS and nonionic surfactants through neonatal rat skin and found that SLS penetrated most rapidly and tended to destroy the stratum corneum. Ethoxylated nonionics penetrated to a lesser extent. Combinations of SLS with nonionics decreased SLS penetration. These studies seem inconsistent with work of Prottey and Ferguson [27]. However, measurements were taken on *neonatal* rat skin at longer time periods (days) where the cumulative damaging effects of SLS potentiate its penetration.

SUMMARY

In this chapter we have summarized what is currently known about the molecular basis of cutaneous skin effects of surfactants, focusing on their action on the stratum corneum. Findings to date suggest that binding to keratin and concomitant protein denaturation resulting in swelling of the membrane are all directly related to induction of cutaneous roughness, tightness, and clinical irritation. Lipid removal potentiates binding and swelling, presumably by exposing new binding sites on the keratin, although lipid removal may also accelerate surfactant penetration into living tissue. Continuous damage resulting from longer exposure to surfactants is required to elicit an inflammatory response manifested as erythema. Because the comparative propensity of the various surfactants to elicit several cutaneous responses (roughness, erythema, and possibly tightness) lines up similarly, we feel that these cutaneous effects are directly related to the initial molecular actions of surfactants described above. The kinetics of the exposure controls the timing of each cutaneous response. Solution properites and structural factors controlling these actions directly need more exploration.

REFERENCES

1. E. R. Cooper and B. Berner, *Surf. Sci. Ser.* (Surfactants in Cosmetics Issue) *16*:Chap. 7, 195 (1984).
2. V. A. DeLeo, L. C. Harber, B. M. Kong, and S. J. DeSalva, *Proc. Soc. Exp. Biol. Med. 184*:477 (1987).
3. E. M. Jackson, *J. Toxicol. Cutaneous Ocul. Toxicol. 34*:347 (1984).
4. M. E. Goldyne, *J. Am. Acad. Dermatol. 10*:659 (1984).
5. M. Kawai and G. Imokawa, *J. Soc. Cosmet. Chem. 35*:147 (1984).
6. G. Imokawa, K. Sumura, and K. Katsumi, *J. Am. Oil Chem. Soc. 52*:484 (1975).
7. G. Imokawa, *J. Soc. Cosmet. Chem. 31*:45 (1980).

8. G. Imokawa, K. Sumura, and M. Katsumi, *J. Jpn. Oil Chem. Soc. 23*:719 (1974).
9. G. Imokawa and T. Takeuchi, *Cosmet. Toiletries 91*:32 (1976).
10. J. Reynolds and S. Herbert, *Biochemistry 6*:937 (1967).
11. G. Imokawa and Y. Mishima, *Contact Dermatitis* 5:357 (1979).
12. U. Zeidler and G. Reese, *Proc. International Federation of the Cosmetic Chemistry Society*, Paris, 1982, p. 229.
13. G. J. Putterman, N. F. Wolejsza, M. A. Wolfram, and K. Laden, *J. Soc. Cosmet. Chem. 28*:521 (1977).
14. L. D. Rhein, C. R. Robbins, K. Fernee, and R. Cantore, *J. Soc. Cosmet. Chem. 37*:125 (1986).
15. E. A. Tavss, E. Eigen, and A. M. Kligman, *J. Soc. Cosmet. Chem. 36*:125 (1985).
16. J. C. Blake-Haskins, D. Scala, L. D. Rhein, and C. R. Robbins, *J. Soc. Cosmet. Chem. 37*:199 (1986).
17. A. M. Kligman and W. M. Wooding, *J. Invest. Dermatol. 49*:78 (1967).
18. K. Schrader, *Z. Felt-, lol-Kosmetick-, Pharma-Waschmitte Ind. III*:41 (1985).
19. C. R. Robbins and K. Fernee, *J. Soc. COsmet. Chem. 34*:21 (1983).
20. M. M. Breuer, *J. Soc. Cosmet. Chem. 30*:41 (1979).
21. L. D. Rhein, F. A. Simion, R. L. Hill, R. H. Cagan, J. Mattai, and H. I. Maibach, *Dermatologica 180*:18 (1990).
22. R. L. Hill and L. D. Rhein, *J. Disp. Sci. Tech. 9*:269 (1988).
23. J. G. Dominguez, F. Balaguer, J. L. Parra, and C. M. Pelejero, *Int. J. Cosmet. Sci. 3*:57 (1981).
24. J. A. Faucher and E. D. Goddard, *J. Soc. Cosmet. Chem. 29*: 323 (1978).
25. J. G. Dominguez, J. L. Parra, M. R. Infante, C. M. Pelejero, F. Balaguer, and T. Sautre, *J. Soc. Cosmet. Chem. 28*:165 (1977).
26. D. Howes, *J. Soc. Cosmet. Chem. 26*:47 (1975).
27. C. Prottey and T. Ferguson, *J. Soc. Cosmet. Chem. 26*:29 (1975).
28. J. A. Faucher, E. D. Goddard, and R. D. Kulkarni, *J. Am. Oil Chem. Soc. 56*:776 (1979).

3

The Influence of Electrolyte at Frog Skin Surfaces: Deductions on the Origin of Physiological Electric Circuits

MAX BENDER Department of Chemistry, Fairleigh Dickinson University, Teaneck, New Jersey

I.	Introduction	52
II.	Experimental Data and Discussion	52
	A. Influence of varying Na, K, and Ca uniformly on both sides of the skin	52
	B. Variation of Na concentration on one side of the skin at a time: application of NaCl gradient	53
	C. Effect of lowered pH	56
	D. Reversal of potential	58
	E. Stimulation and depression of the potential: effect of polyvalent cations	58
	F. Anions	67
III.	Indications of Diffuse Double-Layer Effects: Electrokinetics	67
IV.	Electrogenics	69
V.	Communication Between Skin Surfaces and Skin Interior: Signaling	72
VI.	Skin Interior Mechanisms Influencing Electrogenics	73
VII.	Concluding Remarks	74
	References	74

I. INTRODUCTION

The electrical potential generated by living frog skin depends greatly on the composition of the bathing solutions with which the skin is in contact. This dependency is very different at the inside surface of the skin than it is on the outside surface. Highlights of a detailed study [1] are presented and conclusions are drawn by relating the results with biological findings in general and with physicochemical interpretations.

II. EXPERIMENTAL DATA AND DISCUSSION

All experimental bathing solutions were modifications of the following Ringer's solution (I) used as a standard of comparison:

NaCl	111.30 mM dm^{-3}
KCl	3.36 mM dm^{-3}
$CaCl_2 \cdot 2H_2O$	2.04 mM dm^{-3}
$NaHCO_3$	1.19 mM dm^{-3}
Dextrose	5.55 mM dm^{-3}

The ionic strength of this solution is 0.122.

A. Influence of Varying Na, K, and Ca Uniformly on Both Sides of the Skin

Table 1 shows the effect on the potential when the NaCl content is altered uniformly on both sides of the skin. Note for Table 1 that at zero NaCl concentration (the Na from the $NaHCO_3$ being 1.19) the potential is depressed appreciably and that at high NaCl concentrations it was wiped out. When test solutions were replaced by standard Ringer's solution, their depressed potentials were reelevated. This reversibility occurred in the case of lowered Na concentrations, and for the higher ones also provided that the test Na concentration was not too high.

Table 2 displays the effect on the potential of altering the KCl concentration uniformly on both sides. Diminished KCl apparently enhanced the potential, and as with Na, excess electrolyte diminished it. It is interesting that in the absence of K, the potential apparently increased.

In the case of calcium chloride, both bathing solutions being changed in the same way, as shown in Table 3, excess electrolyte had a drastic effect on the potential, similar to Na and K. Decreased calcium and even its elimination resulted in a small decrease in the potential.

TABLE 1 Effect of NaCl Concentration Change in Solution I: Same Solution on Both Sides of Membrane

		E (mV)				
			In test solution			
$M_{NaCl}/10^{-3}$ (mol dm^{-3})	Ionic strength	Solution I pretest	45 min	90 min	Solution I after-test	Time in test solution (min)
0	0.011	60	25	7	50[a]	120
11.0	0.022	44	42	42	48	90
55.6	0.066	68	60	60	68	110
111.3[b]	0.122	95	—	—	95[c]	—
222.3	0.233	90	40	40[d]	87	80
556.5	0.567	44	0	−0.25[e]	[f]	135
1113.0	1.124	87	0	−0.30[e]	[g]	135

[a]Not very stable.
[b]Control.
[c]Steady plateau over 4-h period.
[d]In test solution for 80 min.
[e]In test solution for 110 min.
[f]25% recovery.
[g]No recovery; E became −0.5 mV, then −0.3 mV after 100 min in solution I (final).
Source: From Ref. 1, reprinted with permission of the American Chemical Society.

According to the above, increased ionic strength in general depresses the potential, and this can be drastic. Other experiments described in ref. [1] bear this out. Parallel with Na, there is partial recovery for higher K and Ca concentrations provided that these are not too great.

B. Variation of Na Concentration on One Side of the Skin at a Time: Application of NaCl Gradient

In succeeding expriments the NaCl on the outside of the skin only was replaced by its osmotic equivalent of sucrose and the potential followed with time. Similar measurements were made with only the

TABLE 2 Effect of KCl Concentration Change in Solution I: Same Solution on Both Sides of Membrane

$M_{KCl}/10^{-3}$ (mol dm^{-3})	Ionic strength	E (mV) Solution I pretest	E (mV) In test solution 60 min	E (mV) Solution I after-test	Time in test solution (min)
0	0.119	83	108	88	65
0.34	0.119	65	78	58	65
1.68	0.120	81	93	85	75
3.36[a]	0.122	95	—	95[b]	—
6.72	0.125	40	20	41	90
16.8	0.135	60	45	58	80
33.6	0.152	95	69	80	85
50.0	0.168	28	11	12	185
275.0	0.393	28	5	—	—
320.0	0.438	41	3	—	—

[a]Control.
[b]Steady plateau over 4-h period.
Source: From Ref. 1, reprinted with permission of the American Chemical Society.

inside NaCl solution replaced by sucrose. This enabled comparison of the effect of Na on one side of the skin versus Na on the other. Also, the overall results of the application of a sodium chloride thermodynamic gradient from inside to outside and outside to inside could thus be ascertained. The results are shown in Fig. 1. Note that adequate Na is needed on the inside for maintenance of the potential but not necessarily on the outside (compared with the case in which a standard Ringer's NaCl content was used on both sides). Note also that there was essentially no difference whether a sodium chloride gradient was applied from the inside to the outside or was not applied at all (i.e., where standard Ringer's NaCl was the same on both sides).

A ^{22}Na tracer was used to observe if sodium traveled across the skin between baths [1]. The tracer was added to the inside and

TABLE 3 Effect of $CaCl_2$ Concentration Change in Solution I: Same Solution on Both Sides of Membrane

		E (mV)			
$M_{CaCl_2}/10^{-3}$ (mol dm^{-3})	Ionic strength	Solution I pretest	In test solution 60 min	Solution I after-test	Time in test solution (min)
0	0.116	50	43	—	150
0.204	0.116	110	95	112	90
1.02	0.119	67	66	69	105
2.04[a]	0.122	95	—	95[b]	—
4.08	0.128	76	52	64[c]	85
10.2	0.146	70	4	13	65
20.4	0.177	116	0	70[c]	55
55.0	0.281	46	1	—	—
65.0	0.311	33	21	—	—
75.0	0.341	29	3	—	—
85.0	0. 371	43	17	—	—
90.0	0.386	58	12	—	—
200.0	0.716	50	−11	15	90

[a]Control.
[b]Steady plateau over 4-h period.
[c]90 min.
Source: From Ref. 1, reprinted with permission of the American Chemical Society.

outside solutions as follows: (1) standard Ringer's solution on both sides; (2) standard Ringers's solution on the outside and no NaCl on the inside; and (3) standard Ringer's solution on the inside with no NaCl on the outside. In none of these experiments was measurable ^{22}Na found in the bathing chamber across the skin from the bath in which it had been placed. Applied NaCl thermodynamic gradients, whether from inside to outside or outside to inside, apparently had no bearing on the potential, in contradistinction to the influence of the actual bath ionic composition in its locale

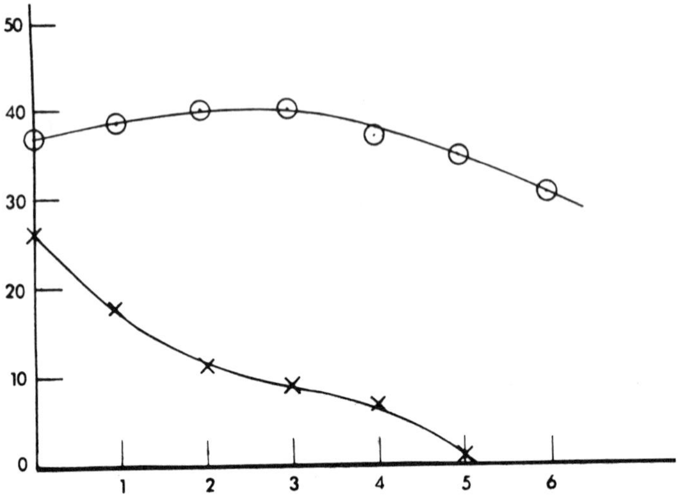

FIG. 1 Frog skin potential under NaCl gradient. Θ, High Na at inside; X, high Na at outside. Abscissa, time (hours); ordinate, potential E (mV). (From Ref. 1, reprinted with permission of the American Chemical Society.)

of inside skin surface or of outside skin surface. Over the duration of these experiments evidence of electric current being carried completely across the skin by ^{22}Na transfer could not be found.

C. Effect of Lowered pH

In addition to a lowering of the potential in an acid medium, measurements of lowered pH (Table 4) show the different effects of H^+ on the inside versus the outside of the skin, and also different degrees of recovery when the test solutions are changed back to standard Ringer's solution. In these tests, skins were first exposed to standard Ringer's solution to achieve a steady-state potential. Then one of the compartment solutions was changed for the same Ringer's solution modified with buffer to the test pH values. There was recovery from the deleterious influence of lowered pH when reverting to standard Ringer's solution provided that the test pH had not been overly lowered, recovery occurring more readily on the outside than on the inside.

TABLE 4 Effect of pH on Membrane Potential (mV)

pH	Compartment	Steady state		0-min test		90-min test		0-min solution I		30-min solution I	
		Test	Control	Test	Control	Test	Control	Test	Control	Test	Control
3.0	Outside	37.4	29.5	23.3	29.5	6.3	27.8	9.9	28.1	27.2	28.2
	Inside	28.8	22.6	19.7	22.8	12.4	21.8	13.1	22.4	13.9	24.4
3.5	Outside	32.1	24.6	35.1	24.8	22.5	25.6	24.0	26.0	37.3	26.2
	Inside	29.3	35.8	28.6	35.4	28.3	33.4	28.9	33.8	30.6	33.5
4.0	Outside	51.7	28.0	52.7	27.8	39.9	27.5	39.3	29.5	38.7	25.3
	Inside	35.6	32.1	35.0	31.8	32.5	25.4	32.5	25.2	33.9	26.5
4.5	Outside	55.5	37.8	54.2	38.0	49.1	35.8	51.3	36.6	50.4	36.6
	Inside	28.2	31.2	28.9	31.3	28.0	30.1	28.4	30.2	25.7	30.6
5.0	Outside	45.3	29.7	43.7	29.9	44.1	29.0	44.9	29.1	37.9	28.6
	Inside	27.7	31.6	27.5	31.4	27.6	30.5	27.2	31.0	28.8	29.2
5.5	Outside	54.9	28.7	56.1	28.9	56.0	27.9	57.3	28.1	45.5	27.8
	Inside	29.1	23.6	29.3	23.6	28.9	22.8	29.8	22.6	30.9	23.0
6.0	Outside	43.0	35.5	42.8	35.1	42.2	35.8	42.3	36.0	37.4	37.1
	Inside	29.8	35.0	30.1	35.0	30.0	31.2	31.0	33.0	31.8	28.5
6.5	Outside	46.1	31.8	46.5	30.0	43.9	26.4	45.1	26.0	43.3	25.0
	Inside	44.9	31.4	44.2	31.5	44.2	32.1	44.5	34.0	42.6	35.1
7.0	Outside	29.7	22.5	30.0	22.6	27.4	20.0	27.7	20.4	26.8	19.4
	Inside	34.9	26.4	35.0	26.4	32.2	22.7	32.3	23.0	31.5	21.6

Source: From Ref. 1, reprinted with permission of the American Chemical Society.

D. Reversal of Potential

Depending on the composition of the compartment solutions, the sign of the electrical potential generated by the frog skin could be reversed. This is evident in Table 1 at high Na^+ values, although these values were very low. There was reversal also with Ca^{2+} at high concentration (Table 3), and the effect was considerably greater than for Na.

Reversal in the sign of the potential occurs when no electrolyte is present (i.e., as with the application of distilled water), the potentials reaching considerable magnitude. (Standard Ringer's potentials are cited as positive based on the positivity of the compartment solution facing the inside surface of the skin.) Results with distilled water are shown in Table 5. The skin was more sensitive to the distilled water if it had not been bathed previously in standard Ringer's solution. Sensitivity to reversal apparently was greater with distilled water in the inside than in the outside compartment, standard Ringer's solution being used for the opposite compartment.

Published results on the sign reversal of frog skin potential includes those of Zadunaisky [2], Martin [3], and Ussing and Windhager [4]. Zadunaisky observed reversal when the sodium was replaced by choline ion in the outside Ringer's solution. Martin obtained a negative value when the outside solution was maintained with choline sulfate Ringer's solution with 2 mM KCl in addition. On treatment of the skin with 10^{-5} M $CuSO_4$ or replacement of the KCl with K_2SO_4, the potential became positive again. In the work of Ussing and Windhager the potential was reversed when Na was replaced by Mg. In their microelectrode impalement experiments measuring the electrical gradient, the direction of the gradient was reversed when Mg was substituted for Na.

E. Stimulation and Depression of the Potential: Effect of Polyvalent Cations

Ussing and Zerahn [5] and Hashida [6] discuss stimulation of frog skin membrane potential by $CuSO_4$. An initial increase of some 18% in potential was found upon application of the nonionic surface-active agent Tergitol 15-s-9 [7,8]. Electrical stimulation has also been found to increase the potential of live frog skin [9–12].

Bender et al. [1] studied the effect of polyvalent cations in general. In Table 6 the result of adding increasing amounts of $AlCl_3$ to standard Ringer's solution (I) on both sides of the membrane is shown. Essentially the potential drops with increased Al concentration. But $LaCl_3$, added to the standard Ringer's solution on both sides, causes a marked increase. This is demonstrated in Fig. 2.

TABLE 5 Effect of Distilled Water on Frog Skin Membrane Potential (mV)

Time (min)	Exptl. 1: steady state in solution I 45 min; then distilled H_2O on both sides	Exptl. 2: direct application of distilled H_2O on both sides	Exptl. 3[a]: solution II[b] on inside distilled H_2O on outside	Expt. 4[a]: solution II[b] on outside distilled H_2O on inside	Exptl. 5: solution II[b] on both sides
0	+35.0	-13.5	-6.5	-15.5	+44.6
15	—	-26.0	-8.5	-16.2	+40.7
30	+31.5	—	-6.5	-16.8	+39.7
45	—	-35.0	-4.0	-18.0	+39.5
60	-10.0	-37.0	—	-17.4	+40.0
90	-13.5	—	—	—	+38.7
105	-16.5	-35.5	—	—	+37.5
120	-19.0	—	—	—	+36.3
145	-20.0	—	—	—	—
150	—	-32.5	—	—	+34.5
200	—	-32.0	—	—	+32.1
210	—	—	—	—	+31.8
225	—	—	—	—	+31.3
240	—	-30.5	—	—	—
255	—	—	—	—	+30.9
285	—	-24.5	—	—	+30.7
315	—	—	—	—	+30.5
330	—	-23.5	—	—	+31.2
360	—	-20.0	-14.5	-22.0	+31.8
390	—	-19.5	—	—	+32.8
420	-	-19.0	—	—	+34.0
450	—	-19.0	—	—	+35.2
480	—	-16.5	—	—	+36.1
525	—	-15.5	—	—	+37.6
575	—	-13.5	—	—	—
590	—	-12.0	—	—	—
600	—	—	—	—	+39.3

[a]Direct application.
[b]Solution II: 56.74 mM NaCl, 0.33 mM KCl, 1.02 mM $CaCl_2$, 1.00 mM $NaHCO_3$; ionic strength = 0.061.
Source: From Ref. 1, reprinted with permission of the American Chemical Society.

TABLE 6 Effect of Including Aluminum in Solution I

		E (mV)				
$M_{AlCl_3}/10^{-3}$ (mol dm^{-3})	Ionic strength	Solution I beginning	Solution I just before test	Beginning of test solution	60 min of test solution	150 min of test solution
0.5	0.125	48.0	32.0	25.5	18.0	21.0[a]
1.0	0.128	42.0	27.0	18.0	10.0	7.0
9.0	0.176	23.5	14.0	4.0	1.5	2.0
13.0	0.200	42.0	27.5	18.5	7.5	3.0
19.0	0.236	59.0	38.5	19.0	7.5	4.0[b]

[a]29.0 mV after 310 min in test solution.
[b]1.0 mV after 260 min in test solution.
Source: From Ref. 1, reprinted with permission of the American Chemical Society.

Tables 7 and 8 give the results for a wide variety of polyvalent cations. Standard Ringer's solution appropriately buffered was used on both sides of the skin. Table 7 shows the results when polyvalent cation was introduced into the outside solution only, up to a concentration of 1 mM. In Table 8 this cation concentration was introduced into the inside solution only.

It can be seen in Table 7 that a considerable variety of polyvalent cations bring about a surge in frog skin potential. Whereas Al acting as the chloride under the experimental conditions portrayed in Table 6 did not bring about this surge, it does when acting as the nitrate under the conditions of Table 7. Comparison with Table 8 indicates that the surge effect happens only with applications of polyvalent cation to the outside surface of the skin. Quite a few of the cations that produce a surge are transition elements.

Table 9 compares the cations that brought about elevation in the potential. The changes are given here as a percentage of the original steady-state values of the potential utilizing standard Ringer's solution on both sides. There is more than a 30% increase with Co, Cu, Cd, Pb, Ce, and Al. Excluding Al, the potential for these cations remained above the original Ringer's steady-state values after 90 min of testing. The effects of Co, Cu, Pb, and Ce remained even after the test solution was replaced with standard Ringer's

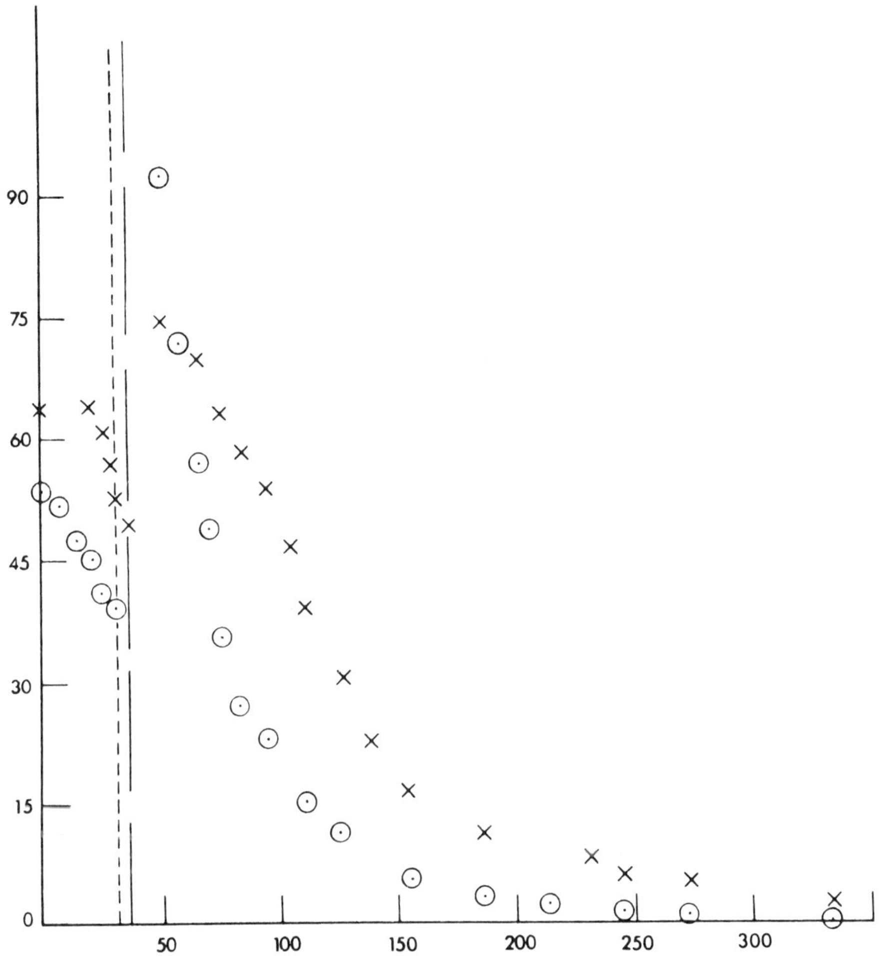

FIG. 2 Influence of La on frog skin membrane potential. X, 20 mM La; ⊙, 35 mM La. Abscissa, time (min); ordinate, potential E (mV). — — —, Solution change to test for 20 mM La; - - -, solution change to test for 35 mM La. (From Ref. 1, reprinted with permission of the American Chemical Society.)

solution. But with Al, although the surge did not last and the potential dropped to below the original Ringer's steady-state values, application of Ringer's after testing not only brought about recovery but produced a steady-state potential greater than that before testing.

TABLE 7 Effect of Polyvalent Cations in Outside Solution on Frog Skin Membrane Potential

Cation salt	E (mV)						Control[a] (% decrease)
	Solution pH	Steady state	Test 0 min	Test 90 min	Solution I 0 min	Solution I 30 min	
$BeSO_4 \cdot H_2O$	5.0	36.6	33.1	21.4	28.8	50.3	9.1
$MgSO_4$	7.5	42.0	41.8	40.9	44.3	42.5	9.6
$Sr(NO_3)_2$	7.5	35.2	35.3	32.7	32.9	32.1	8.1
$BaCl_2 \cdot 2H_2O$	7.5	29.3	29.5	27.3	27.6	26.1	20.0
$ScCl_3 \cdot 6H_2O$	5.0	45.3	45.1	29.1	36.6	37.7	19.8
$YCl_3 \cdot 6H_2O$	6.0	49.2	59.3	46.7	47.3	47.3	1.1
$La(NO_3)_3 \cdot 6H_2O$	6.5	44.8	55.2	43.9	46.3	46.2	6.6
$Ce(NO_3)_3 \cdot 6H_2O$	6.0	43.5	63.0	56.0	55.1	57.5	8.9
$Cr(NO_3)_3 \cdot 9H_2O$	5.0	40.8	48.5	21.5	35.9	40.4	-20.9

$MnCl_2 \cdot 4H_2O$	6.5	35.2	41.4	39.4	37.9	34.7	8.6
$FeSO_4 \cdot 7H_2O$	5.0	36.6	38.6	33.2	35.4	33.4	3.9
$CoCl_2 \cdot 6H_2O$	6.5	38.8	55.2	53.8	52.3	43.9	1.1
$NiCl_2 \cdot 6H_2O$	7.5	64.1	78.8	76.3	72.7	67.4	0.0
$CuSO_4 \cdot 5H_2O$	7.5	49.3	83.1	68.1	68.4	69.6	2.7
$Zn(NO_3)_2 \cdot 6H_2O$	6.5	57.2	71.3	65.0	69.4	62.3	-1.8
$Cd(NO_3)_2 \cdot 4H_2O$	7.5	30.5	45.8	43.2	37.3	26.7	6.7
$HgCl_2$	7.5	32.8	5.1	2.1	2.0	1.6	1.4
$Al(NO_3)_3 \cdot 9H_2O$	5.0	30.7	40.3	24.9	32.3	38.8	10.3
$Pb(C_2H_3O_2)_2 \cdot 3H_2O$	5.5	23.1	30.7	25.5	25.5	25.8	14.1
$Th(NO_3)_4 \cdot 4H_2O$	5.0	71.9	73.2	50.3	54.5	70.4	-0.3
$UO_2(C_2H_3O_2)_2 \cdot 2H_2O$	5.0	44.5	18.1	12.1	16.9	23.5	1.8

[a]Percentage decrease in E 120 min after steady state.
Source: From Ref. 1, reprinted with permission of the American Chemical Society.

TABLE 8 Effect of Polyvalent Cations in Inside Solution on Frog Skin Membrane Potential

	E (mV)						
Cation salt	Solution pH	Steady state	Test 0 min	Test 90 min	Solution I 0 min	Solution I 30 min	Control[a] (% decrease)
$BeSO_4 \cdot H_2O$	5.0	29.7	29.2	25.8	27.2	27.3	4.7
$MgSO_4$	7.5	32.0	31.8	30.7	30.4	30.6	11.0
$Sr(NO_3)_2$	7.5	31.5	31.7	29.4	29.4	28.7	6.5
$BaCl_2 \cdot 2H_2O$	7.5	27.8	28.2	25.7	26.5	26.1	11.3
$ScCl_3 \cdot 6H_2O$	5.0	30.8	30.7	30.7	31.3	31.3	3.4
$YCl_3 \cdot 6H_2O$	6.0	30.7	30.7	30.2	30.5	32.7	-3.8
$La(NO_3)_3 \cdot 6H_2O$	6.5	48.3	49.0	46.5	42.8	45.9	8.8
$Ce(NO_3)_3 \cdot 6H_2O$	6.0	24.6	25.2	23.6	25.9	25.8	9.8
$Cr(NO_3)_3 \cdot 9H_2O$	5.0	25.5	27.7	23.7	25.9	25.8	-14.7

$MnCl_2 \cdot 4H_2O$	6.5	30.0	30.3	30.1	30.9	28.9	5.2
$FeSO_4 \cdot 7H_2O$	5.0	34.7	34.3	32.0	32.6	31.3	0.8
$CoCl_2 \cdot 6H_2O$	6.5	28.4	31.8	28.9	31.6	27.1	6.6
$NiCl_2 \cdot 6H_2O$	7.5	39.3	39.4	35.9	36.0	36.5	3.0
$CuSO_4 \cdot 5H_2O$	7.5	63.0	63.2	62.0	60.4	58.0	2.4
$Zn(NO_3)_2 \cdot 6H_2O$	6.5	42.6	43.2	41.6	42.1	46.4	0.7
$Cd(NO_3)_2 \cdot 4H_2O$	7.5	28.9	29.6	30.3	31.5	29.0	7.7
$HgCl_2$	7.5	37.7	37.0	8.1	7.9	3.8	6.9
$Al(NO_3)_3 \cdot 9H_2O$	5.0	27.5	27.5	26.0	26.0	28.2	10.7
$Pb(C_2H_3O_2)_2 \cdot 3H_2O$	5.5	35.0	34.5	31.1	32.0	32.8	14.7
$Th(NO_3)_4 \cdot 4H_2O$	5.0	42.6	43.0	39.6	42.0	45.6	5.0
$UO_2(C_2H_3O_2)_2 \cdot 2H_2O$	5.0	32.9	31.9	23.4	30.1	31.9	11.3

[a]Percentage decrease in E 120 min after steady state.

Source: From Ref. 1, reprinted with permission of the American Chemical Society.

TABLE 9 Comparison of Cations that Stimulate Frog Skin Potential[a]

Cation	Spike 0 min	90-min test	30-min in solution I after test	Control (over period of experiment)
Mn	17.7	11.9	-1.4	-8.6
Co	42.3	38.6	13.2	-1.1
Ni	22.9	19.0	5.1	0.0
Cu	68.6	38.1	41.2	-2.7
Zn	24.6	13.6	8.9	1.8
Cd	50.0	41.6	-12.4	-6.7
Pb	33.0	10.4	11.7	-14.1
Fe	5.5	-9.3	-8.7	-3.9
Ce	44.8	28.7	32.2	-8.9
La	23.2	-2.0	3.1	-6.6
Cr	18.8	-47.2	-1.0	20.9
Al	31.3	-18.9	26.3	-10.3
Y	20.5	-5.1	-3.9	-1.1
Th	1.8	-30.0	-2.1	0.3

[a]Percent change in steady-state E (+ = increase; - = decrease).
Source: From Ref. 1, reprinted with permission of the American Chemical Society.

Cd influenced the skin such that after the surge there was a decrease in the potential following 30 min in Ringer's post-test. With surges of less than 30% (except for Cr and Th), a progressive decrease took place toward the steady-state values upon exposure to Ringer's and remained after testing. In the Cr and Th experiments, there was an appreciable drop in the potential after testing, but recovery occurred in 30 min.

The addition of Mg, Sr, and Ba, all elements of the group IIA of the periodic table, brought about little change. This was found (for Ca values, see Table 3). The toxic elements Be, Hg, and UO_2 [13–15] depress the potential as applied from the outside solution. Be and UO_2 produced little effect on the inside, while Hg, the most poisonous, was deleterious, although not as seriously as when applied to the outside. In terms of recovery, there was none with Hg; good recovery of Be membranes on application of Ringer's, even surpassing the original steady state; and partial recovery with UO_2. Sc falls between the depressants and the category of elements that affected little change in potential. There was some degree of poisoning due to the Sc, although the membrane did not show this at first. Some recovery occurred when Ringer's solution was applied.

F. Anions

From the data presented above some inferences can be drawn regarding the role of anions in influencing the electrical potential generated by live frog skin. For instance, the difference observed in the behavior of Al as described in the experimental results listed in Table 6 contrasted with those in Tables 7 and 8 could perhaps be that of the Cl^- compared to NO_3^-. But under the circumstances of the inside solution experimentation (Table 8), where no surges were observed, the special attributes of nitrate, sulfate, and acetate are not apparent.

Cl^- being the anion for which, as described above, an NaCl thermodynamic gradient was applied, it may be concluded that a Cl^- gradient in one direction or the other across the skin was essentially of no consequence. These experiments also indicate that reduction of Cl^- concentration on the outside from that of standard Ringer's to a much lower value has little effect. Considering still other experiments outlined in ref. 1, different concentrations of Cl^- in the inside bathing solution had little, if any, effect on the measured potential.

III. INDICATIONS OF DIFFUSE DOUBLE-LAYER EFFECTS: ELECTROKINETICS

Basically the very strong influence of both species and concentration of the bathing solution ions acting through an interface (physiological in these experiments) as described above suggests that Gouy–Chapman electrical diffuse double-layer effects are important in understanding the phenomena observed. This assumption is supported by considerable evidence in the literature. Tables 1, 3, and 5 show that there is an optimum range of ionic strength in which the potential is positive. The fact that negativity is achieved at either ionic strength extreme indicates that isoelectric conditions (i.e., zero potential at the change of sign points, presuming a plot of the data) are attainable at both high and low ionic strengths.

High sensitivity to solution concentration and the sign of charge reversal at low concentration are aspects of diffuse double-layer phenomena [16]. The reversal of sign of the potential in distilled water, considering the drastic change in bathing solution ionic strength and the absence of key (physiological ions), could have its basis in a radical change in the sorption of potential-determining ions.

At the higher ionic strengths Ca^{2+} brings about reversal (Table 3) much more effectively than does Na^+ (Table 1), as the Schulze–Hardy principle predicts [16]. Regarding this contrast of results using divalent rather than monovalent cations, it should be mentioned that in substituting $MgSo_4$ for Na_2SO_4 on the outside (Na_2SO_4 Ringer's solution being used on the inside), Ussing and Windhager [4] both lowered and reversed the potential.

Diffuse double-layer theory affords an explanation of the decreased potential that occurs with increased ionic strength as being caused by the swamping of the net charge of the double layer. As to the reversibility of recovery of the electrical potential upon removal of given insults, there is a parallel with flocculation–deflocculation diffuse double-layer phenomena studied in ref. 16. It was shown in this work that deflocculation of ZnS suspensions was easier the less drastic the treatment in bringing about their flocculation.

The lowering of the potential under increased acidic conditions (Table 4) may be due at least partially to H^+ acting as a potential-determining ion [17,18]. The special effects observed for polyvalent cations and the results obtained with cationic, anionic, and nonionic (surface-active) agents discussed in refs. 7 and 8 should have some foundation in electrical double-layer activity.

Diffuse double-layer theory helps explain Ferreira's observations that the short-circuit current is progressively different in the presence of SO_4^{2-} versus Cl^- versus Br^- versus I^- [19]. Mullins [20] found an appreciable difference tween Cl^- and I^- physiological transport despite their equivalent conductances in water being essentially the same. In the work of Curran et al. [21] it is observed that Ca decreases active Na transport, and hormone increases it, through their effects on the permeability of the outward-facing membrane of the cells, where the potential is generated. Reference is also made to Cereijido et al. [22], where measurements of Na permeability at different outside concentrations did not conform with the rates expected passively.

Lindley and Hoshiko [23] arranged different cations and anions in an order according to how well they replaced Na, K, and Cl for maintenance of a given level of frog skin electrical potential. The sequence of these ions was found to be parallel to the Hofmeister (lyotropic) series. An anionic "Hofmeister" series was shown by Singer and Civan [24] for toad urinary bladders in terms of transepithelial potential, short-circuit current, and mucosal-to-serosal flux, and Kidder et al. [25] observed a selectivity arrangement of Li, Na, and K.

Influx–outflux shunt experiments carried out by Mandel and Curran [26,27] might be explained through electrocapillarity [28]. The reduced but different active transport (permeability) rates in Ouabain-poisoned skin of Cl, urea, K, and mannitol were seen to be not only dependent on the applied electrical potential but also on the external salts concentration as well. There was a degree of reversibility in the measurements. Large changes were observed upon depolarizing the potential, the outside ionic concentrations directly influencing the depolarization potential. In other experiments, Mandel and Curran [29] found the degree of Na active current to be independent of high outside Na concentrations at increased applied potential, indicating electrocapillarity swamping effects.

Electrocapillarity effects are also indicated in Candia's work [12], where applied electrical potential was seen to influence the "passive unidirectional" Na outflux in frog skin. He found that the resistance of the frog skin changed with the potential applied. Alternating current applied across the skin brought about an increase in the potential generated by the skin. Candia's explanation included a consideration of the presence of a double lattice of fixed charges in which Na was mobile, the lattice acting as a rectifying structure. Experiments described in ref. 1 show a decrease in the resistivity of frog skin with increased applied voltage. Ions are held in the diffuse double layer with different degrees of tenacity [17,18]. Thus, the more strongly held ions will begin to move through "channels" when the voltage becomes sufficiently great. In their experiments in which the potential was reversed by replacing Na with Mg, Ussing and Windhager [4] found that the electrical resistance of the frog skin increased. This could possibly be the result of the influence of the divalent Mg in appropriate double-layer areas involved with the electrical conductance.

IV. ELECTROGENICS

The conclusions of the experimental work described above [1] are not in keeping with the description of Na transport required for the Koefoed-Johnson and Ussing model and its modifications [30]. Overall thermodynamic concentration gradient explanations of the electricity generated in frog skin appear inadequate.

Besides the evidence from direct application of Na and Cl gradients, it is significant that the lower curve of Fig. 1 for an applied gradient is a fair match for the data when there is no concentration gradient (1 mM NaCl was on both sides of the membrane). At the same time the upper curve for applied gradient is comparable to the data obtained when standard Ringer's solution is present on both sides. Cross-membrane gradients of H^+, of the various polyvalent cations considered, and of large cations and anions [1] showed little, if any, effect on frog skin electrical potential magnitudes.

Important features are the ionic species and their concentrations. Their influence is markedly different on one side of the skin than on the other. These surface interactions are essentially independent of each other. The lack of evidence of net transfer of material amounts of ions across the skin between bathing solutions according to transport and sorption experiments [1] is as expected.

Mandel and Curran [29] report that when they applied a steep Na gradient from the inside bathing solution to the outside solution against inward Na transport (short-circuit current) they actually found that this short-circuit current increased in the range of 70%. Candia [12] observed much of the Na outflux not to be a function of electrochemical gradient.

In different papers [31–34] it is shown that the epithelial membrane keeps an electrochemical balance without Na in transport. In working with live frogs in freshly kept deionized water, McAfee [35] observed that they did not lose Na to the water. The Na gradient from the concentration level in the frogs to zero in the water was ineffective for any passive transport of the Na. Active (physiologically alive "pumped") Na transport does occur as illustrated by Krogh [36], his frogs taking up NaCl from the surrounding medium, with concentrations as low as 10^{-5} M.

When Rotunno et al. [37] and Cereijido and Rotunno [31] bathed the outside of the skin in low-Na solution, the Na in the skin was retained despite the concentration gradient. The separate results of research by Cereijido et al. [22], Klahr and Bricker [38], Ussing [39], and Bricker et al. [40] pointed to active Na transport.

In various experiments [4,41–45] the electrical potential of frog skin was measured with microelectrodes, these electrodes being inserted to different depths of penetration. While the electrical potential profiles obtained should be interpreted carefully considering the electrode–electrolyte (KCl) impact at the electrode–skin interface in progressive positions of the electrode, results indicated that centers active in generation of the overall potential exist practically everywhere in the whole skin. Figure 3 is an example of such measurements on two different skins.

These jagged potential profiles, with their areas of negativity, potential depression, plateaus, and peaks, are not readily interpretable from overall thermodynamic gradients. Understandably, propagated, oriented, electrically charged particles in skin that is still alive after having been excised should account for at least part, if not all, of the measured potential. Metabolism is occurring and physiological matter, especially protein, is changing in chemistry,

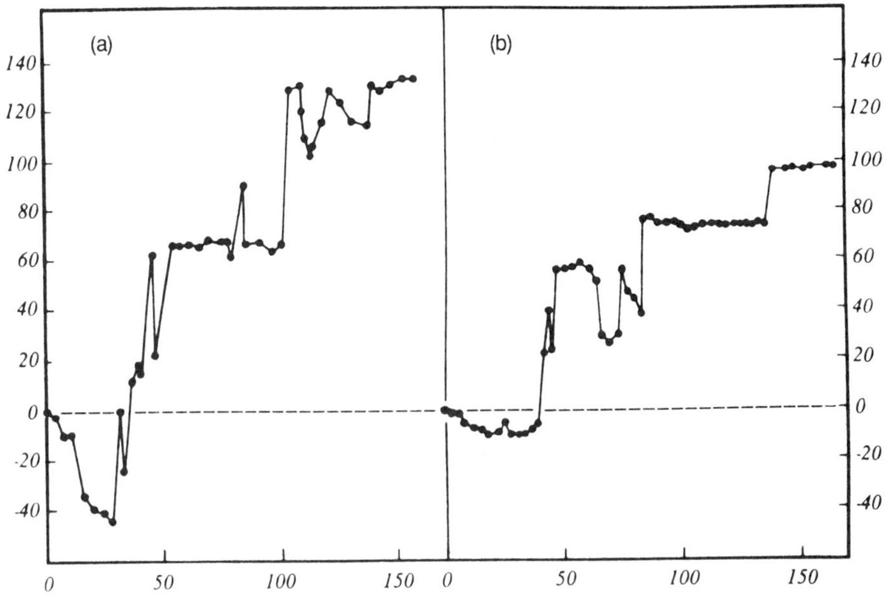

FIG. 3 Frog skin potential profiles. Abscissa, distance from outside surface (meters $\times 10^{-6}$); ordinate, potential (volts $\times 10^{-3}$). (From ref. 43, reprinted with permission of Blackwell Scientific Publications Ltd.)

form, and location. There should be life in the frog skin network, with axoplasmic organelles and other particles in transport, all such particles and molecules being electrically charged (see Chapter 15).

An outline of living processes of this nature in human skin is given by Friberg et al. in Chapter 1. The translocation manifestations of presumably electrically charged particles include the arrival of material from the inner part of the body to the hypodermis, dermis, and epidermis. Progressive changes in cell shape and structure take place, causing complete cells in the stratum basale to become lifeless protein aggregates in the stratum corneum. Fibrous and structural proteins, including actin and tubulin (with its microtubules), are synthesized, and filaments build up during travel to skin locations in line with their necessary functions.

The description of the mitotic spindle by McIntosh and McDonald [46] illustrates more "organized dynamics," which should be considered as taking place in the live skin. This is also interpretable as a source of electric potential. In mitosis the functional orderly procedure of directed motion leaves little to chaotic chance. DNA threads condense after replication to chromosomes that move to the spindle equator and splitting of the cell is brought about with microtubule fibers forming, growing, and organizing in the manner essential to the operation. The newly formed cells translocate.

In the face of all this active generation of electric current and potential by the charged moieties in motion, a thought or two on the overall electrical circuitry is in order. Charged particles should be moving in all directions depending on their function in the skin, but there is a net resultant flow. In Fig. 3, for instance, the downside of peaks could indicate a resultant electrogenic flow in the direction opposite to that overall. The measuring instrument receives some of this electrical flow, while the remainder of the flow shunts back through the skin, depending on the skin's (inanimate) conductivity as based on electrons, protons, and relatively unbound ("unharnessed") charged particles, including ions. Considering the extreme difference in constitution and in function of the inside surface of the skin versus the outside surface (Chapter 1), the observed dissimilarity in influence of a given bathing solution on the two surfaces is understandable.

V. COMMUNICATION BETWEEN SKIN SURFACES AND SKIN INTERIOR: SIGNALING

In light of the fact that only skin surface exposure to bathing solution brings about highly significant effects [1]—where the electrical potential is changed with Na^+, K^+, Ca^{2+}, and H^+ concentrations; drops severely due to the poisonous action of Be, Sc, Hg(II), and UO_2; surges due to the introduction of La or other substances; reverses in sign; and shows recovery when "insults" are removed—there must be rapid communication with the skin's interior electrogenerating centers. Presumably, the network of nerves in the (now excised) skin must play a part. In Chapter 15 attention is given to signaling and rapid message relay, utilizing examples from the multitude of papers in the literature.

Significantly, in this vein of signaling, the 1989 Nobel prize in medicine was awarded to the virologists J. M. Bishop and H. E. Varmus for insight into the complicated signal systems governing cell growth. Signaling takes place in mitosis as described above. In experiments on lymphokine messenger RNA stability, Lindsten et al. [47] have demonstrated that simuli with mitogen or antigen at a T-cell surface alter gene expression by inducing specific changes in the RNA degradation. When Regehr et al. [48] used electrical activation on the synapses of hippocampal pyramidyl nerve cells, they found Ca accumulating in the cells. In their use of the fluorescent calcium indicator fluo-3, Cornell-Bell et al. [49] show how hippocampal astrocytes are promptly induced by glutamate to develop waves of increased free cytoplasmic calcium.

Thus, according to the nature of the bathing solution components and depending on their concentrations, different kinds of messages presumably are forwarded. They vary in intensity, penetrating the skin to different degrees in different pathways. Under such circumstances the extreme effects of poisons and the recovery of electrical potential by replacement of insults in the bathing solution are communicated.

VI. SKIN INTERIOR MECHANISMS INFLUENCING ELECTROGENICS

There are several possible ways in which the contributions by electrogenic regions to the overall potential may become altered as a result of being signaled. First, the rate of locomotion of particles may be affected. Their direction can be influenced. Then there is the magnitude and sign of charge of the electrical diffuse double layer associated with the moving particles and also of the medium through or along which they travel.

The electrical charge on the medium can be sufficiently great to pose as a barrier that restricts travel or brings about translocation in an opposite direction. With passive ionic diffusion experiments in synthetic membranes [17,18,50,51], the results indicated that electrolyte conditions could cause barriers to diffusion flow, in the form of severely increased membrane charge, the highly interactive double layers impeding passage through them of dissolved electrolyte. Under live conditions of active flow-propagated charged particles, the medium through which they travel should be opposite in sign of charge, the isoelectric point being unlikely. When conditions change such that charges on moving particles and on the medium through which they travel become the same, there is restriction and even channel closing. Reversal of travel is a possibility, especially if there is an insult at the bathing solution interface that provokes living (intelligent) particles to travel away rather than toward (see Chapter 15). Changes in magnitude and sign of charge would follow from changes in molecular conformation, degree of ionization, and extent and nature of (ionic) preferential sorption.

Also for consideration are initiated ionic fluxes. For instance, K^+ flow is brought on by protein kinase C enzyme translocation, the result of neural firing [52]. The Ca accumulations mentioned above as researched by Regehr et al. [48] and Cornell-Bell et al. [49] are additional examples. Also, there is the phosphorylation according to Alkon [52] of protein due to CAM (calcium-activated) kinase II activated at postsynaptic sites. Such changes in ionic environment and protein constitution should influence interfacial charges. The kind of binding and degree of binding of the ions are altered as they shift locale and they change in concentration. Their (ionic) activity coefficients change. Phosphorylation presumably alters the charge situation of the protein. Thus the observed surges described above due to La and the other multivalents introduced in the bathing solutions should be understandable in this light.

VII. CONCLUDING REMARKS

In closing, a view is presented that points to the origin of electrical circuits in a given complete physiological system. The electrogenics are the result of live active propulsion of electrically charged particles. They may be self-propelled and/or the presumably oppositely charged pathway (i.e., the microtubules along which they travel) propels them. Circuits may be completed via the flow of passive charged particles, including ions, protons, and electrons.

Upon disruption of a segment of the physiological system as in the excision of frog skin, that segment remains electrogenic, the circuitry now including the bathing solution and the instrumentation that measures potential, current, and other factors. In the electrical potential of the circuitry being influenced by given bathing solution ions and their concentrations, and allowing for appropriate signaling to the skin interior, it is concluded that the charges on active translocating particles and on their (active) pathways throughout the skin are in the nature of electrical diffuse double layers. Thus the potentials generated are electrokinetic in nature: the fall potential when particles are propelled through given medium, and the streaming potential when liquid is propelled through the medium. Bathing solution changes thus affect the electrokinetics and thereby the potential generated. It being demonstrated that signals can bring about interior changes in concentration and distribution of given ions [i.e., Ca (Regehr et al. [48] and Cornell-Bell et al. [49])], the tremendous impact of the bathing solution changes on the electrical potential is understandable.

REFERENCES

1. M. Bender, K. Hillman, L. Tirri, M. Graben, D. Fenster, and M. Rosensaft. The influence of the composition of the bathing solutions on frog skin membrane potential, in *Bioelectrochemistry: Ions, Surfaces, Membranes*, Advances in Chemistry

Series 188 (M. Blank, ed.), American Chemical Society, Washington, D.C., 1980, pp. 411–443.
2. J. A. Zadunaisky, *Am. J. Physiol.* *207*(5):1010 (1964).
3. D. W. Martin, *J. Cell. Comp. Physiol.* *63*:245 (1964).
4. H. H. Ussing and E. E. Windhager, *Acta Physiol. Scand.* *61*:484 (1964).
5. H. H. Ussing and K. Zerahn, *Acta Physiol. Scand.* *23*:110 (1951).
6. K. Hashida, *J. Biochem.* (*Tokyo*) *1*:21 (1922).
7. G. K. Rennie, J. C. Hill, E. D. Goddard, and R. K. Kulkarni, *J. Pharm. Pharmacol.* *27*:363 (1975).
8. R. D. Kulkarni and E. D. Goddard, Abstract E9, *49th National Colloid Symposium*, June 1975.
9. T. C. Barnes, *Am. J. Physiol.* *130*:557 (1940).
10. A. Finkelstein, *J. Gen. Physiol.* *47*(3):545 (1964).
11. A. Finkelstein, *J. Gen. Physiol.* *43*(3):545 (1964).
12. O. A. Candia, Abstracts, *11th Annual Meeting of the Biophysical Society*, 1967, 52, TB10, p. 52.
13. J. Schubert, *Sci. Am.* *199*:27 (1958).
14. C. D. Harrington, *Uranium Production Technology*, Van Nostrand, New York, 1959.
15. *Merck Index*, 8th ed., Merck & Co., Rahway, N.J., 1968.
16. M. Bender, *J. Phys. Chem.* *57*:466 (1953).
17. M. Bender, J. K. Moon, J. Stine, A. Fried, R. Klein, and R. Bonjouklian, *J. Chem. Soc. Faraday Trans. 1* *71*:491 (1975).
18. M. Bender, B. Khazai, and T. E. Dougherty, *J. Colloid Interface Sci.* *63*(2):346 (1978).
19. K. T. G. Ferreira, *Biochim. Biophys. Acta* *150*:587 (1968).
20. L. J. Mullins, *Am. J. Physiol.* *194*:369 (1958).
21. P. F. Curran, F. C. Herrera, and W. J. Flanigan, *J. Gen. Physiol.* *46*:1011 (1963).
22. M. Cereijido, F. C. Herrera, W. J. Flanigan, and P. F. Curran, *J. Gen. Physiol.* *47*:879 (1964).
23. B. D. Lindley and T. Hoshiko, *J. Gen. Physiol.* *47*:749 (1964).
24. I. Singer and M. M. Civan, *Am. J. Physiol.* *221*:1019 (1971).
25. G. W. Kidder, M. Cereijido, and P. F. Curran, *Am. J. Physiol.* *207*:935 (1964).
26. L. J. Mandel and P. F. Curran, *J. Gen. Physiol.* *59*:503 (1972).
27. L. J. Mandel and P. F. Curran, *Biochim. Biophys. Acta* *282*:258 (1972).
28. A. W. Adamson, *Physical Chemistry of Surfaces*, 4th ed., Wiley-Interscience, New York, 1982.
29. L. J. Mandel and P. F. Curran, *J. Gen. Physiol.*, *62*(1):1 (1973).

30. V. Koefoed-Johnson and H. H. Ussing, *Acta Physiol. Scand.* *42*:298 (1958).
31. M. Cereijido and C. A. Rotunno, *J. Gen. Physiol.* *51*:2805 (1968).
32. P. F. Curran and M. Cereijido, *J. Gen. Physiol.* *48*:1011 (1965).
33. E. G. Huf, N. S. Doss, and J. P. Wills, *J. Gen. Physiol.* *41*:397 (1957).
34. N. Levinsky and W. Sawyer, *J. Gen. Physiol.* *36*:607 (1953).
35. R. D. McAfee, *Science* *178*:183 (1972).
36. A. Krogh, *Skand. Arch. Physiol.* *76*:60 (1937).
37. C. A. Rotunno, M. I. Pouchan, and M. Cereijido, *Nature* *210*: 597 (1966).
38. S. Klahr and N. S. Bricker, *J. Clin. Invest.* *43*(5):922 (1964).
39. H. H. Ussing, *Acta Physiol. Scand.* *17*:1 (1949).
40. N. S. Bricker, T. U. L. Biber, and H. H. Ussing, *J. Clin. Invest.* *42*:88 (1963).
41. M. Cereijido and P. F. Curran, *J. Gen. Physiol.* *48*(4):543 (1965).
42. H. H. Ussing, *Ann. N.Y. Acad. Sci.* *137*(2):543 (1966).
43. L. Engbaek and T. Hoshiko, *Acta Physiol. Scand.* *39*:348 (1957).
44. H. H. Ussing, Transport of electrolytes and water across epithelia, in *The Harvey Lectures*, Ser. 59, Academic Press, New York, 1965.
45. F. M. Snell and T. K. Showdhury, Intracellular transport, in *Symposium of the International Society for Cell Biology*, Vol. 5 (K. B. Warren, ed.), Academic Press, New York, 1966, p. 141.
46. J. R. McIntosh and K. L. McDonald, *Sci. Am.* Oct 1989, p. 48.
47. T. Lindsten, C. H. June, J. A. Ledbetter, G. Stella, and C. B. Thompson, *Science* *244*:339 (1989),
48. W. A. Regehr, J. A. Connor, and D. W. Tank, *Nature* *341*: 533 (1989).
49. A. H. Cornell-Bell, S. M. Finkbeiner, M. S. Cooper, and S. J. Smith, *Science* *247*:470 (1990).
50. M. Bender, Water and ions in bological systems, in *Proc. 4th International Conference*, Bucharest, Romania, May 24–28, 1987. Birkhäuser, Stuttgart, West Germany, 1987, p. 415.
51. M. Bender, *Interfacial Phenomena in Biotechnology and Materials Processing*, Elsevier, Amsterdam, 1988, p. 105.
52. D. L. Alkon, *Sci. Am.* *261*:42 (1989).

4

Interfacial Phenomena with Reference to the Oral Cavity

PER-OLOF J. GLANTZ and ROLF W. ATTSTRÖM Faculty of Odontology, Lund University, Malmö, Sweden

ANNE E. MEYER Center for Biosurfaces, State University of New York, Buffalo, New York

ROBERT E. BAIER Faculty of Health Sciences, State University of New York, Buffalo, New York

I. Introduction 77
II. Surface Properties of Enamel and Dentine 78
III. Adhesion in Restorative Dentistry 79
IV. Oral Hygiene Aspects 79
V. Oral Film Formation 80
VI. Oral Fouling 85
VII. Surface and Colloid Chemical Aspects on Saliva 91
References 102

I. INTRODUCTION

In few parts of the body do adhesive interactions play such important roles in the causation of pathological processes and in the restoration of damage produced by such processes as in the oral cavity. For instance, dental plaque is attached to the surfaces of teeth by forces of adhesion with subsequent growth, in the plaque, of harboring microorganisms, some of which will sooner or later attack the tooth surfaces and/or the periodontium.

Modern restorative dental treatment also utilizes the adhesive properties of restorative materials and of tooth substances and saliva. Thus the fundamental reason why a modern composite restoration remains retained in its cavity is because the total forces of adhesion between the restoration and the tooth surface is greater than the combined forces striving to remove the restoration. Furthermore, the retention of a complete maxillary denture to the underlying mucosa at least partially depends on the strength of the adhesion between the denture base, the saliva, and the denture-bearing mucosa.

II. SURFACE PROPERTIES OF ENAMEL AND DENTINE

Surface chemical studies such as contact angle determinations on extracted teeth show that plane-ground but otherwise untreated human enamel and dentine surfaces are covered with low-energy materials of similar wettability [1,2]. Thus in their basic surface chemical behavior there are only minute differences between these two types of tooth materials, creating a good example of the fact that in a heterogeneous system such as a tooth, the composition of the surface does not necessarily reflect the composition of the bulk. Certain differences can be observed between tooth contact angles for different test liquids, indicating that the surface free energy of both enamel and dentine is also built up by polar forces and hydrogen bonds.

The presence on teeth of thin organic surface films with relatively low adhesiveness probably explains why it is so comparatively difficult to cause things to adhere to natural tooth surfaces (i.e., to teeth that have not been etched or otherwise primed). It probably also explains why until recently the only clinically useful method of retaining fillings, crowns, and coatings on teeth has been to obtain mechanical retention of the restorative material in micro- or macroscopic undercuts in the tooth itself or on its surface [3,4].

The effects of acids on the adhesiveness of enamel and dentine are therefore of great interest. In studies of the influence of various acid solutions on the wettability of enamel, it can be shown that acid solutions have effects on the low-energy surfaces of tooth surfaces [2,5–7]. Sometimes they even seem to cause exposure of the mineral phase, onto which, as already mentioned, comparatively much stronger adhesive contacts and thus better retention can be obtained. (For reviews of the mechanisms involved, see refs. 8, 9, and 10.) This acid influence evidently varies in intensity and importance, depending on the nature and concentration of the acid solutions as well as on which tooth structure is treated [1,7,11].

A human enamel surface becomes hydrophilic after rather short exposures even to weaker acids such as lactic acid, whereas on human dentine, with its higher organic material content, it takes stronger acid solutions and much longer exposure times before true hydrophilic surfaces can be obtained. This is then one of the reasons why acid etching techniques, which are successful on enamel, often fail to work on dentine.

It is possible to demonstrate the presence of certain reparative processes on enamel and dentine after mild acid attacks. Thus if the tooth surface has not been made fully hydrophilic, it slowly restores its previous degree of hydrophobicity. The presence of this process of repair is another indication that enamel and dentine can produce the low-energy materials normally covering their surfaces.

III. ADHESION IN RESTORATIVE DENTISTRY

As has been mentioned, bonding of restorative resins to enamel retention is obtained primarily through mechanical interlocking by tags penetrating the etched surface, whereas the formation of tags in the dentine surface only gives rise to mechanical bonds of low strength. This interlocking must therefore be supplemented with bonding of a chemical nature in order to reach clinically useful levels. Most attempts so far to bond restorative resins to dentine have involved the application of bifunctional molecules containing methacrylate groups.

The first research along this line was reported by Buonocore and co-workers [12], who used an adhesive molecule with methacrylate groups. An adhesive of this type has found commercial use and has been reported to mediate an initial bond strength to dentine of about 3 MPa. The strength, however, dropped to 2 MPa as a consequence of water storage. Most modern dentine adhesives consist of di- or multifunctional molecules with groups capable of reacting with dentine components such as calcium or collagen. Further, they also contain methacrylate groups to bond to the actual restorative resin. It is likely that the bonding mechanisms are of a complex nature and that there is a mutual dissolution and mixing process between the adhesive and dentine components that gives rise to a true transition zone between the tooth and the resonous restoration. For reviews of materials and techniques claimed to cause or enhance adhesion of restorative resins to dentine, see, for example, Brauer and Huget [13], Causton and Johnson [14], and Asmussen and Munksgaard [15]. Recently promising in vivo and in vitro adhesion data have been presented for the fairly recently introduced so-called glassionomer cements [16].

IV. ORAL HYGIENE ASPECTS

The presence of low-energy surfaces on teeth is probably essential for the maintenance of good oral health. Bearing the high adhesiveness of hydroxy apatite in mind and also the high number of pathogenic or potentially pathogenic microorganisms in saliva, we should be grateful to possess the ability to form organic tooth surfaces with relatively low adhesiveness.

In clinical studies on humans it has been demonstrated that the differences in critical surface tension observed for different types of tooth surfaces has relevance for in vivo adhesion of dental plaque (Fig. 1) and cells [1]. Independently, Baier [17,18] and Dexter [19] have described a similar relationship between this surface chemical descriptor—critical surface tension—and the number of retained cells on various substrata. These investigators identified the least adhesive range of critical surface tension for solids to be entered around 25 mN·m (dyn/cm). Results by Baier and DePalma [20], Schankenraad et al. [21], and Christersson et al. [22] have extended this relation to include the similar influence of critical surface tension values on general cell—surface interactions.

Fluoridation of teeth is a type of surface treatment known to have effects on the frequency of caries. Studies of the possible influence of fluorine on the adhesiveness of tooth surfaces demonstrate that certain fluorides can decrease the adhesiveness of enamel surfaces [23—25].

V. ORAL FILM FORMATION

If we leave the adhesive properites of the isolated tooth and discuss what happens to a surface when it is exposed in the oral cavity, it is well known that after some time teeth in the oral cavity are covered by organic material generally known as pellicle [26]. This organic film is further obstructing or preventing adhesive contacts with tooth surfaces. To understand the importance of cleansing and isolating procedures in clinical situations where adhesive contacts with tooth surfaces are aimed at, it is therefore essential to know the rate at which the pellicle is formed in vivo.

In vivo studies of oral films have provided detailed information through use of the range of surface physicochemical analytical techniques shown in Fig. 2 [27,28]. The primary materials used for the adsorption of these films were silica test pieces and germanium prisms treated to give different types of surface chemical characteristics, representing both normal tooth surfaces and a range of restorative materials.

The results of these studies show basically that the formation of oral films proceeds at high speed and is of a certain qualitative selectivity. That is, the layers deposited are similar to one another but obviously different from the bulk salivary pool.

Additional information has been obtained by exposing test plates in both protected, stagnant regions of the oral cavity and in regions

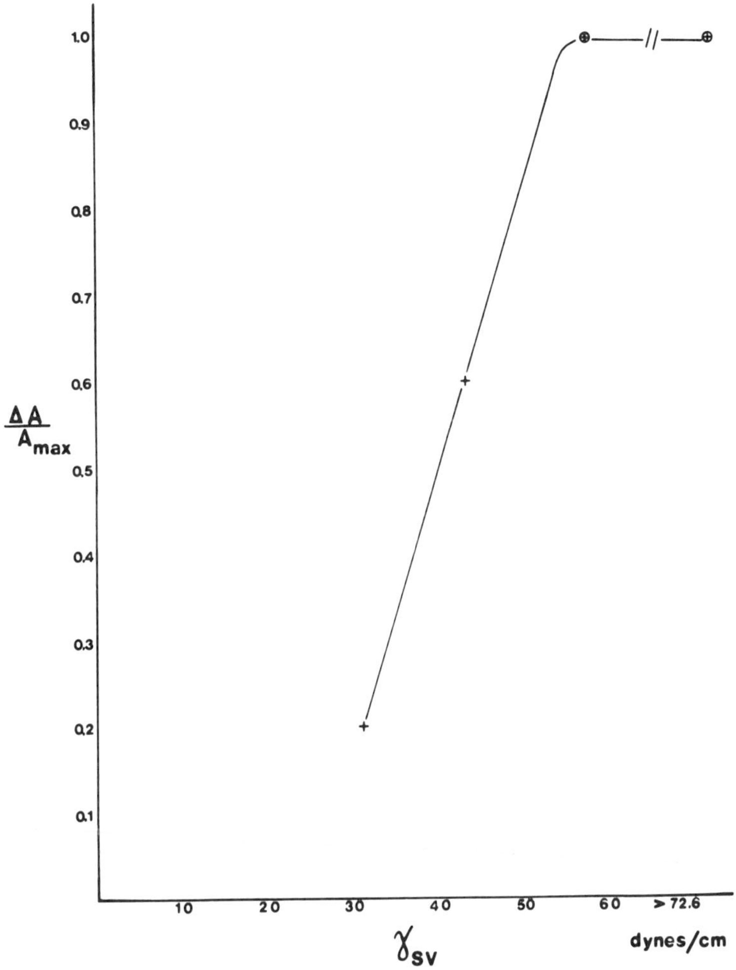

FIG. 1 Relation between the free surface energy of solids (γ_{SV}) and their plaque-adhering property per unit of area (ΔA), expressed in relation to materials with maximum plaque adhering properties (A_{max}).

exposed to considerable shear forces. Test prisms left free in the vestibular sulcus, exposed to high shear, acquired the same type and amounts of films on their surfaces as when kept in special holders shielding them from shear forces. When the prisms with their

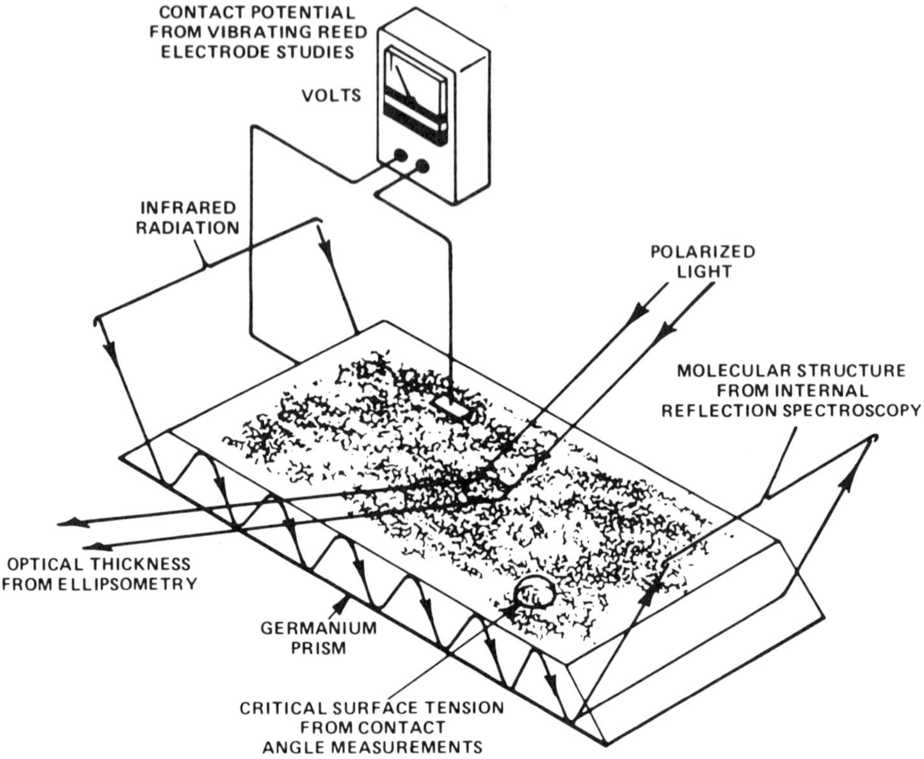

FIG. 2 Physicochemical analytical methods applied to the characterization of oral films.

adsorbed films were not rinsed free of excess clinging saliva after removal from the oral cavity, the dried films were thicker by about 10 to 40 nm. Furthermore, in these no-rinse experiments, no differences could be detected between the thicknesses of films formed on high-energy and low-energy surfaces. This is clearly not the case when the intraorally exposed prisms are rinsed with water; then the retained films are relatively thicker on low-energy than on high-energy surfaces, at the same time as roughly the same amount of material seems to be adsorbed on both high- and low-energy surfaces. This in turn suggests that a more native configuration or coagulum exists for the molecules adsorbed on the low-energy surfaces. The results of recent studies also indicate that on high-energy materials, more potential protein sites can be surface-associated and also that the packing system is more complicated, with greater film densities on high-energy surfaces. This result corresponds with those from the previously mentioned experiments of growing plaque on various types of solids under clinical conditions [1].

A third observation made was that on test pieces which have been exposed in a stagnating system, particles such as microorganisms or biological debris adhered to the film-coated prisms much faster than when the prisms were exposed to shearing forces by the movement of oral mucosa or by saliva. In Fig. 3 a holder is shown that allows stagnation of the saliva between the vertically positioned test prisms and the buccal surfaces of the first lower premolar teeth.

As a further step in ongoing surface chemical studies in the oral cavity, a method has been developed to perform intraoral wetting studies and thereby make it possible to calculate the clinical adhesiveness of tooth surfaces and restorative dental materials. These studies showed that, in vivo, the critical surface tensions for all tested surfaces and subjects were in the bioadhesive range 32 to

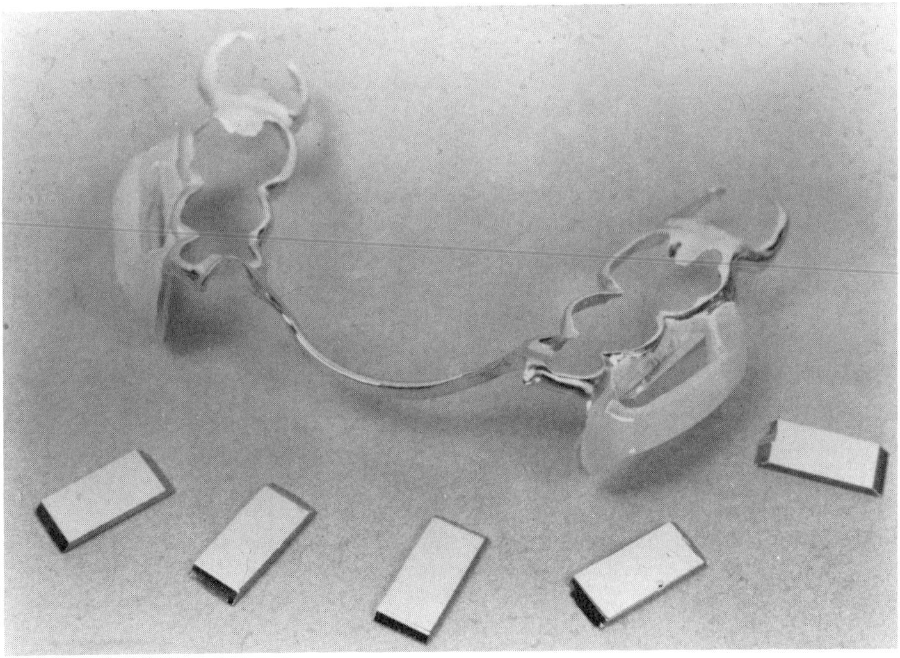

FIG. 3 Ge prisms and intraoral holder for such prisms used to collect salivary films in oral areas of moderate stagnation.

50 dyn/cm. The formation of organic films on solid surfaces in the oral cavity thus brings these to a common state of bioadhesiveness. The formation of organic films was also observed to have a masking effect on minor surface irregularities. Generally speaking, these results support the previously mentioned findings from the prism studies: that the speed of oral film formation is extremely high and that film qualities have a high degree of uniformity in composition on different kinds of surfaces despite some small variations from one human being to another.

The oral cavity is exposed deliberately or accidentally to a wide range of substances, some of which are or may be toxic, abrasive, and/or infected. Yet the teeth, especially, are expected to withstand repeated and sometimes prolonged attacks from such substances without showing mechanical or chemical failure. Tooth enamel and dentine are the two most highly differentiated hard tissues that exist in the human body. In many ways they act as the hard tissue counterparts to the brain and central nervous system, which are our most highly differentiated soft tissues. Enamel has no ability, and dentine a very restricted ability, to regenerate after having obtained macroscopic defects. Therefore, it must be of special importance for anyone with teeth to try and prevent these precious tissues from being destroyed. The ability of saliva to form oral films, both quickly and completely, probably provides an important step in the processes of preserving good oral health and, consequently, to a certain extent also good general health.

The only normally appearing clinical actions so far found to alter the surface chemical characteristics of the films are drops in pH. In in vivo studies, it has even been noted that the pH decrease created by chewing only a couple of pieces of orange is enough to increase the wettabilities of the films. Bearing in mind the general protective actions of oral films discussed previously, it is understandable that they are pH sensitive; if not, they would have protective effects for masticated foodstuffs even after passage to the stomach, and thus have delaying influences on digestive processes.

What types of actions or "driving forces" are responsible for the spontaneous formation of the oral biological films? This question cannot be answered precisely given our present state of knowledge. Bearing in mind, however, that in saliva and other biological fluids there are organic macromolecules with hydrophilic, hydrophobic, and ionically charged groups such as sulfates, phosphates, amines, and carbonyls, it is not impossible that due to the comparatively high spreading pressures of such macromolecules, they cover the surfaces of all solids present by mechanisms explained thermodynamically on the basis of dehydration and favorable entropy shifts. Much more work is required to elucidate this process.

VI. ORAL FOULING

Previously reported studies on various kinds of natural and restored organic as well as inorganic surfaces, including clinical wetting studies, indicate that salivary films form with similar speed on any type of surface. It is therefore likely that any solid entering the oral cavity or being present there is provided with a coating of salivary origin. Examples of such solids are foodstuffs, prosthetic appliances, dental instruments, and microorganisms. In a series of studies, Olson et al. [29] and Olson and Glantz [30] have used particle microelectrophoresis to study oral streptococci. Among other things, they found that these cocci behaved much like any other colloid particle and had zeta potentials which (for their signs and magnitudes of charge) were dependent on the composition and properties of the surrounding media.

Under clinical conditions, in the oral cavity, it is therefore likely that microorganisms are also surrounded by the same or a similar type of biological film as teeth and the artificial materials studied. A so-called fuzzy coat has been reported to surround oral microorganisms when studied using transmission electron microscopes. It is a fair suggestion that this coating, at least partially, is the biofilm mentioned. Simonsson and Glantz [31] have recently shown that the colloidal stability of suspensions of parotid saliva and a plaque-forming microorganism (*Streptococcus sanguis*) was lower for slow plaque formers than for fast ones, pointing to a possible controlling mechanism in saliva.

Bearing the general presence of salivary films in mind, the first adhesive events that occur when a biological particle such as a microorganism comes close to a solid surface such as a tooth or a restoration are probably controlled primarily by the nature of the adsorbed or associated films on the solid and the microorganism. Initially, there is therefore probably limited effect of the inherent surface chemical properties of either the naked solid or microorganism.

If oral microorganisms, teeth restorations, and the oral mucosa are all covered with salivary films affecting adhesion, it must be extremely difficult to create in vitro adhesive systems simulating intraoral conditions. Even if, for example, microorganisms and tooth materials in such a system are covered primarily with the true salivary films, it takes only a slight alteration in the composition, the pH, or the ionic strength of the liquid phase of an in vitro system to alter the adhesive properties of the film or even to desorb it, especially from surfaces such as a tooth—or any other biological surface, for that matter. After desorption or alteration has taken place, adhesive events in the in vitro situation may well proceed along routes that have no or very little relevance to the in vivo situation.

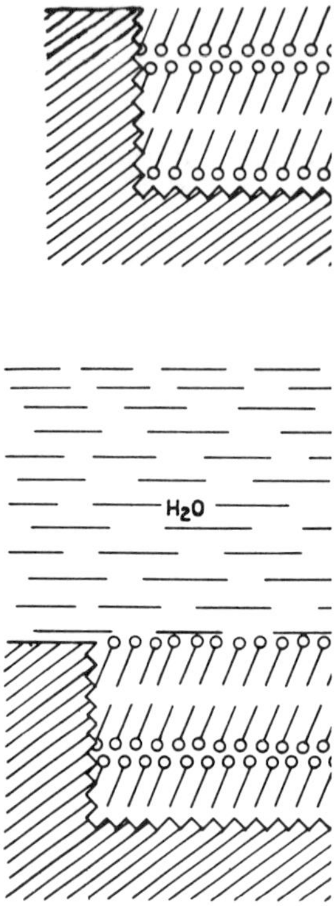

FIG. 4 Oriented lipid multilayers exposing methyl groups toward air (top) and carboxyl groups toward water (bottom).

In an in vitro system this has been demonstrated by Larsson and Glantz [32], who prepared lipid multilayers according to the well-known Langmuir–Blodgett technique. In these experiments, a solid was moved up and down through a condensed monolayer of behenic acid on water, thus transferring the monolayer to the solid in such ways that the carboxyl groups formed the outer surface when the solid exposed its interface toward water, and the methyl end groups formed the surface toward air (Fig. 4). When identical amounts of an oral strain of *S. sanguis* were added to these two types of surfaces, a drastic difference in adhesion of the microorganisms was observed. Methyl group surfaces showed no significant visual difference in bacterial adhesiveness compared to ordinary metal surfaces, whereas no visually noticeable adhesion at all was observed on the carboxyl group surfaces. It was concluded that the obstruction of adhesion in the presence of this type of lipid multilayer with carboxyl group surfaces might be due to electrostatic repulsion, as the charge density was very high in these experiments, corresponding to up to one electron per unit area of about 23 Å^2. These observed effects were not limited only to the strain of *S. sanguis*, but could also be demonstrated when *Escherichia coli* was used.

When the effects of ions were tested in these experiments, a few colonies could be observed on the carboxyl group surfaces at as low a value as 0.5% (w/w) sodium chloride. At a concentration of 2% sodium chloride, there were no longer any significant differences in initial attachment on methyl group surfaces compared to carboxyl group surfaces. These observations are in agreement with the expected effects of salts on the ionic double layer according to the DLVO theory of colloidal adhesion. Similar results have been presented by Cowan et al. [33] and Weerkamp et al. [34] using a system with hydroxyapatite or powdered enamel and dentine.

In saliva, the number of microorganisms present is generally measured to be about 10^8 per milliliter. If physical-chemical interactions take place primarily upon collision between microorganisms and solid surfaces, the particle mobility is so high that only a few minutes should be needed to cover the surface with a microorganism monolayer in this actual system. Especially as human whole saliva is thixotropic, or shows rheodestruction, moderate shearing of the system should promote such reactions. On surfaces exposed to shear in the oral cavity, however, it has been reported to take up to hours for significant numbers of microorganisms to become attached, while the process seems to be much quicker at stagnation. This finding indicates that the process of bacterial attachment to oral solid surfaces has a general chemically nonspecific background mechanism.

If high concentrations of cocci-shaped and cocci-sized polymeric particles of nonbiological origin are brought close to a tooth surface without direct exposure to saliva, they adhere to it in very much the same fashion as has been described to happen for streptococci. Figure 5 demonstrates the results of a clinical experiment by Glantz et al. [35] in which a deciduous lower front tooth was first brushed with a toothpaste containing large amounts of poly(methyl methacrylate) (PMMA) particles (Bofors Tandkräm). These particles have a diameter of about 0.7 μm, which is similar to that of most streptococci. After brushing, the tooth was extracted, then rinsed with water and prepared for SEM study.

The adsorption pattern for these particles, which by themselves do not possess any appreciable ability to form specific biochemical bonds with the pellicle-covered tooth, has a striking similarity to that reported for many types of oral streptococci on solid surfaces. The almost monolayer arrangement is obvious and so is the close

FIG. 5 Scanning electron micrograph showing incisal surface of deciduous anterior tooth that had been brushed with a toothpaste containing poly(methyl methacrylate) particles (diameter about 0.7 μm) and then rinsed in running tap water.

packing of the particles. Even after a 1 min alcohol wash in an ultrasonic bath, followed by a 10-min treatment with a sodium hypochlorite solution, some of these particles still adhere to the surface, which indicates that they must have comparatively strong attachment to the pellicle-covered enamel.

Another important observation is that on some clean solid surfaces placed in the oral cavity, the majority of the first adhering microorganisms retained after rinsing are rod-shaped (Fig. 6). This might be explained on the basis of surface-area considerations, provided that the work of adhesion per unit area of the microorganism–solid interfaces is comparatively low.

Another reason why certain microorganisms are found more often than others on solid surfaces could be that out of the random population that first adheres nonspecifically to a clean or protein-coated surface, they are the ones that have an ability to alter the environment in their own favor. For instance, metabolism and growth could form additional bonds over the interface (resulting, especially, from extracellular exudates). Variations in the ability to metabolize on film-covered solid surfaces and not in primary adhesion to such surfaces could thus be the important etiological factor for common oral diseases. This theory is supported by the findings illustrated in

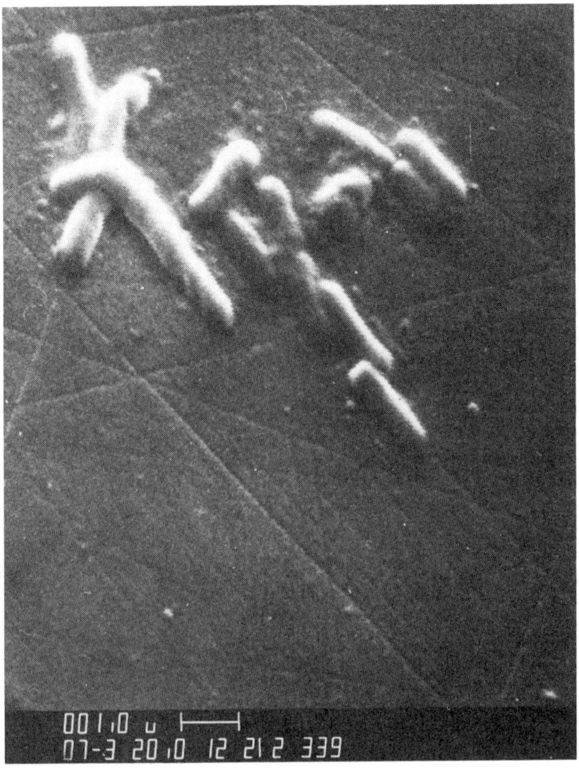

FIG. 6 Scanning electron micrograph showing adhering, mostly rod-shaped microorganisms after a 2-h intraoral exposure of detergent-washed Ge prism.

Figs. 7 and 8. As can be seen around some of the coccoid particle groups shown in Fig. 7, a halo can be observed. This halo suggests that inside its periphery, a biochemical transformation and/or addition of material might have taken place as a consequence of microbiological metabolism. In higher-magnification photos, the halo-surrounded groups of particles were clearly seen to be covered by an extracellular layer (Fig. 8). Since, on the very same test plates, other particles of debris of nonmicrobial origin were found not to have such coverage, it was concluded that this observation was not an experimental artifact, but that the microorganisms in Figs. 7 and 8 had surrounded themselves and the area next to them with a film of sufficient cohesive characteristics to resist removal by rinsing with distilled water. Polysaccharides are typical of substances that

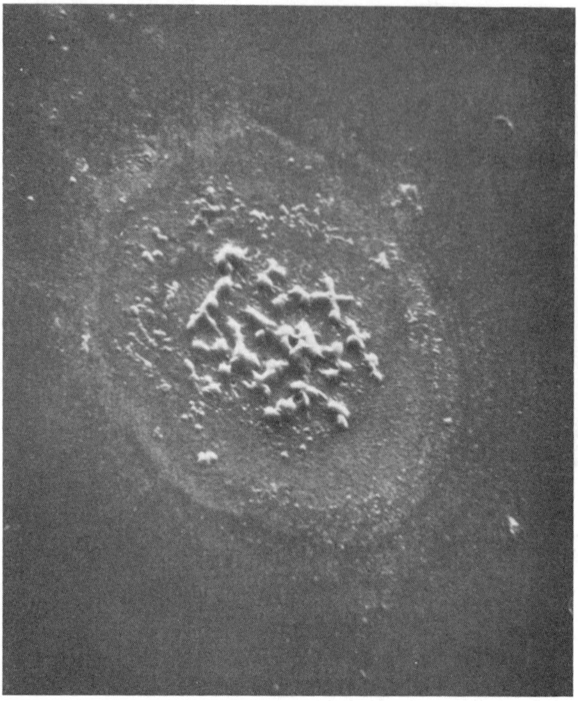

FIG. 7 Scanning electron micrograph (original magnification 50×) showing particles, judged to be microorganisms, adhering to film-covered Ge prism at the end of a 4-h unprovoked intraoral exposure. Note the halo surrounding the particles at an approximate distance of 20 μm.

could give such film characteristics. In areas thus covered, the adhesive properties could very well be altered, thereby perhaps facilitating or hindering local adherence of newly arriving microorganisms. It should also be noted that the film shown in Figs. 7 and 8 was air-dried. In its original oral environment, it is likely that the film material occupied a considerably larger volume.

One can, of course, speculate that acid-producing oral microorganisms are able to induce caries—at least in part—because in the oral cavity they provide the conditions for film breakdown that is physiologically not supposed to take place until the films have reached the stomach. Also, although we do think that generally nonspecific types of interactions are involved during primary

FIG. 8 Scanning electron micrograph (original magnification 200×) of the same group of particles as in Fig. 7. Note the coating on the mostly sphere-shaped particles, judged to be microorganisms.

biological fouling of teeth and oral solid surfaces, we also believe in the later appearance of biologically specific adhesive events (such as lectin—receptor interactions or enzyme—substrate reactions) in the oral cavity. Certain variations in general physicochemical interactions could thus cause the appearance of dramatic biological alterations, for example, and here again considerably more effort is required to define these conditions.

VII. SURFACE AND COLLOID CHEMICAL ASPECTS ON SALIVA

Apart from its film-forming capacity, control of the oral microflora, and wear protection, saliva has many important functions, such as enzymatic activity, buffering, and control of water balance. This range of salivary functions is so wide that it cannot be suitably explained by considering saliva as a simple solution; instead, saliva should be regarded as having an internal structure and to behave like a weak gel, for example [36].

As such, the presence of salivary structural components has been suspected since around the turn of this century [37] and a variety of descriptions of salivary structural features have been presented [38,39]. However, until recently they have not included a structural model for this low-viscosity, non-Newtonian body fluid. In theory, it should be possible to clarify the nature of structural elements in fluids such as saliva by the use of rheology. Detailed characterization has, however, been retarded by the fact that no sufficiently accurate instrument has yet been developed for precision rheological measurements of fresh whole saliva. This is probably due to the fact that saliva is secreted at a limited volume rate, is biochemically unstable, and contains multiple, relatively large parcles such as epithelial cells.

Bearing in mind the failure to obtain consistent rheologic data, alternative approaches have been developed. A method has recently been presented [40–42] for the preservation of intrinsic saliva structures by instantaneous immersion of samples directly into flasks of liquid nitrogen (-185.8°C). Shortly after flash freezing and retrieval from the nitrogen, solid pieces of saliva are handled in the same way as is any other frozen piece of tissue (i.e., they are cryogenically sectioned, fixed, stained, and examined by both light and electron microscopy).

Since the first successful experiment a couple of years ago, this freeze sectioning method has been used for the routine examination of both whole saliva and saliva fractions donated by a range of subjectively healthy volunteers found to be free of clinically detectable oral diseases [41–43].

The results of these studies, so far, point uniformly in the same direction; that human saliva is a highly structured secretional product which perhaps is better described as a tissue than a fluid. On the light-microscopic level, its main structural feature is the continuous filamentous network illustrated in Fig. 9. This network was better observed in fractions of parotid saliva than in submandibular/sublingual ones. It is not markedly affected by water dilution, suggesting that excess amounts of network-forming material exist at the time of secretion.

Figure 10 shows that when particulate biological materials such as desquamated epithelial cells and microorganisms are observed in the whole saliva sections, they are usually aggregated and preferentially retained inside the intrinsic salivary network structures. The saliva filaments appear to have undergone local expansion to accommodate this extraneous particulate material.

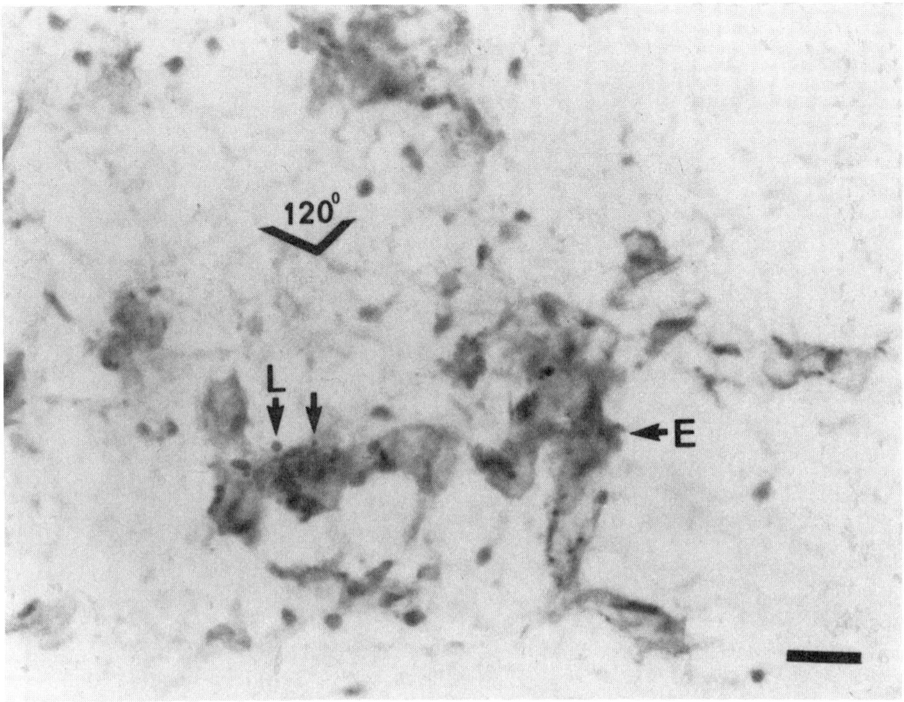

FIG. 9 Oil-red-O-stained and hematoxylin-counterstained section of whole saliva from a 20-year-old subjectively healthy female donor. The section shows the presence of a network created by seemingly amorphous structural components of varying dimensions. Many of the network components meet at approximately 120° angles. The figure also shows the presence of epithelial cells (E) in contact with the network and multiple lipoid droplets (L) often in association with the epithelial cells. Length of bar: 1 mm.

When immunocytochemical staining techniques were applied to the whole saliva sections, strong staining patterns—for epidermal growth factor (EGF), nerve growth factor (NGF), and lysozyme—were seen preferentially located inside the network structures. Gram staining routinely revealed uneven distributions of the microorganisms, most frequently observed in large aggregates predominantly positioned inside the network structure elements (Fig. 11).

Finally, when frozen saliva sections were stained for lipid material, multiple, dispersed lipoid droplets were consistently demonstrated,

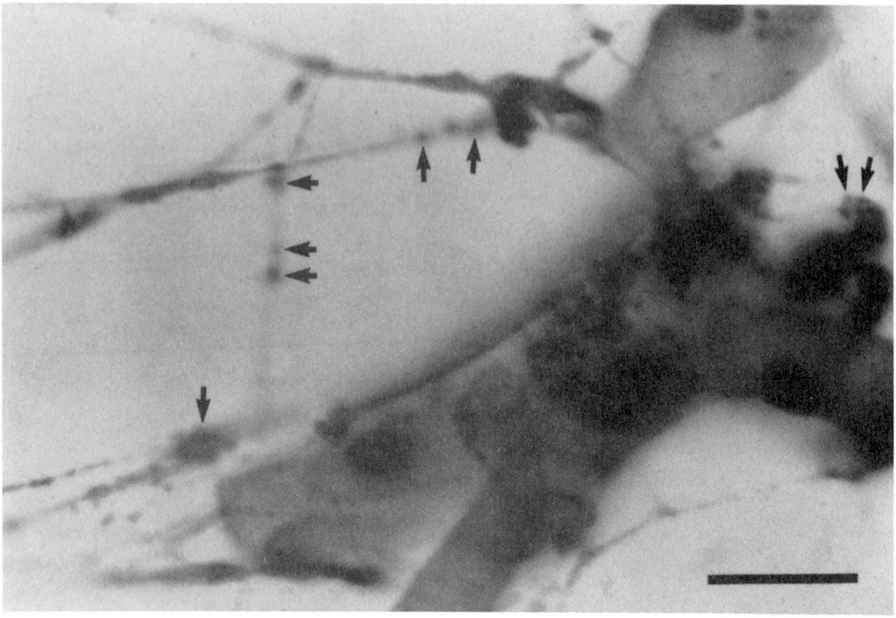

FIG. 10 Hematoxylin-eosin-stained section of a whole saliva from a 51-year-old subjectively healthy male donor. Salivary structures are seen associated with epithelial cells and microorganism-like particles (arrows). Length of bar: 10 μm.

often in association with epithelial cells. These droplets appear in sections of both whole saliva and saliva fractions and have diameters of up to about 10 μm (Fig. 9).

Electron microscopic studies performed so far [42,44] have been mainly of the transmission (TEM) type, supported by a number of examinations using scanning electron microscopy with energy-dispersive x-ray analysis (SEM/EDAX). Generally speaking, the results of electron microscopy support those derived from morphologic analysis, at the same time that they provide information about saliva ultrastructure. The presence of a fine, continuous network structure is verified, and TEM further reveals observable minor network structurs, with diameters as small as 0.1 μm (Figs. 12a and b).

In thin sections (80 to 100 nm), TEM shows the network to have reticulated interior structures of its own. For components with diameters above 1 μm, an outer zone surrounded an inner, more electron dense core. This internal core contained granules of varying

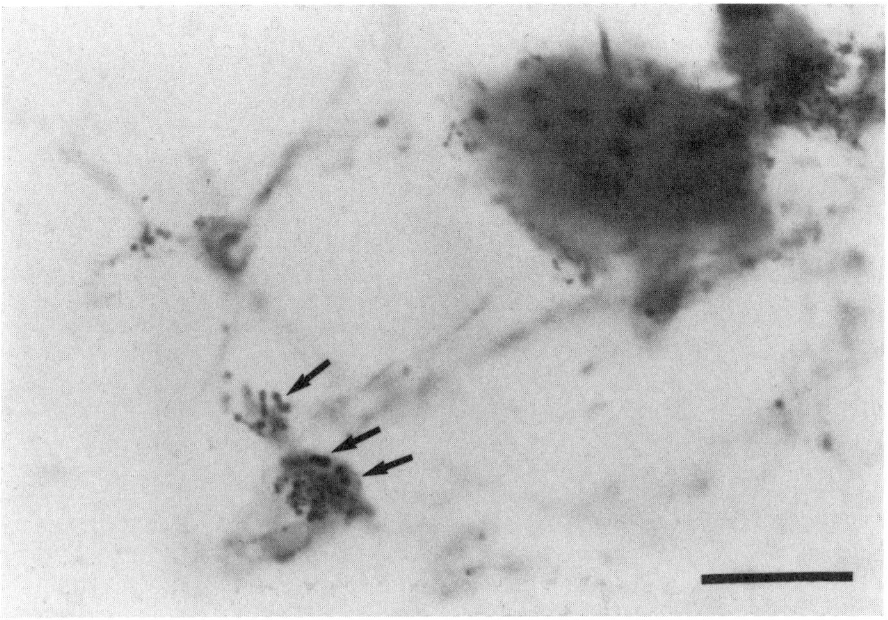

FIG. 11 Gram-stained section of whole saliva from a subjectively healthy 32-year-old female donor. Gram-positive microorganism-like particles (arrows) and an epithelial cell are located inside a salivary network structure. Length of bar: 10 μm.

shapes and electron densities (Fig. 13). Similar filamentous structures with interior granules were observed in pure submandibular/sublingual saliva fractions (Fig. 14), and these sections did stain for EGF and NGF activities within the core regions (Fig. 15).

Transmission electron microscopy also verified the light-microscopic observation that microorganisms are preferentially located inside the salivary network structures (Fig. 16). In particular, we repeatedly observed filaments that seemed to have expanded to incorporate foreign particles. This phenomenon seemed to have occurred as the result of a spontaneous engulfment of material inside the network structures themselves (Fig. 17).

Based on these data the following idealized model of following four saliva structures is proposed:

1. The continuous water-based electrolyte acts mainly as a diluent for water-soluble substances and maintains the water balance of the oral mucosa and the mineral balance of the teeth.

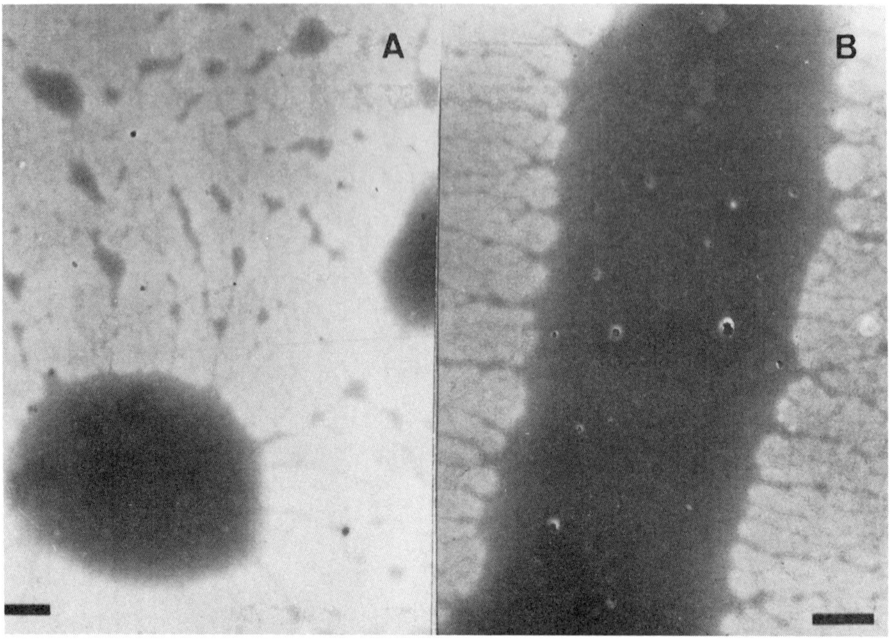

FIG. 12 Transmission electron micrographs of paraformaldehyde-vapor-fixed, approximately 10-μm-thick sections of whole saliva from a subjectively healthy 48-year-old male donor. The pictures show a probable cross section (A) and a probable longitudinal section (B) of major structural components with somewhat varying electron densities. From these components fine network structures are extending. Especially in the cross section, some of the fine components reunite into larger pools of material at a distance from the major component. Length of bars: 1 μm.

2. The scaffold-like continuous network structure, based on the overall staining pattern observed, is believed to comprise higher-molecular-weight salivary glycoproteins. Bearing in mind the comparatively high spreading pressures of such macromolecules, they should be implicated in virtually every type of oral adhesive event taking place in the presence of saliva [45]. They should also serve as partial diffusion barriers and wear protectors for oral surfaces. In addition, they could participate in the buffering activity of saliva, together with certain components of the electrolyte.

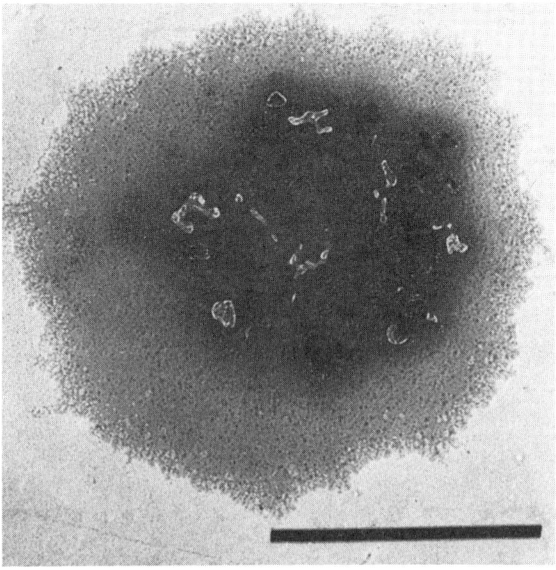

FIG. 13 Transmission electron micrograph of a 90- to 100-nm-thick section of whole saliva from a subjectively healthy 51-year-old male donor. The picture shows the probable cross section of a major structural component with a comparatively thick foamlike outer zone and an inner, more electron dense, also foamlike core exhibiting granules with different shapes, sizes, and electron densities. Length of bar: 1 μm.

3. The network core structures within the surface-active glycoproteinaceous filaments are likely to contain higher concentrations of other types of macromolecules, such as less water-soluble proteins. They are then, for example, able to perform specific biochemical functions of saliva, requiring a protected environment. Our results show these inner-core components to be associated with the previously documented ability of saliva to aggregate epithelial cells and microorganisms [46,47].

Our results also illustrate how the salivary network fibers engulf these cells through some kind of acellular ingestion process and then both collect and concentrate them in enlarged structural segments. Thie "particle passivation" system is important for the control of oral microflora and oral hygiene, but might also provide wear protection of teeth through the viscoelastic micro-cushions thus created.

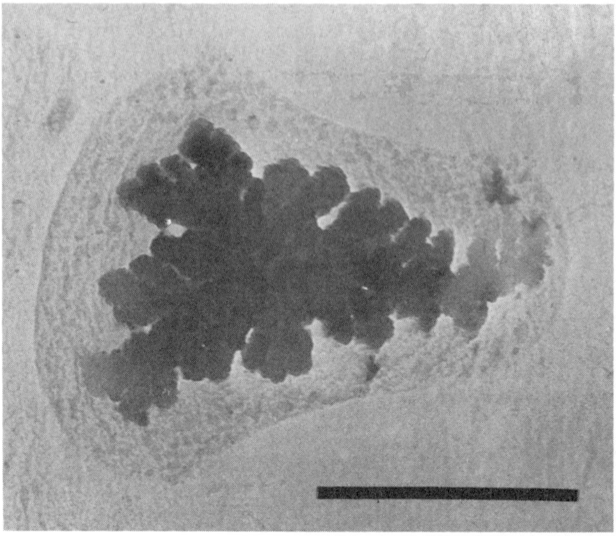

FIG. 14 Transmission electron micrograph of a 90- to 100-nm thick section of submandibular–sublingual saliva from a subjectively healthy 32-year-old female donor. The picture, a probable cross section of a major structural component, shows an outer less-electron-dense zone surrounding more-electron-dense core structures with varying shapes. Length of bar: 1 μm.

The system may also include an interior transport mechanism, details of which remain to be elucidated.

4. The dispersed droplets of lipoid material, with varying dimensions and distribution, are associated predominantly with water-insoluble foreign substances and desquamated epithelial cells.

There is a high probability that saliva also has other minor structures. Generally speaking, however, it is essential to emphasize that the major network elements can explain not only the simultaneous performance in whole saliva of a range of highly specific biological functions, but in fact in principle are a necessity for several reactions which are not mutually compatible when executed in simple salt solutions.

When considering this proposed model it is particularly interesting to note that most of the specialized biological functions seem to be

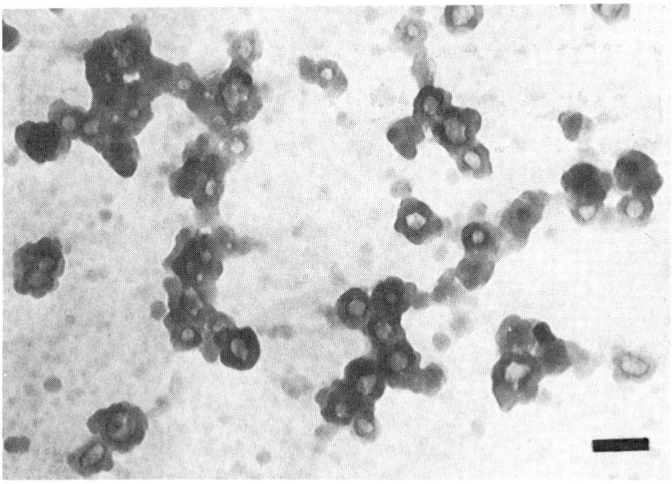

FIG. 15 Transmission electron micrograph of EGF antibody-treated whole saliva specimen from a subjectively healthy 51-year-old male donor. The spherical granular-like reaction structures are composed of an electron dense periphery with an electron-lucent central core, and are arranged in chains. Length of bar: 0.1 μm.

located in relatively protected and possibly different environments inside the fibrous network structures. In this manner, our model provides a rational explanation for many established biological observations. As an example may be mentioned the comparatively rapid healing of oral wounds and lesions that takes place despite the presence of high concentrations of viable pathogenic or potentially pathogenic salivary microorganisms. This now becomes an understandable consequence of the acellular ingestion of oral microorganisms described and a mechanism for local delivery of EGF and NGF active substances in wounds and lesions. The instinctive sucking and licking of wounds on the skin also receives a rational justification. Suggested gastric cytoprotective functions of EGF could very well also be associated with the continuous swallowing of saliva. Finally, it is reasonable to assume that the major principles discussed here for the salivary filaments and functions observed are also valid for other secretory products and it appears possible that these products may display similar structures and functional roles.

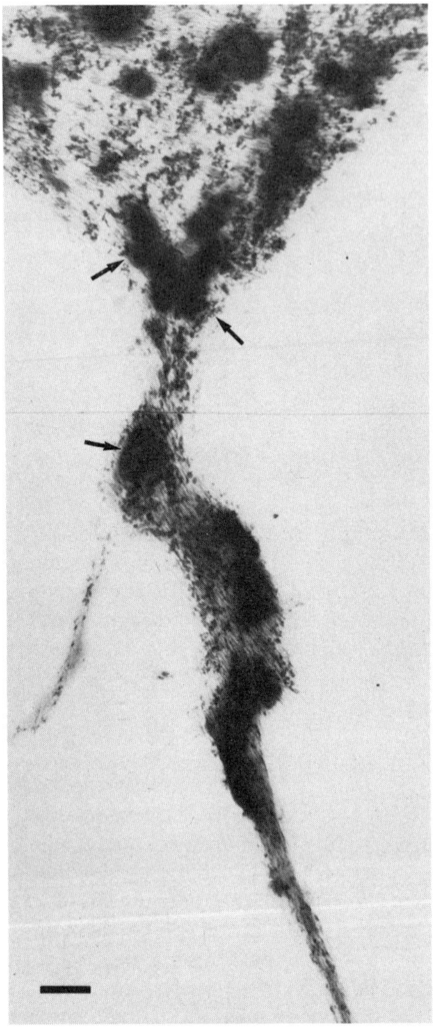

FIG. 16 Transmission electron micrographs of a paraformaldehyde-vapor-fixed and OsO_4 postfixed, approximately 10-μm-thick section of whole saliva from a subjectively healthy 51-year-old male donor. The picture shows a branched major structural component with integrated structures, including multiple microorganism-like particles (arrows). Length of bar: 1 μm.

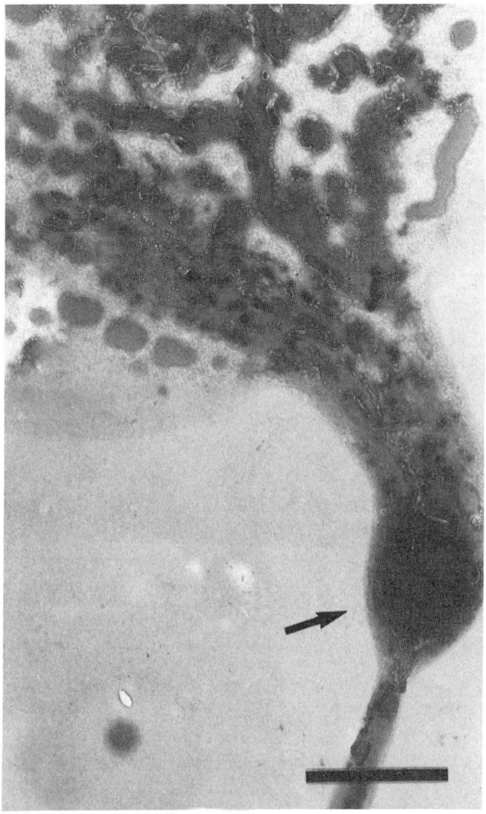

FIG. 17 Transmission electron micrograph of a 90- to 100-nm-thick section of whole saliva from a subjectively healthy 51-year-old male donor. The picture shows a bulbous structural extension of a finer, denser, filament. The swollen segment has a thin, less electron dense background structure with multiple granules of varying size, shape, and dimensions. Of special interest is the large, electron dense, microorganism-like particle (arrow) surrounded by the widening portion of the filament, and the ensemble of smaller granules located between this particle and the expanded interior zone of the enlarged component. Length of bar: 1 μm.

REFERENCES

1. P.-O. Glantz, *Odontol. Revy 20* (Suppl. 17) (1969).
2. P.-O. Glantz, *J. Colloid Interface Sci. 37*)2):281–290 (1971).
3. D. C. Smith, *Br. Dent. J. 124*:381–384 (1968).
4. J. W. McLean and A. D. Wilson, *Aust. Dent. J. 22*(2):120–127 (1977).
5. K. C. Uy and R. Chang, in *Public Health Publication 1494*, U.S. Department of Health, Education, and Welfare, Washington, D.C., 1966.
6. G. V. Newman and L. H. Sharpe, *J. N.J. Dent. Soc. 37*:289–291 (1966).
7. H. Retief, *J. Oral Rehab. 1*:265–284 (1974).
8. M. G. Buonocore, in *Adhesion in Biological Systems* (R. S. Manly, ed.), Academic Press, New York, 1970, pp. 225–254.
9. R. L. Bowen, in *Dental Adhesive Materials*, Proc. symposium held Nov. 8–9, 1973, Hunter Bellevue School of Nursing, New York (G. T. Ward and E. O. Woolridge, eds.), Prestige Graphic Services, New York, 1974, pp. 205–221.
10. G. Brauer, Chemishe Analyse von Methacrylat-Kunststoffen, *Dtsch. Zahnaerztl. Z. 30*(10):672–679 (1975).
11. G. Brauer and J. D. Termini, *J. Dent. Res. 51*:151–160 (1972).
12. M. G. Buonocore W. Wileman, and F. Brudevold, *J. Dent. Res. 35*:846–851 (1956).
13. G. M. Brauer and F. F. Huget, in *The Chemistry of Biosurfaces* (M. L. Hair, ed.), Marcel Dekker, New York, 1972, pp. 731–800.
14. B. E. Causton and N. W. Johnson, *Br. Dent. J. 152*(1):9–11 (1982).
15. E. Asmussen and E. C. Munksgaard, *Scand. J. Dent. Res. 91*(2):153–155 (1983).
16. M. J. Tyas and D. R. Beech, *Aust. Dent. J. 30*(4):260–264 (1985).
17. R. E. Baier, in *Adhesion in Biological Systems* (R. S. Manly, ed.), Academic Press, New York, 1970, pp. 15–48.
18. R. E. Baier, in *Surface Chemistry and Dental Integuments* (A. Lasslo and R. P. Quintana, eds.). Charles C Thomas, Springfield, Ill., 1973, pp. 337–391.
19. S. C. Dexter, *J. Colloid Interface Sci. 70*(2):346–354 (1979).
20. R. E. Baier and V. A. DePalma, in *Management of Arterial Occlusive Disease* (W. A. Dale, ed.), Year Book Medical Publishers, Chicago, 1971, pp. 147–163.
21. J. M. Schakenraad, H. J. Busscher, C. R. H. Wildevuur, and J. Arends, *J. Biomed. Mater. Res. 20*:773–784 (1986).
22. C. E. Christersson, R. G. Dunford, P.-O. Glantz, and R. E. Baier, *Scand. J. Dent. Res. 97*:247–256 (1989).

23. P.-O. Glantz, in *Berichte vom VI. Internationalen Kongress fur grenzflächenactive Stoffe. Chemie, physikalische Chemie und Anwendungstechnik der grenzflächenaktiven Stoffe*, Carl Hauser Verlag, München, 1973, pp. 747–758.
24. van Pelt et al., in *Surface and Colloid Phenomena in the Oral Cavity: Methodological Aspects* (R. M. Frank and S. A. Leach, eds.), IRL Press, London, 1981, pp. 111–118.
25. Perdok et al., *Clin. Prev. Dent. 10*(5):3–9 (1983).
26. T. Sönju and G. Rölla, *Caries Res. 7*:30–38 (1973).
27. R. E. Baier and P.-O. Glantz, *Acta Odontol Scand. 36*:389–301 (1978).
28. P.-O. Glantz, R. E. Baier, and D. W. Goupil, *Acta Odontol. Scand. 39*:169–177 (1981).
29. J. Olson, P.-O. Glantz, and B. Krasse, *Arch. Oral Biol. 21*: 605–609 (1976).
30. J. Olson and P.-O. Glantz, *Arch. Oral Biol. 22*:461–466 (1977).
31. T. Simonsson and P.-O. Glantz, *Acta Odontol Scand. 46*:195–197 (1988).
32. K. Larsson and P.-O. Glantz, *Acta Odontol. Scand. 39*:79–82 (1981).
33. M. M. Cowan, K. G. Taylor, and R. J. Doyle, *Infect. Immun. 55*(7):1552–1557 (1987).
34. Weerkamp et al., *J. Dent. Res. 67*(12):1483–1487 (1988).
35. Glantz, Granath, and Holma, in Fundamentals and Applications of Surface Phenomena Associated with Fouling and Cleaning in Food Processing (B. Hallström and C. Trägårel, eds.), Lund University and University of Wisconsin, 1981, pp. 229–230.
36. W. H. Schwarz, *J. Dent. Res. 66*:660–664 (special issue) (1987).
37. G. Fano, *Arch. Fisiol. 5*:365–370 (1908).
38. N. Colombini, *Boll. Soc. Ital. Biol. Sper. 14*:453–456 (1939).
39. H. Hoepfner, *Kolloid Z. 105*:216–223 (1943).
40. P.-O. Glantz, C. D. Vaughan, and J. R. Natiella, *J. Dent. Res. 67* (special issue A) (1988).
41. P.-O. Glantz, J. R. Natiella, C. D. Vaughan, A. E. Meyer, and R. E. Baier, *Acta Odontol. Scand. 47*:17–24 (1989).
42. P.-O. Glantz, S. M. Wirth, R. E. Baier, and J. E. Wirth, *Acta Odontol. Scand. 47*:7–15 (1989).
43. P.-O. Glantz, S. E. Friberg, S. M. Wirth, and R. E. Baier, *Acta Odontol Scand. 47*:111-116 (1988).
44. P.-O. Glantz, M. A. Meenaghan, K. H. Hyun and S. M. Wirth, *Acta Odontol. Scand. 47*(5):287–292 (1989).
45. P.-O. Glantz and R. E. Baier, *J. Adhesion 20*:227–244 (1986).
46. R. J. Gibbons and M. Nygaard, *Arch. Oral Biol. 15*:1397–1400 (1970).
47. D. I. Hay, R. J. Gibbons, and D. M. Spinell, *Caries Res. 5*: 111–123 (1971).

5

Biomaterial Surface Domains: Bacterial Versus Tissue Cells

ANTHONY G. GRISTINA*, PAUL T. NAYLOR†, and TONI L. MEADOR††
Department of Orthopedic Surgery, Wake Forest University Medical Center, Winston-Salem, North Carolina

QUENTIN N. MYRVIK* Department of Microbiology and Immunology, Wake Forest University Medical Center, Winston-Salem, North Carolina

I.	Introduction	106
II.	Bacteria in Biomaterial Sepsis	107
III.	Adhesion Mechanisms	109
	A. Extracellular polysaccharide production	111
	B. Microcolony formation	113
	C. Glycoproteinaceous conditioning films	115
	D. Tissue adhesion and integration	117
IV.	Characteristics of Commonly Used Biomaterials	117
	A. Metals	118
	B. Polymers	119
	C. Methyl methacrylate	120
V.	Biological Substrata	121
VI.	Clinical Experience	121
	A. Total artificial heart	121
	B. Ocular implants and hydrophilic polymers	122
	C. Host defense response	122
	D. The race for the surface	124
VII.	The Future	125
	References	126

Current affiliations:
*Musculoskeletal Sciences Research Institute, Herndon, Virginia.
†Orthopedic Surgery, Inc., Knoxville, Tennessee.
††Bowman Gray School of Medicine, Winston-Salem, North Carolina.

I. INTRODUCTION

Recent medical and technological advances have made possible the use of manufactured materials (metals, alloys, and polymers) as tissue substitutes in autografts and allografts. Artificial joints and prosthetic vascular grafts, contact lenses, and drug delivery devices are in common use. Although many problems concerning design, function and strength, and rejection have been solved, two major impediments to the extended use of implanted biomaterials are yet to be overcome: infection centered about the biomaterial, and the lack of successful integration of biomaterial surfaces by tissue cells [1].

Implanted materials are subject to infection because bacteria naturally adhere to surfaces as a survival mechanism, whether the surface is metallic, polymeric, or biologic. Upon implantation, biomaterial substrata are coated with a conditioning film derived from host elements in the surrounding fluid (adhesive proteins such as laminin, fibronectin, and collagen). Bacteria may then adhere to elements in this film via receptor–ligand interactions, fimbrial attachment, or possibly chemical bonding. Some species of bacteria produce an exopolysaccharide "slime" which surrounds developing microcolonies, which may optimize nutrition and growth and provide protection from antagonists in the surrounding environment.

For reasons not yet clear, biomaterial-centered infections are highly resistant to host defense mechanisms and antibiotic treatment. The presence of biomaterials appear to disrupt normal macrophage oxidative response, motility, and phagocytosis. The foreign body effect, that is, the increased susceptibility to and morbidity of infection in the presence of a foreign body, is a well-known concept in medicine. However, the exact mechanisms through which the presence of a biomaterial inhibits host defense response are not known. Curiously, even though an infecting bacterial species may be identified, antibiotic treatment which would normally kill suspended populations of the same species is ineffective against the biomaterial-adherent populations.

The second obstacle to the expanded use of biomaterials is the failure of some implants to integrate with tissue cells (bone, vascular, musculotendinous, connective). In practice, many polymer implants are inflammatory and have proven to be more adhesive for bacteria than for tissue cells. Once an implant surface is colonized by bacteria, it will essentially never be colonized by tissue cells and will remain an infectious center, largely because of the inflammatory process at the tissue–implant interface.

Conceptually, it appears that bacterial and tissue cells compete for domination of biomaterial surfaces in a "race for the surface." Bacteria are highly adaptive organisms and naturally adhere to surfaces as a survival mechanism. Tissue cells are less adaptive than bacteria. Successful integration to biomaterial surfaces by tissue cells is defined as the development of a homeostatic interaction between outer membrane macromolecules and the biomaterial surface [2–4]. The response of tissue cells to a biomaterial implant is affected by host defense reaction to a foreign body and by the so-called adhesive proteins and their cellular receptors (integrins), which are vital to intercellular communication and differentiation [5].

Finally, and significantly, compatibility between a biomaterial implant surface and constituents of the surrounding host environment is based on cellular, molecular, and atomic interfacial reactions. Although the physical and chemical characteristics of the bulk phase of commonly used biomaterials may be known, these same characteristics at the material surface may be significantly different due to the rearrangement and relaxation of atomic and geometric structures. Furthermore, implantation of biomaterials into a liquid biologic environment introduces additional chemical and physical factors which are also relatively unknown.

II. BACTERIA IN BIOMATERIAL SEPSIS

Both coagulase-positive and coagulase-negative staphylococci have been isolated from biomaterial-centered infections. In vitro studies have indicated that coagulase-negative staphylococci preferentially colonize polymers, and metals are predominantly colonized by coagulase-positive staphylococci, although both species may be found colonizing both types of surfaces [3,6,7]. Recent studies on the binding of coagulase-negative and coagulase-positive staphylococci to biomaterials have shown that hydrocarbon polymers bind more organisms per unit area than do metals [8]. This difference is intrinsic to the biomaterial and is consistent when slime- and nonslime-producing organisms are compared.

Generally, well-known pathogens such as *Staphylococcus aureus*, *Pseudomonas aeruginosa*, and *Escherichia coli* are commonly involved in biomaterial-centered sepsis. However, when bacteria adhere to surfaces, phenotypic changes occur which appear to affect virulence. These changes may help explain a unique aspect of biomaterial-centered infection, the transformation of normally commensal, autochthonous bacteria to apparently virulent pathogens [9]. For example, although *S. epidermis* is normally a common nonpathogenic saprophyte of human skin, it has emerged as a serious causative organism in biomaterial-related and nosocomial infections, especially in immunosuppressed patients [10–15]. Several strains of *S. epidermidis* produce the extracellular polysaccharide slime believed to be associated with virulence [9,16–18]. *P. aeruginosa* is the most frequently identified cause of corneal keratitis of extended-wear contact lenses [19].

Despite the predominance of these organisms in biomaterial-centered infections, they are by no means the only pathogens involved. Many substratum-induced infections are polymicrobial (Fig. 1). Several species of bacteria may be isolated from within biofilm-enclosed colonies (a fact that often makes treatment more difficult) [20]. Studies of adult osteomyelitis have revealed that more than two-thirds of the infections involve multiple pathogens, including *Pseudomonas, Enterococcus, Streptococcus, Bacillus,* and *Proteus* species [18,21].

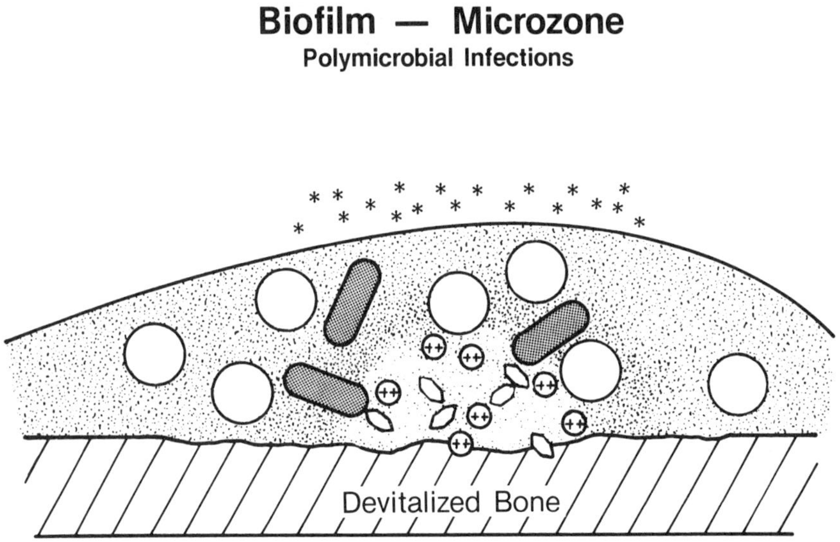

FIG. 1 In osteomyelitis, polymicrobial colonization with aerobic and anaerobic bacteria occurs. A microzone may be formed within which climatic qualities enhance metabolite and ion concentrations. [Reprinted with permission from A. Gristina et al., Microbial adhesion and the pathogenesis of biomaterial-centered infections, in *Orthopaedic Infection: Diagnosis and Treatment* (R. B. Gustila, R. P. Gruninger, and D. T. Tsukayama, eds.), W. B. Saunders Co., Philadelphia, 1989, pp. 3–25.] Bacteria (aerobic, anaerobic); Biofilm; Microzone; ⊕ Ions (Ca^{++}, Mg^{++}), Metabolites; * Antagonists and host defenses.

III. ADHESION MECHANISMS

Fundamental principals of colloid chemistry and theoretical physics may be applied to interactions between tissue and biomaterial surfaces and are essentially the same for bacteria or tissue cells. The surface characteristics of a biomaterial may be modified by chemical activities that occur during implantation or in biologic environments, when previously sterile biomaterials are coated by debris, blood, and components of the surrounding serum and matrix.

Immersed objects have a tendency to attract suspended particles, including microbes, to their surfaces [22]. Derjaguin and Landau [23] and Verwey and Overbeek [24] formulated a theory (DLVO theory) which states that two positions of thermodynamic stability exist near a submerged surface in a colloidal solution (Fig. 2). At long range, a variety of nonspecific weak forces, including gravitation, chemotaxis, and surface tension, attract particles to surfaces. At shorter ranges, however, London–van der Waals forces and steric configurations repel particles from the surface [22], whereas hydrophobic, covalent, hydrogen, and ionic bonding may attract particles [25–27]. In addition, repulsion is increased at short range because the surfaces of both eukaryocytes and prokaryocytes and most substrata are negatively charged.

As a result of these counterbalancing attractive and repulsive forces, a particle will first be positioned at a *secondary minimum* just off the surface (1 to 10 nm) [22, 28]. This concentration of particles at the secondary minimum is called *adsorption* and is a reversible process [22]. However, repulsion decreases rapidly with distance and may be overcome by short-range attractive forces. This short-range binding occurs at the *primary minimum* (less than 1 nm from the surface). Adhesion at the primary minimum occurs only when the potential energy barrier between the contacting surfaces is overcome [28]. Particles held in this position are essentially irreversibly bound to the surface. The process whereby particles leave the secondary minimum and enter the primary minimum is called *adherence* [22].

Under natural conditions, this sequence of events may be affected by environmental factors, such as the complex biochemical nature and intricate topography of the glycocalyx surrounding the eukaryotic cell wall [22]. The action of antibacterial enzymes and antibodies, and mechanical forces such as coughing, peristalsis, excretion, ciliary action, mucous entrapment, and desquamation of colonized cells may greatly increase or diminish microbial attachment [22]. Although cell and biomaterial surfaces charges are both usually negative, isoelectric points of proteins in a liquid environment may be affected by changes in pH that occur with tissue damage and with degradation of the biomaterial [29–31].

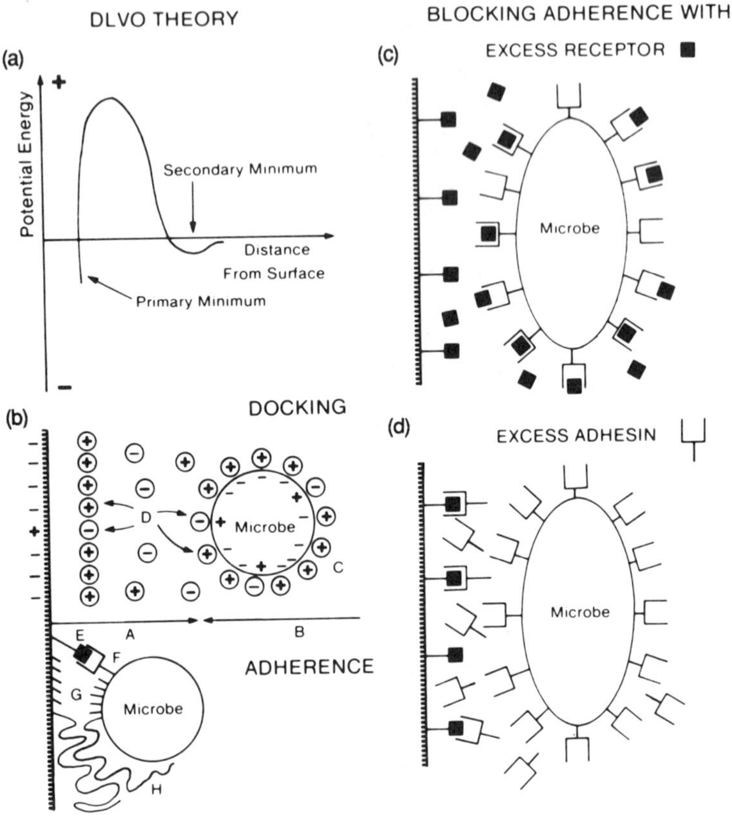

FIG. 2 Panel (a) illustrates the DLVO theory showing two points of low potential energy, one close to the surface (primary minimum) and another just off the surface (secondary minimum), separated by a high-energy barrier. The precise shape and amplitude of this curve can vary considerably with local conditions. Panel (b) shows the forces of repulsion (A) (London–van der Waals repulsion, repulsion by like charges) and the forces of attraction (gravitation, London–van der Waals attraction, surface tension), which hold the particles at the secondary minimum. The letter "C" indicates the charge field surrounding the negatively charged particle (microbe), and "D" indicates the mosaic charges set up by local charge variation and fluctuation. Particles may cross the energy barrier, separating the secondary minimum from the primary minimum by several means: Adhesions (F) such as fibrils or fimbriae may bridge the gap and interact with specific molecular structures or receptors (E) on the surface; the hydrophobic surfaces of microbes promote close contact with lyophilic regions of cell surfaces (G); and microbially

Bacterial attachment to surfaces may also be mediated nonspecifically through fibrils of extracellular polysaccharide [32] or specifically via proteinaceous organelles (fimbriae, pili, curli [33]) which react with surface receptors over distances greater than 20 nm [33–35]. Fimbriae may also interact nonspecifically (by charge or hydrophobic interaction) with inorganic substratum elements [35].

The sequence of attachment, therefore, is: encounter, initial attachment, and adhesion (Fig. 3). Reversible, nonspecific adhesion is dependent on the general and relatively long-range physical characteristics of the bacterium, the fluid interface, and the substratum. Specific, irreversible adhesion, which occurs after initial attachment, is probably time dependent, involving protein adhesion–receptor interactions and synthesis of carbohydrate polymers. Clearly, the sequence of these processes may vary in vivo.

A. Extracellular Polysaccharide Production

Shortly after attachment and adhesion (3 to 4 h or less), if environmental conditions (temperature, nutrients, antagonists, and cation balance) are favorable, bacteria may produce or increase the rate of production and extracellular polysaccharide polymers. The composition and function of the exopolysaccharide "slime" varies depending on bacterial species, growth phase, and nutrient conditions.

Exopolysaccharides produced by coagulase-negative staphylococci are usually composed of neutral monosaccharides, amino sugars, uronic acid, and polyols [16,36–38]. Studies of slime produced by *S. epidermidis* have revealed only the monomeric carbohydrate moieties and several amino acids, including glucose, galactose, mannose, glycerol, hexosamine, glycine, alanine, and phenylalanine [16,19, 39,40].

With a rich supply of nutrients, maximum biofilm development may occur within 1 to 3 days [41]. The extracellular polysaccharide slime serves a variety of purposes (Table 1). The slime is polyanionic and acts as an ion-exchange resin, concentrating nutrients and limiting the penetration of charged molecules such as cationic biocides [43]. Functional results of this characteristic include enhanced nutrition, moderated susceptibility to phagocytosis, antibody

---

FIG. 2 (Continued)
produced polymeric compounds or slime are adsorbed by the surface and coat the microbe (H), gluing the organism onto the surface. Adherence can be blocked by flooding the system with excess receptor [panel (C)] or with excess adhesion [panel (D)]. (From ref. 22; reprinted with permission.)

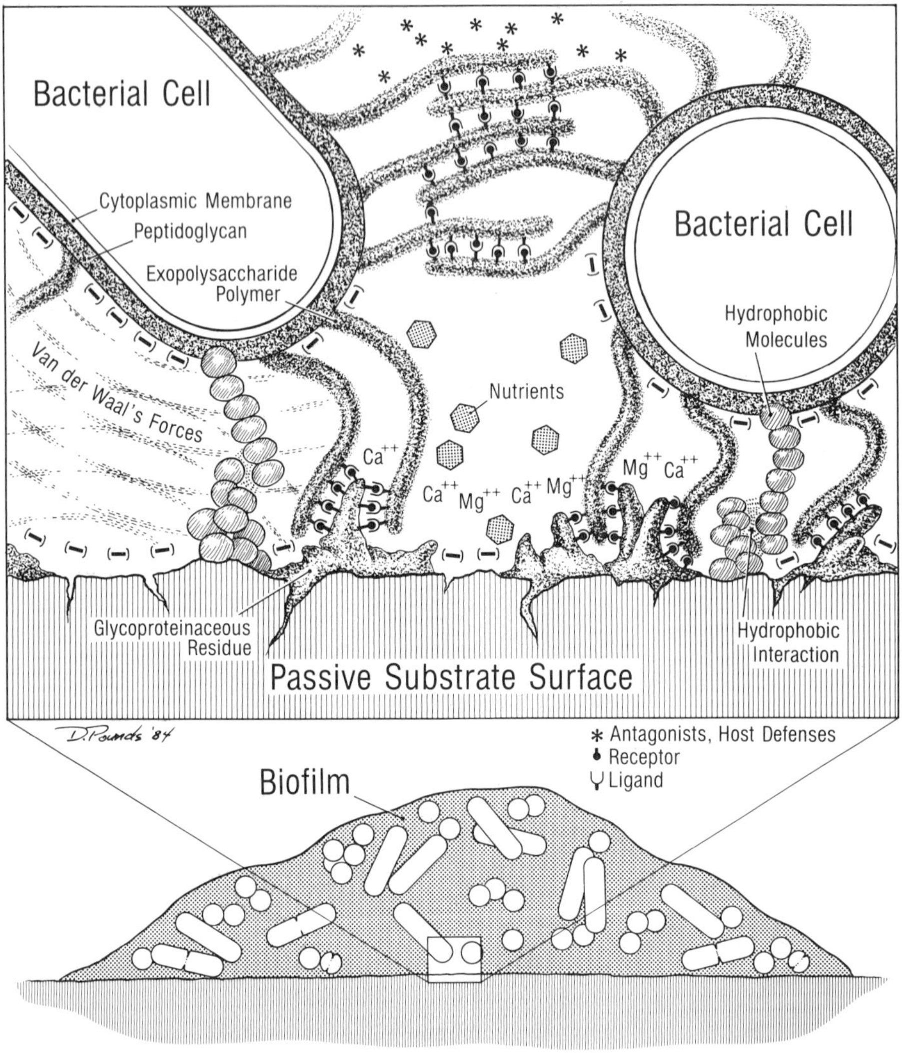

FIG. 3 Mechanisms of bacterial adherence. At specific distances the initial repelling forces between like charges on the surfaces of bacteria and substrate are overcome by van der Waals forces and hydrophobic interactions occur between molecules. Under appropriate conditions there is extensive development of exopolysaccharide polymers, allowing ligand—receptor interaction and pertinacious bonding of the bacteria to the substrate. (From ref. 18; reprinted with permission. Copyright 1985 by AAAS.)

TABLE 1 Biological Properties of Extracellular Slime Substance of *S. epidermidis*

1. Adhesion-like property
   - Adhesion to mammalian cells (?)
   - Attachment to polymer surfaces (?)
   - Intercell adherence (cell-cluster formation)
2. Interference with host defense mechanisms
   - Inhibition of T-cell blastogenesis
   - Inhibition of B-cell blastogenesis
   - Inhibition of immunoglobulin production
   - Enhancement of PMN adherence
   - Inhibition of PMN chemotaxis
   - Stimulation of PMN degranulation
   - Inhibition of bacterial opsonization
   - Inhibition of PMN chemiluminescence
   - Inhibition of *S. epidermidis* uptake by PMN
3. Enhancement of *S. epidermidis* virulence in mice
4. Interference with the action of antistaphylococcal antibiotics
5. Heparin-like property

Source: Ref. 42; reprinted with permission. Copyright 1988 by British Society for Antimicrobial Chemotherapy.

response, and increased virulence [17,18,39,42,44–48]. Characklis and Cooksey have also proposed that the biofilm may act as a site of increased intercellular communication and genetic transfer [49].

B. Microcolony Formation

Bacterial colonies may form as individual bacteria adhere to each other through polymer-to-polymer or polymer-to-surface interactions. These colonies may contain a single species or be polymicrobial [1]. The microenvironment that is created within a biofilm-enclosed colony appears to be an important factor in subsequent bacterial growth, propagation, and pathogenesis through the concentration of charged particles, molecules, and nutrients.

Most organisms within a mature microcolony are not attached to a two-dimensional surface but are fixed within the three-dimensional structure of the biofilm [41]. Bacterial species in consortia have shown higher metabolic activities than those of free organisms, a characteristic that may be related to cross-feeding and the development of suitable microenvironments within the adherent biofilm [50].

Many different species of bacteria may thrive within the biofilm, with significant interaction and even interdependency between organisms [41], such as the growth and activity of anaerobic species within a biofilm which is itself located in a highly aerobic bulk environment [48,51,52]. For example, the formation of dental caries involves a succession of microbes. The primary colonizing aerobic species grow on salivary glycoproteins. Facultative and anaerobic organisms then use the metabolic products of the primary colonizer and produce deleterious organic acids [41]. In oil pipelines, hydrocarbon-degrading bacteria create the anaerobic conditions and produce the necessary nutrients for sulfate-reducing bacteria, which are the principal cause of anaerobic corrosion [53].

Under normal conditions, host defense mechanisms are available to limit nutrients necessary for bacterial propagation. For example, iron is an essential element in bacterial growth and virulence [54–56]. However, free ionic iron is not usually available in the extracellular matrix of the human host [55]. In serum, the concentration of free iron is kept at a low level through binding by the protein transferrin [57,58]. Similarly, lactoferrin restricts the availability of iron on mucosal surfaces [54,55,59]. Consequently, successful pathogenic replication in an animal host may be restricted, based on the availability of iron and other nutrients [60]. The presence of biomaterials acts to disrupt these mechanisms: first, by providing a surface for bacterial adhesion, and second, through corrosion of the biomaterial surface, which provides nutrients for bacterial propagation and growth.

Both metallic and polymeric biomaterials are subject to corrosion and/or biodegradation upon implantation in aqueous environments. Corrosion of implanted metals may release cations (Fe^{3+}, Mg^{2+}, Cu^{2+}, Mn^{2+}, Zn^{2+}, K^{+}, Ni^{2+}) that can serve as coenzymes for protein and sugar metabolism and DNA replication [61,62]. Some microorganisms have shown an ability to mediate a significant amount of metallic corrosion [63]. For example, sulfate-reducing bacteria facilitate biocorrosion in anaerobic environments via depolarization of cathodes [63]. Bacteria may use trace ions such as Mg^{2+} and Ca^{2+} to stabilize (via acidic groups) exopolysaccharides in a gel, which enhances cell-to-cell and cell-to-surface adhesion and increases resistance to antagonists [3,47]. Some synthetic polymers, such as polyurethane and methyl methacrylate, contain ester bonds that may be hydrolyzed by staphylococci [64].

The formation of slime-enclosed microcolonies allows bacteria to compete with host proteins for iron and other nutrients. Bacteria may accumulate ions such as iron by localization of metabolites and siderophores [65] and by sequestering it from binding by transferrin or lactoferrin. The release of free iron from the biomaterial surface may also serve to saturate the iron-binding proteins. The sequence of inflammation, corrosion, reduced pH, increased iron concentration, and saturation of iron-binding proteins may result in increased virulence and inhibition of macrophage function [55–57,60,66].

Biofilm development involves not only phases of attachment, growth, and polysaccharide production, but also a phase of detachment [41]. Portions of the slime-enclosed microcolony may eventually be detached secondary to hemodynamic forces or trauma. Detachment may also be a normal feature of a dynamic equilibrium resulting, for example, from nutrient or oxygen depletion arising within the film [67]. These pathogenic inocula may then serve as a source of secondary infection and hematogenous septic emboli.

C. Glycoproteinaceous Conditioning Films

Implanted biomaterials may be rapidly coated by constituents of the serum and surrounding matrix, including fibronectin, osteonectin, vitronectin, albumin, fibrinogen, laminin, and collagen, and covalently bound, short-chain oligosaccharides [3,29,68–76]. Bacterial and tissue cells may then adhere to constituents of this film. Eukaryocytes have been shown to adhere to surfaces coated with fibronectin and collagen [69,70]. Certain strains of *S. aureus*, *S. epidermidis*, and *E. coli* have receptors or sets of receptors for fibronectin and collagen epitopes [71,73,74,76]. Studies have shown that pathogenic bacteria have greater numbers of conditioning film protein receptors than do a similar strain of nonpathogenic bacteria [69].

The interaction between conditioning film proteins and tissue and bacterial cells also depends on the physical state and quantity of bound proteins on the biomaterial surface. In vitro studies of fibronectin-coated biomaterials demonstrate that a low concentration of bound fibronectin enhances bacterial adherence, while a high concentration of bound fibronectin inhibits binding by the same bacteria [73–75]. Bacterial adhesion may also be decreased by the presence of albumin [76]. Thus conditioning protein molecules may play a variety of roles in bacterial adhesion, depending on their concentration and environmental conditions. Therefore, the role of fibronectin in the pathogenesis of infection remains to be established.

Collagen is a common constituent of the extracellular matrix and conditioning film and a major component of cartilage [77]. Cartilage is acellular and appears to promote colonization by bacteria. Infected rabbit joints examined under a microscope revealed that *S. aureus* predominantly colonized cartilaginous, not synovial, surfaces (Fig. 4). Biochemical studies have revealed that certain strains of *S. aureus* have collagen receptors on their surfaces [78]. The gradual destruction of cartilage that is seen in arthritis may be a consequence of the actions of host-originated inflammatory products and collagenase, which some strains of *S. aureus* can produce [79,80]. It may be possible to block bacterial adhesion to collagen by using a synthetic peptide $(\text{pro-gly-pro})_n$ that is believed to be similar to the collagen epitope [78].

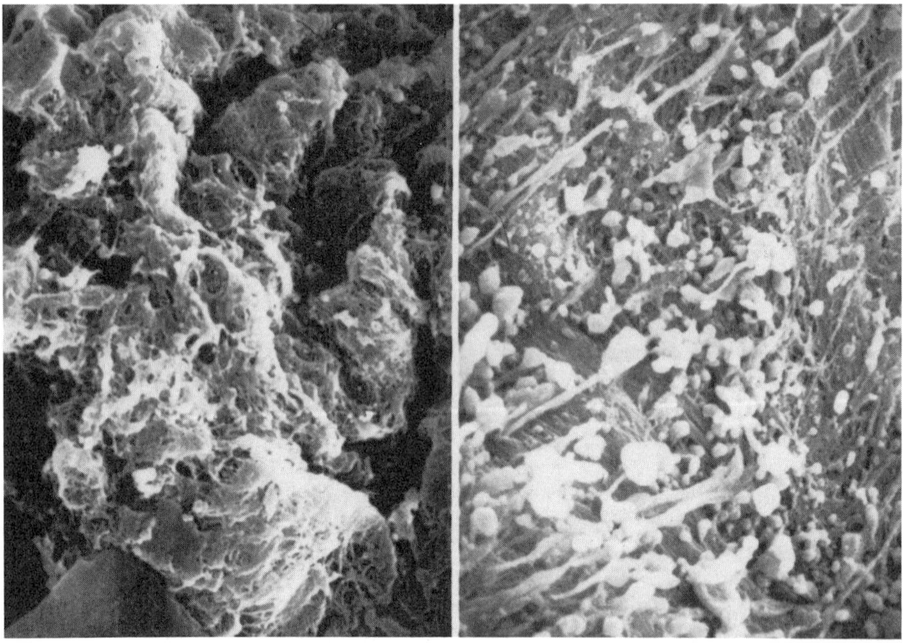

FIG. 4 Specimens of human synovium (a) and cartilage (b) from a clinically septic knee. Rare and isolated cocci were found on synovium. Abundant cocci are intermixed with inflammatory cells and debris on cartilage. [From ref. 77; reprinted with permission. Copyright 1988 by Butterworth & Company (Publishers) Ltd.]

The initial amount of surface free energy and the state of cleanliness of an implanted material may dramatically influence the adsorption of conditioning films and the subsequent adhesion of tissue and bacterial cells [29,81]. High surface energy appears to favor bacterial and tissue cell adhesion.

D. Tissue Adhesion and Integration

Integration may be defined as the functional assimilation of a transplanted tissue or tissue substitute by surrounding tissue without an intervening fibroinflammatory layer. Successful tissue integration of any biomaterial requires long-term physicochemical and biomechanical compatibility. Disruption of cell matrices and the glycocalyx, cell necrosis, and the presence of an inflammatory layer increases susceptibility to bacterial colonization.

Tissue cells are more highly specialized and less adaptive than bacteria, but like bacteria, they interact with other cells or biomaterials via adhesive glycoproteins (Fig. 5) [5,82,83]. Endothelia, epithelia, and fibroblasts have receptors (integrins) for the adhesive proteins (fibronectin, collagen, laminin, and fibrinogen) which mediate integration [5,82,83]. Skin, muscle, tendon, bone grafts, and organ transplants may be integrated, depending on the presence of adhesive surfaces/proteins and the viability of the graft.

The integration of autografts depends on mediation by adhesive proteins and cell survival without an intervening inflammatory layer or "non-self"-stimulated response. Allografts are more difficult to integrate. Allografts are immunogenic and usually induce an inflammatory response before cellular integration occurs.

IV. CHARACTERISTICS OF COMMONLY USED BIOMATERIALS

The principal metals used as tissue substitutes are titanium, titanium alloy, Vitallium, and stainless steel. In general, laboratory studies indicate that all metals behave similarly with regard to adhesion and colonization by bacteria, but their response may vary depending on the specific clinical setting.

The elemental composition and atomic geometry of the surface (0 to 10 Å) of a metal alloy may be significantly different from its bulk phase composition because of segregation of specific elements at the surface [4,84,85]. A thin (10 to 200 Å) oxide layer forms almost immediately on metal alloys upon exposure to the atmosphere. This oxide layer represents the true interactive surface between metallic implants and constituents of the surrounding milieu [2]. The stoichiometry of these surface oxides is variable and their integrity

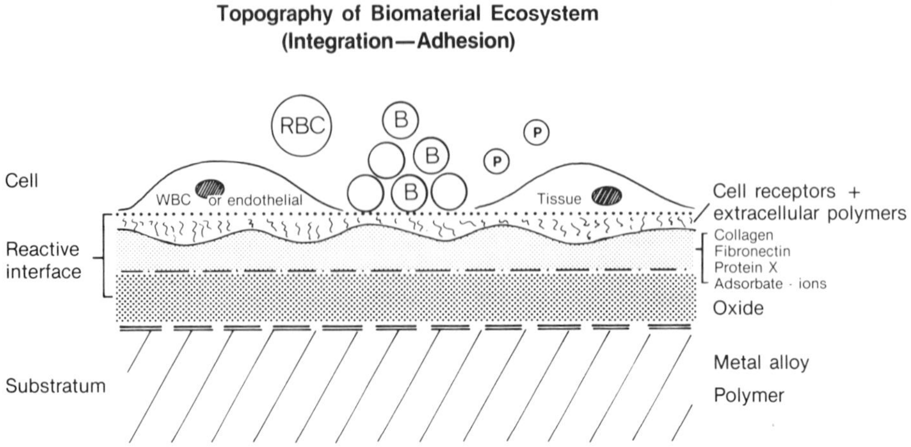

FIG. 5 A biomaterial substratum interacts with available cells via a reactive interface, the passivated surface (metallic oxide or polymer surface). This surface may be coated by a host-derived, aggregated lycoprotein mosaic. Extracellular polymers, cell receptors, and outer membrane glycoproteins then interact with constituents of this conditioning film. RBC, red blood cell; B, bacteria; P, platelets. [Reprinted with permission from A. G. Gristina et al., The race for the surface: Microbes, tissue cells, and biomaterials. In *Molecular Mechanisms of Adhesion* (L. Switalski, M. Hook, and E. H. Beachey, eds.), Springer-Verlag, New York, 1989, pp. 177–211.]

may be disrupted by grain boundaries and pinhole defects. The surface geometry and energy states of nonmetallic crystalline or amorphous polymers are also subject to rearrangement and relaxation, depending on their molecular composition or crystal structure and size.

Surface diffusion also plays a role in catalytic activity [86]. Due to the acqusition of energy from thermal vitrations in the underlying lattice, molecules adsorbed to clean surfaces diffuse about freely [86], and interact more frequently on a surface than in free solution. Adsorption at specific sites may also lower the activation energy for specific chemical reactions, which may then proceed at a lower temperature [86].

A. Metals

Metal surfaces are created by planar cuts through crystalline structures. The geometric arrangement of metal atoms at the exposed surface plane, and thus the number of unsatisfied bonds, depends on surface cut orientation [85,86]. Most metals have moderate-to-high surface energies (more than 40 dyn/cm), which are believed to be adhesive for tissue cells [29] and potentially capable of catalyzing chemical reactions [85,86].

1. Titanium

Pure titanium is an inert material with high corrosion resistance due to a tightly adherent oxide layer composed of several oxides, such as TiO, TiO_2, and Ti_2O_2, of which TiO_2 is the most common [87].

Titanium alloy (Ti6Al4V: 90% titanium, 6% aluminum, and 4% vanadium) is used for cementless total joint replacements that are designed for fixation by bony ingrowth (Fig. 6). Fiber titanium has mechanical properties near those of trabecular bone [88]. Pure titanium and Ti6Al4V are considered to be biocompatible and nontoxic [31]. However, aluminum may be neurotoxic and its long-term effects in vitro require further investigation [89].

New bone begins to form around a titanium implant within a few days of insertion. Apparently, the titanium surface is osteoconductive, as direct deposition of bone onto titanium implants has been observed [87]. Thus, in living bone, titanium may be less heavily colonized by bacteria than are other metals, which are separated from bone by a fibrous and sometimes inflammatory tissue sheath.

B. Polymers

Polymers are macromolecules composed of repeating units of monomers in which carbon forms the backbone, although oxygen, nitrogen, silicon, and sulfur may also be present. Most medical polymers are composed of amorphous zones, which confer toughness, and crystalline zones, which confer rigidity. The degree of crystallinity affects the rate of water sorption, which in turn affects the rate of biodegradation [29].

In vivo, the behavior of polymers is determined by molecular orientation, geometric isomerism, conformation, and configuration [90]. The stereochemistry of the polymer may influence its hydrophobicity and biological interactions [83,84]. The stereoregularity of the polymer has a strong influence on density and the rate and extent of solubility.

The histological response to polymers depends on the biocompatibility of the components, and degradation by heat, oxidation, mechanical perturbation, hydrolysis, enzymatic action, or electromagnetic radiation [91–93]. Polyesters may be broken down to eventually produce pyruvate, which can enter the Krebs cycle [94]. In general, polymers have surface energies in the low and intermediate

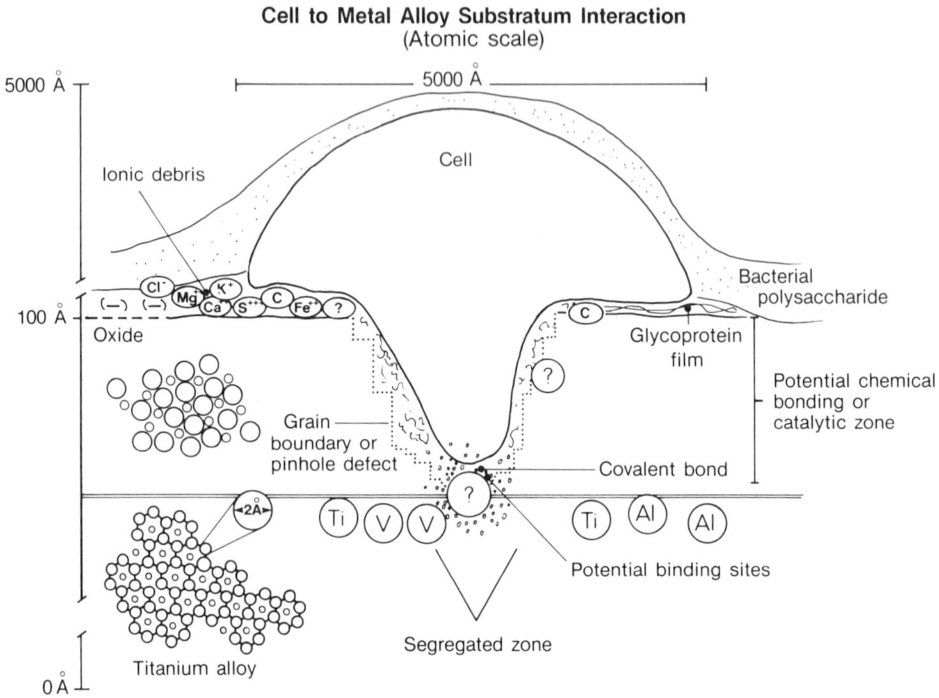

FIG. 6 Schematic representation of the possible interaction between cells and a metal alloy biomaterial surface at any atomic level. Cell and surface perimeters are exaggerated because of the nonlinear scale. A cell (bacterial or tissue cell) is shown at an irregularity in the oxide layer of Ti6Al4V. Atomic geometry and electronic state direct the accumulation of ionic contaminants, resulting in catalytic processes. Ionic debris (ovals encircling dots) may include Cl^-, Mg^{2+}, K^+, Ca^{2+}, S^{3+}, C, and Fe^{2+}. Unsatisfied or dangling bonds provide potential binding sites. (From ref. 1; reprinted with permission. Copyright 1987 by AAAS.)

range. Therefore, tissue integration with polymers is poor and often characterized by an inflammatory inferface [95].

C. Methyl Methacrylate

Methyl methacrylate is used as fixation material for joint replacements. It is noncrystalline and porous and is not well integrated by tissue cells, due to the extensive formation of connective tissue, which interferes with the mechanical interlock between cement and bone. Goldring et al. have reported that a synovial-like membrane forms at the bone–cement interface in loose total hip replacements [96]. Methyl methacrylate may actually provoke an inflammatory response [95], inhibit host defenses [97], and provide an excellent substratum for bacterial colonization [98].

V. BIOLOGICAL SUBSTRATA

It is virtually impossible to implant a biomaterial device without causing tissue damage. Traumatized soft tissue is rich in nutrients and receptors which may be used in bacterial colonization. Endothelial cells are surrounded by a well-developed glycocalyx. Disruption of this extramembranous polysaccharide margin exposes fibronectin and other receptors for bacterial adhesion [99–101]. Endothelial damage may also be a factor in site localization by trauma or by septic emboli in osteomyelitis.

Gram-positive bacteria, *S. aureus*, *Streptococcus pyogenes*, *Str. mutans*, and *Str. sanguis*, which are common causes of bacterial endocarditis, may bind to fibronectin, fibrin, or platelets [22,71,73, 102]. Conditioning of plastic valves and traumatized natural valves by fibronectin, fibrin, and platelet vegetations may be an initial step in bacterial colonization. Bridging by platelets allows bacterial binding to damaged tissues [6,101,102].

Bone is a composite of calcium hydroxyapatite crystals in a collagen matrix. Some species of staphylococci have receptors for collagen and may easily colonize devitalized bone, and may metabolize proteinaceous and mineral constituents [3,18,103]. Bone allografts, which represent a large mass of dead bone and cartilage, have a clinically demonstrated infection rate of 5 to 14% [104].

VI. CLINICAL EXPERIENCE

A. Total Artificial Heart

The development and use of artificial hearts is a milestone in medicine. They are constructed from a variety of materials, including polyurethane, poly(vinyl chloride), Dacron, pyrolytic carbon, graphite, and titanium [105]. The complex structure of the artifical heart requires compatibility between materials and between the biomaterial surface, surrounding tissue and blood elements. Hemodynamic interactions within the device create fluid eddies and cause damage to tissue and blood elements, conditions favoring the formation of clotting cascades and adhesion by bacteria. Potential receptor sites for bacteria are exposed at the juncture between the natural vessel and the total artificial heart. The transcutaneous drive lines required to furnish power for the artificial heart provide a pathway for contamination.

Total artificial heart patients manifest classic features of biomaterial-centered infections [106–109]. The clinical course of the two patients in these studies was typical of resistant biomaterial-centered infections, including intermittent septicemia, distant abscess formation, and organ failure. Polymicrobial colonization was typical in each patient, although a single species appeared numerically dominant. The infections, which included nonpathogens or uncommon organisms, were resistant to treatment and were only partially suppressed by antibiotics.

Tissue necrosis and death adjacent to biomaterial surfaces of the artificial heart created an ideal environment for bacterial colonization [108]. Characteristically, bacteria colonized body cavities adjacent to the device, the drive lines, interior hemodynamic surfaces in contact with blood elements, exterior surfaces traversing tissue planes, and great vessels [107]. Therefore, the infections that occurred may have resulted from direct contamination or by hematogenous spread.

B. Ocular Implants and Hydrophilic Polymers

The use of extended-wear contact lenses is dramatically increasing the frequency of bacterial corneal ulceration. The contact lens itself provides a suitable substratum for bacterial colonization and a source of subsequent inocula to compromised epithelial cells [19, 110].

Recent studies of extended-wear contact lenses in vitro have demonstrated biofilm-mediated adhesion by *P. aeruginosa* and *S. epidermidis* [19,110]. The *Pseudomonas* appears to be held in a position via a glue-like mechanism. Alternatively, *S. epidermidis* appears to achieve direct contact with the undersurface of the lens, allowing access to and a possible chemical bonding to the damaged corneal epithelial surface. In contrast, *Pseudomonas* is rarely observed in direct contact with the lens.

C. Host Defense Response

Host defense or immune mechanisms are invariably disrupted by the implantation of biomaterials and tissue transplants. The mechanisms of this disruption are unknown. Early concepts suggested that tissue damage caused by a foreign-body increased susceptibility to infection, activated host defenses, and stimulated the release of inflammatory mediators, including oxygen radicals and lysosomal enzymes.

Bacteria then amplified tissue damage by serving as inflammatory agents. Other possible mechanisms include the influence of the extracellular polysaccharide slime and the "preemption" of the macrophage oxidative response.

Extracellular slime produced by *S. epidermidis* has been reported to inhibit T-cell and B-cell blastogenesis, polymorphonuclear leukocyte chemotaxis, opsonization, and chemiluminescence, and to enhance the virulence of *S. epidermidis* in mice [41]. Johnson et al. showed that phagocytosis by polymorphonuclear leukocytes of *S. epidermidis* grown on a plastic surface decreased over time [111]. As slime production by *S. epidermidis* is time dependent, these results support the theory that excess extracellular slime impairs phagocytosis by polymorphonuclear leukocytes. Conversely, the presence of foreign bodies also enhances infection by non-slime-producing organisms (e.g., *S. aureus*); therefore, the presence of slime alone cannot fully explain the impairment of host defense mechanisms.

Related studies have shown that after organisms adhere to biomaterials their resistance to antibiotic therapy increases, and that this is specific to the type of biomaterial [112,113]. Organisms adhering to some hydrocarbon polymers are extremely resistant to antibiotic treatment. When the same organisms adhere to metals, their resistance to antibiotic therapy is greatly reduced.

Early theories proposed that the mechanism of this increased antibiotic resistance was that the biofilm formed a "diffusion barrier" [41]. However, more recent studies have discounted this mechanism and have shown that bacteria within biofilms have a decreased metabolic rate and undergo phenotypic changes that may influence resistance and virulence [9,114–116]. It has also been shown that significantly higher levels of antibiotics are required to kill surface-adherent bacteria than the same species of bacteria in suspension [112]. Therefore, bacteria on surfaces or within microcolonies are probably physiologically different from free-floating organisms. Consequently, their response to antibiotics may be altered, depending on the specific chemistry of the cell and the mechanism of action of the antibiotic.

The interaction between professional phagocytes and a foreign body may somehow exhaust or preempt killing mechanisms. Klock and Bainton [117] observed that granulocytes exposed to nylon fibers retained their chemotactic response but had an impaired bactericidal response. Studies have shown that neutrophils inside Teflon cages placed in peritoneal cavities exhibited decreased phagocytic and bactericidal potential and reduced superoxide production [118,119]. A similar pattern was noted when normal rabbit alveolar macrophages were exposed to poly(methyl methacrylate) spheres [39]. From these results it would appear that foreign body surfaces trigger a "slow burst" and "preempt" a second burst if the phagocyte subsequently encounters an infectious organism. It is almost certain that the killing capacity of such preempted phagocytes is impaired.

D. The Race for the Surface

The molecular mechanisms of bacterial adhesion and tissue cell integration of biomaterial surfaces are interrelated and similar. Therefore, an implant surface may be conceptualized as the object of a "race" between macromolecules, bacteria, and tissue cells (Fig. 7). Because of organic adsorbates and the nature of surfaces, potential colonizing cells first encounter a highly altered, reactive interface. If tissue cells arrive first at a biomaterial surface and integration is achieved, bacterial adhesion may be prevented by the presence of extracapsular polysaccharides and functional host defense mechanisms.

However, bacteria are relatively ubiquitous in biologic environments and have well-established mechanisms for adhering to inanimate substrata and certain cell surfaces. Tissue cells are invariably

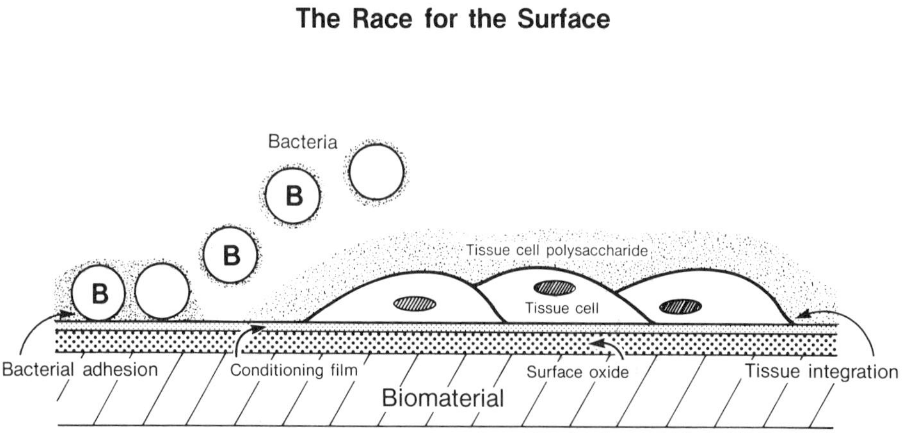

FIG. 7 At the instant of insertion, a biomaterial represents a ready surface for colonization. Macromolecules, bacteria, and tissue cells compete for domains at the reactive surface interface. The destiny of an implant will be determined by the conditioning macromolecules and cells that dominate its surfaces. (From ref. 1; reprinted with permission. Copyright 1987 by AAAS.)

unable to displace pioneer bacterial colonies, and biomaterials used currently generally do not facilitate tissue integration. Tissue cells are highly specialized, and in effect, for most surfaces, their colonization potential is low compared to that of the more adaptive bacterial cells. Therefore, implanted biomaterial surfaces may primarily be the "eminent domain" of organic and inorganic moieties and bacteria rather than of tissue cells. In summary, the "race for the surface" is a contest between tissue cells and bacteria for domination of an available biomaterial surface.

VII. THE FUTURE

The key to solving the problem of biomaterial-centered infection may be the cireation of idealized biomaterial surfaces that will "direct" adhesion by tissue cells and prevent adhesion by bacteria. Advanced instrumentation such as ion scattering spectroscopy, Auger electron spectroscopy, electron spectroscopy for chemical analysis, low-energy electron diffraction, scanning tunneling microscopy, secondary ion mass spectrometry, atom scattering techniques, electron-stimulated desorption, and photon-stimulated desorption may be used to characterize the atomic and molecular phenomena that occur at biomaterial surfaces [4,84,85,120,121]. This information may then be used to study molecular transformations and cellular responses at the surface in a liquid biologic environment.

Surface modification techniques such as heavy ion implantation, chemical vapor deposition, and vacuum evaporation may be used to create quantum states or energy levels that will decrease adhesion by organic or ionic moieties present in conditioning films or on adhesive bacterial cell surfaces. Alternatively, adhesive zones or margins may be created to enhance biocompatibility and tissue integration. Surface area or texture may be altered to provide or prevent mechanical interdigitation.

"Precolonization" of prosthetic vascular grafts with endothelial cells has been attempted, with encouraging results that indicate a lower infection rate [122–125]. Similar encouraging results have been achieved indirectly in experiments with artificial skin (epidermal and combined dermal layers) [126,127].

The use of appropriate antibiotics may be effective in the period immediately following implantation, before bacteria have a chance to adhere (Fig. 8). The selection and method of delivery (topical, layered, or eluted from the surface) of an antibiotic should be based on its ability to interfere with bacterial polymer synthesis and adhesive mechanisms, or to penetrate biofilms. Analogs that compete with adhesions, lectins, ligands, or receptors may be developed based on detailed knowledge of the molecular structure of receptors.

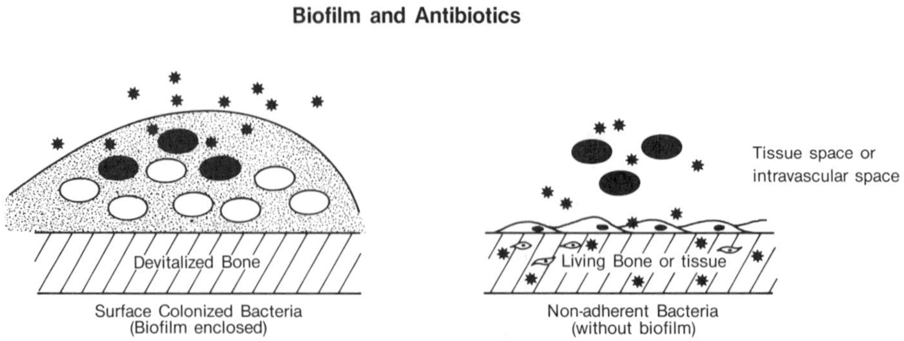

FIG. 8 Antibiotic effectives are blocked by biofilm about bacteria in organized infection on the surface of a sequestrum or biomaterial substratum. Free-floating bacteria, not protected by biofilm, are more susceptible to standard antibiotic levels that reach organisms via normal circulation or tissue. [Reprinted with permission from A. G. Gristina et al., Microbial adhesion and the pathogenesis of biomaterial-centered infections, in *Orthopaedic Infection: Diagnosis and Treatment* (R. B. Gustilo, R. P. Gruninger, and D. T. Tsukayama, eds.), W. B. Saunders Co., Philadelphia, 1989, pp. 3–25.] ○ Live bacteria; ● Dead bacteria; ✱ Antibiotic; Living tissue cell (osteoblast, endothelium, periosteum); Biofilm.

Modern technology has made possible the creation and use of manufactured materials as substitutes for damaged natural host tissues and organs, thereby relieving pain and extending survival. The same technology must now be used to defeat the highly adaptive mechanisms by which bacteria use manufactured materials for survival, and to encourage the sensitive and normal processes of host tissue cell biomolecular communication in the race for the surface.

REFERENCES

1. A. G. Gristina, Biomaterial-centered infection: microbial adhesion versus tissue integration, *Science 237*:1588–1595 (1987).

2. T. Albrektsson, The response of bone to titanium implants, *CRC Crit. Rev. Biocompat. 1*:53–84 (1985).
3. A. G. Gristina, C. D. Hobgood, and E. Barth, Biomaterial specificity, molecular mechanisms and clinical relevance of *S. epidermidis* and *S. aureus* infections in surgery, in *Pathogenesis and Clinical Significance of Coagulase-Negative Staphylococci* (G. Pulverer, P. G. Quie, and G. Peters, eds.), Gustav Fischer, Stuttgart, West Germany, 1987, pp. 143–157].
4. B. Kasemo and J. Lausmaa, Surface science aspects on inorganic biomaterials, *CRC Crit. Rev. Biocompat. 2*:335–380 (1986).
5. E. Ruoslahti and M. D. Pierschbacher, New perspectives in cell adhesion: RGD and integrins, *Science, 238*:491–497 (1987).
6. G. D. Christensen, W. A. Simpson, and E. H. Beachey, Adhesion of bacteria to animal tissues: complex mechanisms, in *Bacterial Adhesion: Mechanisms and Physiologic Significance* (D. C. Savage and M. Fletcher, eds.), Plenum Press, New York, 1985, pp. 279–305.
7. B. Sugarman and E. J. Young, eds. *Infections Associated with Prosthetic Devices*, CRC Press, Boca Raton, Fla., 1984.
8. P. Naylor, L. Webb, R. Jennings, and A. Gristina, Resistant infection and internal fixation devices, Abstract, *Orthopaedic Trauma Association Annual Meeting*, Dallas, Oct. 27–29, 1988, p. 21.
9. G. D. Christensen, L. M. Baddour, and A. Simpson, Phenotypic variation of *Staphylococcus epidermidis* slime production in vitro and in vivo, *Infect. Immun., 55*:2870–2877 (1987).
10. K. Henderson, Infections associated with prostheses, shunts, and implants, in *Infectious Disease: Diagnosis and Management* (T. T. Yoshikawa, A. W. Chow, and L. B. Guze, eds.), Wiley, New York, 1980, pp. 274–285.
11. S. H. Dougherty and R. L. Simmons, Infections in bionic man: the pathobiology of infections in prosthetic devices, part 2, *Curr. Prob. Surg. 19*:269–318 (1982).
12. D. F. Bandyk, G. A. Berni, B. L. Thiele, and J. B. Towne, Aortofemoral graft infection due to *Staphylococcus epidermidis*, *Arch. Surg. 119*:102–107 (1984).
13. H. Masur and W. D. Johnson, Prosthetic valve endocarditis, *J. Thorac. Cardiovasc. Surg. 80*:31–37 (1980).
14. F. D. Lowy and S. M. Hammer, *Staphylococcus epidermidis* infections, *Ann. Intern. Med. 99*:834–839 (1983).
15. D.J. Winston, D. V. Dudnick, M. Chapin, G. H. Winston, R. P. Gale, and W. J. Martin. Coagulase-negative staphylococci bacteremia in patients receiving immunosuppressive therapy, *Arch. Intern. Med. 143*:32–36 (1983).

16. I. Ichiman and K. Yoshida, The relationship of capsular type of *Staphylococcus epidermidis* to virulence and induction of resistance in the mouse, *J. Appl. Bacteriol.*, *51*:229–241 (1981).
17. J. Dankert, A. H. Hogt, and J. Feijen, Biomedical polymers: bacterial adhesion, colonization, and infection, *CRC Crit. Rev. Biocompat.* *2*:219–301 (1986).
18. A. G. Gristina, M. Oga, L. X. Webb, and C. Hobgood, Bacterial colonization in the pathogenesis of osteomyelitis, *Science* *228*:990–993 (1985).
19. M. M. Slusher, Q. N. Myrvik, J. C. Lewis, and A. G. Gristina, Extended-wear lenses, biofilm, and bacterial adhesion, *Arch. Ophthalmol.* *105*:110–115 (1987).
20. A. G. Gristina, L. X. Webb, and E. Barth, Microbial adhesion, biomaterials and man, in *Infection in the Orthopaedic Patient* (R. Coombs and R. H. Fitzgerald, eds.), Butterworths, London, 1989, pp. 30–42.
21. G. Cierny, L. Couch, and J. Mader. Adjunctive local antibiotics in the management of contaminated orthopaedic wounds, in *Final Program of the AAOS 53rd Annual Meeting*, American Academy of Orthopaedic Surgeons, Park Ridge, Ill., 1986, p. 86.
22. G. D. Christensen, W. A. Simpson, and E. H. Beachey, Microbial adherence in infection, in *Principles and Practice of Infectious Diseases*, 2nd ed. (G. L. Mandell, R. G. Douglas, Jr., and J. E. Bennett, eds.), Wiley, New York, 1985, pp. 6–22.
23. B. V. Derjaguin and L. Landau, Theory of the stability of strongly charged lyophobic sols and of the adhesion of strongly charged particles in solutions of electrolytes, *Acta Physiochim. URSS* *14*:633–656 (1941).
24. E. J. W. Verwey and J. T. G. Overbeek, *Theory of the Stability of Lyophobic Colloids*, Elsevier, London, 1948.
25. R. G. Greig and M. N. Jones, *Biosystems* *9*:43–55 (1977).
26. A. S. G. Curtis, Cell adhesion, in *Progress in Biophysics and Molecular Biology*, Vol. 27 (J. A. V. Butler and D. Noble, eds.), Pergamon Press, Elmsford, N.Y., 1973, pp. 317–386.
27. R. M. Pashley, P. M. McGuiggan, B. W. Ninham, and D. F. Evans, Attractive forces between uncharged hydrophobic surfaces: direct measurements in aqueous solutions, *Science 229*: 1088–1089 (1984).
28. P. J. Watt and M. E. Ward, Adherence of *Neisseria gonorrhoeae* and other *Neisseria* species to mammalian cells, in *Bacterial Adherence: Receptors and Recognition*, Ser. B, Vol. 6 (E. H. Beachey, ed.), Chapman & Hall, London, 1980, pp. 253–288.
29. R. E. Baier, A. E. Meyer, J. R. Natiella, R. R. Natiella, and J. M. Carter, Surface properties determine bioadhesive outcomes: methods and results, *J. Biomed. Mater. Res.*, *18*:337–355 (1984).

30. J. Lyklema, Interfacial electrochemistry of surfaces with biomedical relevance, in *Surface and Interfacial Aspects of Biomedical Polymers*, Vol. 1, *Surface Chemistry and Physics* (J. Andrade, ed.), Plenum Press, New York, 1985, pp. 293–336.
31. D. C. Mears, *Materials and Orthopedic Surgery*, Williams & Wilkins, Baltimore, 1979.
32. A. Olsen, A. Jonsson, and S. Normark, Fibronectin binding mediated by a novel class of surface organelles on *Escherichia coli, Nature 338*:652–655 (1989).
33. J. W. Costerton and R. T. Irvin, The bacterial glycocalyx in nature and disease, *Annu. Rev. Microbiol. 35*:299–324 (1981).
34. T. A. Stenstrom and S. Kjelleberg, Fimbriae mediated nonspecific adhesion of *Salmonella typhimurium* to mineral particles, *Arch. Microbiol. 143*:6–10 (1985).
35. G. W. Jones and R. E. Isaacson. Proteinaceous bacterial adhesions and their receptors, *CRC Crit. Rev. Microbiol. 10*:229–260 (1984).
36. G. Peters, F. Schumacher-Perdreau, B. Jansen, M. Bey, and G. Pulverer, Biology of *Staphylococcus epidermidis* Extracellular slime, in *Pathogenesis and Clinical Significance of Coagulase-Negative Staphylococci* (G. Pulverer, P. G. Quie, and G. Peters, eds.), Gustav Fischer, Stuttgart, West Germany, 1987, pp. 15–32.
37. B. K. Brandley and R. L. Schnarr, Cell-surface carbohydrates in cell recognition and response, *J. Leokocyte Biol. 40*:97–111 (1986).
38. I. W. Sutherland, Bacterial polysaccharides: their nature and production, in *Surface Carbohydrates of the Prokaryotic Cell* (I. W. Sutherland, ed.), Academic Press, London, 1978, pp. 27–96.
39. Q. N. Myrvik, W. Wagner, E. Barth, P. Wood, and A. G. Gristina, Effects of extracellular slime produced by *Staphylococcus epidermidis* on oxidative responses of rabbit alveolar macrophages, *J. Invest. Surg.* (in press).
40. A. G. Gristina, E. Barth, L. X. Webb, and Q. N. Myrvik, Microbial adhesion, molecular mechanisms and therapeutic strategies in biomaterial- and compromised tissue-centered infections, in *34th ECCM Symposium: Foreign Body-Related Infections* (P. K. Peterson and A. Fleer, eds.), Excerpta Medica, Amsterdam, 1987, pp. 26–35.
41. W. A. Hamilton, Biofilms: microbial interactions and metabolic activities, *Symp. Soc. Gen. Microbiol. 41*:361–385 (1987).
42. G. Peters, New considerations in the pathogenesis of coagulase-negative staphylococcal foreign body infections, *J. Antimicrob. Chemother. 21*(Suppl. C): 139–148 (1988).

43. J. W. Costerton and E. S. Lasen, The inherent biocide resistance of corrosion-causing biofilm bacteria, *Mater. Perf. 23*: 13–17 (1984).
44. S. Schwartzmann and J. R. Boring III, Antiphagocytic effect of slime from a mucoid strain of *Pseudomonas aeruginosa*, *Infect. Immun. 3*:762–767 (1971).
45. M. N. Gwynn, L. T. Webb, and G. N. Rolinson, Regrowth of *Pseudomonas aeruginosa* and other bacteria after the bactericidal action of carbenicillin and other β-lactam antibiotics, *J. Infect. Dis. 144*:263–269 (1981).
46. J. C. Nickel, J. B. Wright, I. Ruseska, T. J. Marrie, C. Whitfield, and J. W. Costerton, Antibiotic resistance of *Pseudomonas aeruginosa* colonizing a urinary catheter in vitro, *Eur. J. Clin. Microbiol. 4*:213–218 (1985).
47. M. Fletcher, Adherence of marine microorganisms to smooth surfaces, in *Bacterial Adherence: Receptors and Recognition*, Series B, Vol. 6 (E. H. Beachey, ed.), Chapman & Hall, London, 1980, pp. 345–374.
48. R. J. Gibbons and J. V. van Houte, Bacterial adherence in oral microbial ecology, *Annu. Rev. Microbiol. 29*:19–44 (1975).
49. W. G. Characklis and K. E. Cooksey, Biofilms and microbial fouling, *Adv. Appl. Microbiol. 29*:93–138 (1983).
50. W. H. Jeffrey and J. H. Paul, Activity measurements of planktonic microbial and microfouling communities in a eutrophic estuary, *Appl. Environ. Microbiol. 51*:157–162 (1986).
51. W. A. Hamilton, Energy sources for microbial growth: an overview, in *Aspects of Microbial Metabolism and Ecology* (G. A. Codd, ed.), Academic Press, London, 1984, pp. 35–57.
52. H. W. Paerl, Influence of attachment on microbial metabolism and growth in aquatic ecosystems, in *Bacterial Adhesion: Mechanisms and Physiologic Significance* (D. C. Savage and M. Fletcher, eds.), Plenum Press, New York, 1985, pp. 363–400.
53. W. A. Hamilton, Sulphate-reducing bacteria and anerobic corrosion, *Annu. Rev. Microbiol. 39*:195–217 (1985).
54. J. J. Bullen, The significance of iron in infection, *Rev. Infect. Dis. 3*:1127–1138 (1981).
55. E. D. Weinberg, Iron and susceptibility to infectious disease, *Science 184*:952–956 (1974).
56. M. Dho and J. P. Lafont, Adhesive properties and iron uptake ability in *Escherichia coli* lethal and nonlethal for chicks, *Avian Dis. 28*:1016–1025 (1984).
57. J. J. Bullen, H. J. Rogers, and E. Griffiths, Bacterial iron metabolism in infection and immunity, in *Microbial Iron Metabolism: A Comprehensive Treatise* (J. B. Beilands, ed.), Academic Press, New York, 1974, pp. 518–551.

58. A. B. Schryvers and L. J. Morris, Identification and characterization of the human lactoferrin-binding protein from *Neisseria meningitidis, Infect. Immun. 56*:1144–1149 (1988).
59. P. L. Masson, J. F. Heremans, and C. H. Dive, An iron-binding protein common to many external secretions, *Clin. Chim. Acta 14*:729–734 (1966).
60. M. R. W. Brown and P. Williams, The influence of environment on envelope properties affecting survival of bacteria in infections, *Annu. Rev. Microbiol. 39*:527–556 (1985).
61. A. L. Lehninger, *Principles of Biochemistry*, Worth, New York, 1985.
62. J. Black, Does corrosion matter? *J. Bone J. Surg. Br. Vol. 70*:517–519 (1988).
63. L. Daniels, N. Belay, B. Rajagopal, and P. J. Weimer, Bacterial methanogenesis and growth from CO_2 with elemental iron as the sole source of electrons, *Science 237*:509–511 (1987).
64. A. Ludwicka, R. Locci, B. Jansen, G. Peters, and G. Pulverer, Microbial colonization of prosthetic devices v. attachment of coagulase-negative staphylococci and "slime"-production on chemically pure synthetic polymers, *Zentralbl. Bakteriol. Mikrobiol. Hyg. Abt I Orig. B 177*:527–532 (1983).
65. S. Sriyoschati and C. D. Cox, Siderophore-mediated iron acquisition from transferrin by *Pseudomonas aeruginosa, Infect. Immun. 52*:885–891 (1986).
66. I. Kochan, J. T. Kvach, and T. I. Wiles, Virulence associated acquisition of iron in mammalian serum by *Escherichia coli, J. Infect. Dis. 135*:623–632 (1977).
67. J. A. Howell and B. Atkinson, Sloughing of microbial film on trickling filters, *Water Res. 18*:307–315 (1976).
68. J. R. W. Govan, Mucoid strains of *Pseudomonas aeruginosa*: the influence of culture medium on the stability of mucus productions, *J. Med. Microbial. 8*:513–522 (1975).
69. R. A. Proctor, Fibronectin: A brief overview of its structure, function, and physiology, *Rev. Infect. Dis. 9*(suppl. 4): S317–S321 (1987).
70. P. B. van Wachem, C. M. Vreriks, T. Beugeling, J. Feijen, A. Bantjes, J. P. Detmers, and W. G. van Aken, The influence of protein adsorption of interactions of cultured human endothelial cells with polymers, *J. Biomed. Mater. Res. 21*: 701–218 (1987).
71. L. M. Switalski, C. Ryden, K. Rubin, A. Ljungh, M. Hook, and T. Wadstrom, Binding of fibronectin to *Staphylococcus* strains, *Infect. Immun. 42*:628–633 (1983).
72. G. M. Vercellotti, J. B. McCarthy, P. Lindolm, P. K. Peterson, H. S. Jacob, and L. T. Furcht, Extracellular matrix

proteins (fibronectin, laminin and type IV collagen) bind and aggregate bacteria, *Am. J. Pathol. 120*:13–21 (1985).
73. M. Hermann, P. E. Vaudaux, D. Pittet, R. Auckenthaler, P. D. Lew, F. Schumacher-Perdreau, G. Peters, and F. A. Waldvogel, Fibronectin, fibrinogen, and laminin act as mediators of adherence of clinical staphylococcal isolates to foreign materials, *J. Infect. Dis. 158*:693–701 (1988).
74. P. E. Vaudaux, F. A. Waldvogel, J. J. Morgenthaler, and U. E. Nydegger, Adsorption of fibronectin onto polymethylmethacrylate and promotion of *Staphylococcus aureus* adherence, *Infect. Immun. 45*:768–774 (1984).
75. S. F. Mohammad, N. S. Topham, G. L. Burns, and D. B. Olsen, Enhanced bacterial adhesion on surfaces pretreated with fibrinogen and fibronectin, *Trans. Am. Soc. Artif. Intern. Organs 34*:573–577 (1988).
76. M. Deyme, A. Baszkin, J. E. Proust, E. Perez, E. Albrecht, and M. M. Boissonnade, Collagen at interfaces. II. Competitive absorption of collagen against albumin and fibrinogen, *J. Biomed. Mater. Res. 21*:321–328 (1987).
77. A. Voytek, A. G. Gristina, E. Barth, Q. Myrvik, L. Switalski, M. Hook, and P. Speziale, Staphylococcal adhesion to collagen in intra-articular sepsis, *Biomaterials 9*:107–110 (1988).
78. P. Speziale, G. Raucci, L. Visai, L. M. Switalski, R. Timpl, and M. Hook, Binding of collagen to *Staphylococcus aureus* Cowan 1, *J. Bacteriol. 167*:77–81 (1986).
79. D. J. McCarthy, *Arthritis and Allied Conditions*, Lea & Febiger, Philadelphia, 1985.
80. R. L. Smith and D. J. Schurman, Comparison of cartilage destruction between infectious and adjuvant arthritis, *J. Orthop. Res. 1*:136–143 (1983).
81. R. E. Baier, Conditioning surfaces to suit the biomedical environment: recent progress, *J. Biomech. Eng. 104*:257–??? (1982).
82. K. R. Gehlsen, L. Dillner, E. Engvall, and E. Ruoslahti, The human laminin receptor is a member of the integrin family of cell adhesion receptors, *Science 241*:1228–1229 (1988).
83. I. I. Singer, S. Scott, D. W. Kawka, D. M. Kazazis, J. Gailit, and E. Ruoslahti, Cell surface distribution of fibronectin and vitronectin receptors depends on substrate composition and extracellular matrix accumulation, *J. Cell Biol. 106*:2171–2182 (1988).
84. R. Tromp, R. J. Hamers, and J. E. Demuth, Quantum states and atomic structure of silicon surfaces, *Science 234*:304–309 (1986).
85. J. R. Noonan and H. L. Davis, Atomic arrangements at metal surfaces, *Science 234*:310–316 (1986).

86. R. Gomer, Surface diffusion, *Sci. Am. 247*(Aug.):98–109 (1982).
87. T. Albrektsson, P.-I. Branemark, H. A. Hansson, and J. Lindstrom, Osseointegrated titanium implants: requirements for ensuring longlasting, direct bone-to-implant anchorage in man, *Acta Orthop. Scand. 52*:155–170 (1981).
88. W. Rostoker, J. O. Galante, and G. Shen, Some mechanical properties of sintered fiber metal composites, *J. Test. Eval. 2*:107–112 (1974).
89. H. W. Wisnewski, Neurotoxicity of aluminum, in *Neurotoxicology* (H. Rozin and R. Shiraki, eds.), Raven Press, New York, 1977, pp. 313–335.
90. J. O. Hollinger and G. C. Battistone, Biodegradable bone repair materials: synthetic polymers and ceramics, *Clin. Orthop. 207*:290–305 (1986).
91. J. D. Andrade, D. E. Gregonis, and L. M. Smith, Polymer surface dynamics, in *Surface and Interfacial Aspects of Biomedical Polymers*, Vol. 1, *Surface Chemistry and Physics* (J. D. Andrade, ed.), Plenum Press, New York, pp. 15–41.
92. R. K. Kulkarni, Brief review of biochemical degradation of polymers, *Polym. Eng. Sci. 227*(Oct.) (1965).
93. G. Odian, *Principles of Polymerization*, 2nd ed., Wiley-Interscience, New York, 1981.
94. J. O. Hollinger, Preliminary report on the osteogenic potential of a biodegradable copolymer of polylactide (PLA) and polyglycolide (PGA), *J. Biomed. Mater. Res. 17*:71–82 (1983).
95. S. C. Kozinn, N. A. Johanson, and P. G. Bullough, The biologic interface between bone and cementless femoral endoprostheses, *J. Arthroplasty 1*:249–259 (1986).
96. S. R. GOldring, A. L. Schiller, M. Roelke, C. M. Rourke, D. A. O'Neill, and W. H. Harris, The synovial-like membrane at the bone–cement interface in loose total hip replacements and its proposed role in bone lysis, *J. Bone J. Surg. Am. Vol. 65*:575–584 (1983).
97. K. L. Spilizewski, R. E. Marchant, J. M. Anderson, and A. Hiltner, In vivo leukocyte interactions with the NHLBI-DTB primary reference materials: polyethylene and silica-free polydimethylsiloxane, *Biomaterials 8*:12–17 (1987).
98. M. Oga, Y. Sugioka, C. D. Hobgood, A. G. Gristina, and Q. N. Myrvik, Surgical biomaterials and differential colonization by *Staphylococcus epidcermidis*, *Biomaterials 9*:285–289 (1988).
99. L. K. Birinyi, E. C. Douville, S. A. Lewis, H. S. Bjornsen, and R. F. Kempczinski, Increased resistance to bacteremic graft infection after endothelial cell seeding, *J. Vasc. Surg. 1*:193–197 (1987).

100. R. J. Hamill, J. M. Vann, and R. A. Proctor, Phagocytosis of *Staphylococcus aureus* by cultured bovine aortic endothelial cells: model for postadherence events in endovascular infections, *Infect. Immun. 54*:833–836 (1986).
101. U. S. Ryan, Metabolic activity of pulmonary endothelium: modulations of structure and function, *Annu. Rev. Physiol. 48*:263–277 (1986).
102. W. P. M. Houdijk, P. G. de Groot, P. F. E. M. Nievelstein, K. S. Sakariassen, and J. J. Sixma, Subendothelial proteins and platelet adhesion: von Willebran factor and fibronectin, not thrombospondin, are involved in platelet adhesion to extracellular matrix of human vascular endothelial cells, *Arteriosclerosis 6*:24–33 (1986).
103. M. Wrangstadh, P. L. Conway, and S. Kjelleberg, The production and release of an extracellular polysaccharide during starvation of a marine *Pseudomonas* sp. and the effect thereof on adhesion, *Arch. Microbiol. 145*:220–227 (1986).
104. R. G. Burwell, G. E. Friedlaender, and H. J. Mankin, Current perpsectives and future directions: the 1983 invitational converence on osteochondral allografts, *Clin. Orthop. 197*: 141–157 (1985).
105. R. A. Fuller and J. J. Rosen, Materials for medicine, *Sci. Am. 255*:118–125 (1986).
106. P. Didisheim, D. B. Olsen, D. J. Farrar, P. M. Portner, B. P. Griffith, D. G. Pennington, and A. G. Gristina, Infections and thromboembolism with implantable cardiovascular devices, *Trans. Am. Soc. Artif. Intern. Organs 35*:54–70 (1989).
107. J. J. Dobbins, G. S. Johnson, C. M. Kunin, and W. C. DeVries, Postmortem microbiological findings of two total artifical heart recipients, *JAMA 259*:865–869 (1988).
108. A. G. Gristina, J. J. Dobbins, B. Giammara, J. C. Lewis, and W. C. DeVries, Biomaterial-centered sepsis and the total artificial heart, *JAMA 259*:870–874 (1988).
109. W. DeVries, The permanent artificial heart: four case reports, *JAMA 249*:849–859 (1988).
110. G. A. Stern, A. Lubniewski, and C. Allen, The interaction between *Pseudomonas aeruginosa* and the corneal epithelium, *Arch. Ophthalmol. 103*:1221–1225 (1985).
111. G. M. Johnson, D. A. Lee, W. E. Regelmann, E. D. Gray, G. Peters, and P. G. Quie, Interference with granulocyte function by *Staphylococcus epidermidis* slime, *Infect. Immun. 54*:13–20 (1986).
112. P. Naylor, R. Jennings, Q. Myrvik, L. Webb, and A. Gristina, Antibiotic sensitivity of biomaterial-adherent *Staphylococcus epidermidis*, *Orthop. Trans. 12*:524–525 (1988).

113. N. K. Sheth, T. R. Franson, and P. G. Sohnle, Influence of bacterial adherence to intravascular catheters on in vitro antibiotic susceptibility, *Lancet 1*:1266–1268 (1985).
114. W. W. Nichols, S. M. Dorrington, M. P. E. Slack, and H. L. Walmsley, Inhibition of tobramycin diffusion by binding to alginate, *Antimicrob. Agents Chemother. 32*:518–523 (1988).
115. W. W. Nichols, J. J. Evans, M. P. E. Slack, and H. L. Walmsley, The penetration of antibiotics into aggregates of mucoid and non-mucoid *Pseudomonas aeruginosa, J. Gen. Microbiol.* (in press).
116. P. Gilbert, M. R. W. Brown, and J. W. Costerton, Inocula for antimicrobial sensitivity testing: a critical review, *J. Antimicrob. Chemother. 20*:147–154 (1987).
117. J. C. Klock and D. F. Bainton, Degranulation and abnormal bactericidal function of granulocytes procured by reversible adhesion to nylon wool, *Blood 48*:149–161 (1976).
118. W. Zimmerli, P. D. Lew, and F. A. Waldvogel, Pathogenesis of foreign body infections: evidence for a local granulocyte defect, *J. Clin. Invest. 73*:1191–1200 (1984).
119. W. Zimmerli, A. Waldvogel, P. Vaudaux, and U. E. Nydegger, Pathogenesis of foreign body infections: description and characteristics of an animal model, *J. Infect. Dis. 146*:487–497 (1982).
120. T. Engel, Low-energy atom scattering from surfaces, *Science 234*:327–333 (1986).
121. B. D. Ratner, Surface characterization of biomaterials by electron spectroscopy for chemical analysis, *Ann. Biomed. Eng. 11*:313–336 (1983).
122. S. T. Boyce, D. J. Christianson, and J. F. Hansbrough. Structure of a collagen-GAG dermal skin substitute optimized by cultured human epidermal keratinocytes, *J. Biomed. Mater. Res. 22*:939–957 (1988).
123. S. T. Boyce and J. F. Hansbrough, Biologic attachment, growth, and differentiation of cultured human epidermal keratinocytes on a graftable collagen and chondroitin-6-sulfate substrate, *Surgery, 103*:421–431 (1988).
124. C. J. Doillon, A. J. Wasserman, R. A. Berg, and F. H. Silver, Behaviour of fibroblasts and epidermal cells cultivated on analogues of extracellular matrix, *Biomaterials 9*:91–96 (1988).
125. L. X. Webb, R. T. Myers, A. R. Cordell, C. D. Hobgood, J. W. Costerton, and A. G. Gristina, Inhibition of bacterial adhesion by antibacterial surface pretreatment of vascular prostheses, *J. Vasc. Surg. 4*:16–21 (1986).
126. S. T. Boyce, D. J. Christianson, and J. F. Hansbrough, Structure of a collagen-GAG dermal skin substitute optimized

by cultured human epidermidal keratinocytes. *J. Biomed. Mater. Res. 22*:939–957 (1988).

127. S. T. Boyce and J. F. Hansbrough, Biologic attachment, growth, and differentiation of cultured human epidermal keratinocytes on a graftable collagen and chondroitin-6-sulfate substrate, *Surgery 103*:421–431 (1988).

6

Proteins in Blood Plasma at Interfaces

LEO VROMAN Department of Chemical Engineering,
Columbia University, New York, New York

I.	Introduction	137
II.	Initial Interactions of Plasma with Solid Surfaces	139
	A. Events that precede coagulation	139
	B. Coagulation	143
	C. Immune system; complement activation	145
III.	Conclusions	146
	References	148

I. INTRODUCTION

Seen from a distance by an engineer, our blood may appear to be a rather Newtonian liquid moved by a pump of rather solid material through a system of pipes, equally solid. Since artificial blood vessels and valves and whole "hearts" are being implanted with partial success, the distant view has survived in efforts that require a very simple foundation to build up a gradual understanding of the relationships between our blood and its natural container. Unfortunately, approaching that relationship from a distance is like trying to understand the line that represents the horizon: The closer we come, the more complex it turns out to be and the farther away our answers appear.

It is perhaps better to start by realizing that blood and its blood vessel wall (the endothelium) have developed a relationship through their joint evolution and through their growth in each living organism that will be disrupted by separating them. In fact, when some blood is taken out of circulation and placed in any container, it undergoes irreversible changes. It will clot unless prevented from doing so by additives that cause different changes. These may be subtle enough to allow its return to the body, as in transfusions; but when observed closely enough, these changes can be detected.

One consequence of this lifelong relationship between blood and endothelium is that we cannot study the blood vessel as if it were merely a container that we wash clear of all remnants of blood and then reexpose, for example, to blood containing a labeled protein, to see if the living wall will adsorb it. This wall is one that reacts to any intervention in its own complex ways. It passes ions and water actively and selectively; certain proteins are embedded in it, others are produced by it and released; and certain cells can pass right through it. The "surface" consists of phospholipids and other organic molecules as well, and as are all cell membranes, it is of near-liquid mobility. Possibly, any approaching protein molecule will find this undefinable interface adapting to any particle's peculiar charge distribution by rearranging these phospholipid and other members of its superficial scenery so that adsorption onto it would be an event that needs to be redefined.

Most if not all artificial surfaces, on the other hand, will adsorb any protein out of solution. If the solution contains only one purified protein, that protein will be adsorbed at a rate that depends on the nature of the interface (in turn being determined by solvent ions and solid material surface structure and texture) on the nature of the protein and its concentration, on flow, and on temperature. At physiological concentrations a protein such as albumin, immunoglobulin G, or fibrinogen—all normally occurring in blood plasma at concentrations of at least several milligrams per milliliter—adsorption out of solution onto a material such as glass or oxidized metal or various polymers occurs within 1 s and reaches near-equilibrium (having formed a more-or-less confluent monomolecular layer) within a minute.

The behavior of a single purified protein at an "unnatural" solid surface even under the most "natural" conditions of dissolved state, concentration, and temperature of the protein barely allows prediction of what it will do under more life-like conditions, where a very large number of proteins will be present in solution. Their physical and biochemical interactions at the interface will then lead, often via very rapid and intricate pathways, to a mosaic-like surface distribution. Eventually, the process may slow down enough to allow measurements—a state easily mislabeled as "equilibrium." In the absence of calcium ions or in the presence of heparin or some other inhibitor of blood coagulation, we can indeed find the same set of proteins present at the interface after an hour of solid—plasma contact that we would have found after only a few minutes of contact. If no additives are used to inhibit coagulation, however, the plasma will clot within a few minutes; that is, it will convert its dissolved fibrinogen first to fibrin monomers and will then polymerize them to form insoluble strands of fibrin.

Depending on the nature of the solid surface, fibrin may form and adhere first on the solid and will then require us to redefine the plasma–solid interface. Again depending on the surface of the material, the entire clot, if formed in the presence of blood platelets, will "retract." Freshly drawn blood placed in a glass test tube, for example, will clot in about 5 minutes and will, within hours, detach itself from the walls and contract by twisting its fibrin network while keeping most red blood cells trapped. The clear liquid that will now separate the retracted clot from the test tube wall is called serum. Serum not only lacks fibrinogen (having been consumed to form fibrin) but contains activated protein molecules of the complement system. Both clotting and complement activation involve serial interactions among nine or more proteins.

The dramatic consequences of placing blood in a glass test tube (i.e., seeing it clot and retract) have misled us for about 100 years into assuming that the first event upon contact is one that involves some protein species to be adsorbed, become activated, transmit its activated state to some other protein species, and thus initiate clotting. This now appears to be incorrect. We shall see that under certain conditions at least, one protein species will be deposited by the plasma first, to be replaced by another protein species, and so on, without immediate involvement of the clotting system. Several of the proteins briefly spending time at the interface may be recognized by blood platelets or white cells. These formed elements, with their specific receptor sites for specific proteins, will then adhere and go through specific changes. Such visible events may be regarded as rapid, gentle, and biologically significant probes for the adsorbed protein species. Thus patterns of platelet and granulocyte distribution deposited by blood on, for example, hemodialyzer surfaces may result from the effects of flow on the underlying but invisible film of proteins and their interactions rather than from a direct effect of flow on the visible formed elements themselves. Only brief descriptions of some sequences of interactions will be given below, with references to overviews that provide more detail.

II. INITIAL INTERACTIONS OF PLASMA WITH SOLID SURFACES

A. Events that Precede Coagulation

While blood normally will clot within a few minutes of contact with glass, in contact with polymers such as polystyrene, polypropylene, or polytetrafluoroethylene (Teflon), it may take up to 30 min to clot. Again our preoccupation with coagulation would mislead us into thinking that such hydrophobic materials would not adsorb proteins. Onto air–liquid and liquid–liquid as well as on liquid–solid interfaces, blood plasma deposits proteins within 1 s [1–3]. One of the first proteins deposited is fibrinogen, but especially on materials with a high critical surface tension and negative surface charge–surfaces that activate clotting–this fibrinogen film is replaced [4–6], often within 30 s, by factor XII and by high-molecular-weight kininogen (HMWK) [7–9]. In narrow spaces [10,11] and in separated flow [12], this displacement may be slower than in other areas.

The possible biological significance of these findings cannot be understood without some knowledge of platelet adhesion. Platelets adhere to artificial materials where they find fibrinogen remaining [13]. They adhere with specific receptors, GPIIb-IIIa, to arginylglycylaspartic acid (RGD) areas of the fibrinogen molecule [14,15]. The attraction of adsorbed versus dissolved fibrinogen is an indication that this protein, perhaps like many others, changes its conformation while being attached to the solid. Added to blood, fibrinogen in solution is not bound by the platelet receptors. For this binding out of solution and for subsequent enhancement of platelet aggregation by fibrinogen, the receptors must be activated first, whereas it appears that such preactivation of receptors is not needed for platelets to attach themselves to fibrinogen that has been adsorbed onto a surface. Perhaps the receptors can be activated by sets of neighboring fibrinogen molecules adsorbed at the material–blood interface, although evidence for this mechanism is still indirect [16]. There is also evidence that in solution, fibrinogen hides certain sites that do become available when fibrinogen is adsorbed onto materials [17].

The "time window" during which whole blood flowing through a device can deposit platelets, which is the period from fibrinogen deposition until its removal by HMWK and factor XII, may thus be narrowed further by the need for adsorbed fibrinogen to expose its attractive sites. Certainly under conditions where the window is wider (in narrow spaces, where amounts of HMWK per surface area are too small to remove fibrinogen efficiently) or delayed (where flow is disturbed), more platelets should be expected to arrive in time for their attachment to the waiting fibrinogen layer. This was found to be the case in specific areas of a stagnation point flow chamber [12]. All of the experimental data described above were gathered on plasma (or whole blood) that had been anticoagulated, either by addition of sodium citrate to bind calcium ions or by addition of an anticoagulant. The sequence of protein interactions, including the depostion of factor XII and HMWK, is not prevented by these additives.

Investigations with serial dilutions of plasma in a simple system of a convex lens resting belly-down on a glass, anodized tantalum, or oxidized silicon surface indicate that fibrinogen is not the first protein deposited by plasma. Plasma deposits a series of proteins, each being displaced by the next to a degree that depends on the width of space, the nature of the substrate, and the dilution of normal plasma used [10]. The sequence from first to last protein present is *albumin, gamma globulins (IgG), fibrinogen* and *fibronectin, factor XII* and *HMWK*. The fact that this sequence is also the one from a protein present in highest concentration to a protein present in the lowest concentration of the series may be coincidental. All efforts so far to reproduce the effect by mixing, for example, traces of albumin with larger amounts of fibrinogen, have failed [18].

Mathematical models of displacement have been proposed. One of these [19] requires that surface adsorption and desorption rate constants depend on surface coverage and on interaction among adsorbed molecules. Another [20] proposes a gradual loss of elutability of adsorbed fibrinogen, caused, for example, by conformational changes.

The behavior of purified IgG in narrow spaces may indicate another mechanism of desorption as well. Injected between a convex lens and a glass or anodized tantalum-coated slide, IgG appears to form a more than monomolecular layer from which excess is rapidly desorbed, except where the space between the lens and the slide surface is within a certain range. For example, an 8-mg solution forms a thick ring where this space is about 10 to 30 μm, while a more dilute solution leaves a correspondingly larger ring. This finding suggests that IgG deposits a multimolecular layer if enough IgG in solution per surface area is available, and will then remove the excess if sufficient IgG has remained in solution to do so. Only where there is barely enough IgG to form the thick layer, after which the solution is so depleted that it cannot remove the excess, will this layer remain [11].

The observations described above can be made with glass slides; adsorbed proteins can then be stained with a concentrated solution of Coomassie Blue. Surfaces with intense interference colors, such as certain preparations of anodized tantalum sputtered glass or silicon crystal wafers onto which silicon oxide had been vapor-deposited, allow direct observation of deposited proteins. In all of these studies, adsorbed proteins can be identified at the interface by exposure to an antiserum or purified antibody to one of the proteins suspected to be present at the interface.

The recording ellipsometer—an instrument that records changes in azimuth and ellipticity of reflected light while adsorption takes place—is quite useful for semiquantitation of these molecular interactions [2, 5,8,19]. For example, an oxidized and hence wettable silicon crystal slide is placed in buffer solution (usually in a sodium barbiturate buffer at pH 7.4 rather than in phosphate buffer since the latter interfers with adsorption) in the ellipsometer and the azimuth and ellipticity of the reflected light are recorded. About 0.1 ml of a solution of 8 mg of IgG per milliliter of buffer is added and the effect of adsorption is recorded. When equilibrium is approached, about 0.1 ml of normal plasma is added. The recording will indicate that the plasma is rapidly depositing matter on top of the IgG layer and will then remove this excess again. If an unoxidized and hence hydrophobic silicon surface had been used as the substrate for IgG adsorption, the layer of matter deposited by plasma is not removed again [18].

The possible physiological significance of such a finding will not be clear without discussing its consequences beyond molecular interactions. In whole heparinized blood, granulocytes adhered where IgG had been preadsorbed onto a hydrophobic rather than onto a hydrophilic (high free surface energy, high critical surface tension) substrate [21]. The complement system may have been involved in at least some of these experiments. Aggregated IgG is known to bind complement components [22]. The peculiar behavior of IgG as described, as well as the indications of marked differences in its orientation and activity toward the complement system and/or granulocyte receptor sites, may result indirectly in effects by narrow spaces and by flow on the deposition of granulocytes, and these effects may alter somewhat in nature from those that determine platelet adhesion to fibrinogen at the same material surface.

Plasma fibronectin, although perhaps not involved in granulocyte adhesion, may affect the adhesion of monocytes, especially in narrow spaces. Between lens and slide, a 10% solution of normal plasma leaves a ring where this space is between 6 and 13 μm wide; higher plasma dilutions leave correspondingly larger and broader rings [11]. Although no deposition of fibronectin by undiluted plasma was found, one may extrapolate and assume that some fibronectin is left by undiluted plasma or whole blood in spaces less than 6 μm wide.

The rates at which fibrinogen is deposited and displaced [4–9] and at which fibronectin is deposited and displaced [23] by plasma in general parallel the extent of their displacement in narrow spaces. This allows us to speculate on the relevance of data obtained with serial dilution in terms of data obtained with graded gaps between surfaces. Any protein ring formed between lens and slide shows by its inner diameter, where the plasma layer was too thin to contain enough of this protein to form a confluent carpet, while its outer diameter shows where the plasma layer was thick enough to contain its next competitor in sufficient amounts per unit of surface area to remove completely the protein under study. Studies with plasma samples lacking fibrinogen, HMWK, or factor XII confirmed this.

For example, where normal plasma created a small ring of IgG surrounded by fibrinogen, congenitally fibrinogen-deficient plasma created a ring of IgG so broad that it included the area where normal plasma would have deposited fibrinogen, and preaddition of fibrinogen rendered the pattern normal. Congenitally HMWK-deficient plasma allowed its fibrinogen to form a ring with normal inside diameter, within which IgG was found, but the outside diameter of this fibrinogen deposit was limited only by the periphery of the area the plasma had occupied [10].

One technique that allows true quantitation of adsorbed protein is the use of labels. Several investigators [4,6,24] have used radiolabeled fibrinogen to show that plasma does not merely mask or distort the adsorbed fibrinogen layer, but actually removes it. This can also be shown simply by rinsing the surface on which plasma had been injected between lens and slide, thus removing the lens, and coating the residual protein film with metal oxide powder, then reexposing to plasma. Where the first layer of plasma had been so thin that it lacked enough HMWK to remove the fibrinogen film, subsequent exposure to oxide suspension caused this area as well as all other adsorbed protein to become coated with the powder. When the thick second blanket of plasma was then applied, it slowly lifted off the oxide where it had rested on fibrinogen, but not where it had rested on HMWK. If the second blanket was of plasma obtained from a patient lacking HMWK, it could not remove the oxide [25].

A more specific tool for identifying adsorbed proteins and their fate is provided by coating various deeply colored powders (ferric and ferrous iron oxide, chromium oxide, ultramarine) first with protein A and then with the desired antibody [26]. Color patterns created by a combination of these preparations poured onto a surface bearing a combination of adsorbed proteins can be lifted off with adhesive tape, leaving the substrate proteins for further study.

B. Coagulation

Two systems of interactions lead to clotting: the intrinsic and extrinsic systems. The intrinsic system acts via proteins present in the plasma only; the extrinsic system, which provides rapid clotting and platelet aggregation to seal off wounds, requires a tissue factor that inserts itself in the sequence of events at a point the intrinsic system would reach only after several minutes. Until about 1960 the extrinsic system, with its ability to seal wounds rapidly, may have been the only one of physiological significance, and in fact the rare "patients" who lack either factor XII or HMWK have been discovered accidentally because their deficiency causes on symptoms. With the introduction of artificial blood vessels and the search for blood-compatible biomaterials, the intrinsic system has gained meaning.

The intrinsic system is activated by contact of the blood plasma with the surface that is introduced. The actual process starts with adsorption of factor XII (Hageman factor, named after a patient who lacked it and thus caused its discovery). A change in conformation of the adsorbed molecule converts it to an active enzyme, factor XIIa, which will act on its substrate, prekallikrein (probably carried to the surface by HMWK), and convert it into an active enzyme, kallikrein, which may in turn modify HMWK in such a way that it and the molecule it carries are more avidly adsorbed onto the clot-promoting solid [27,28]. Some molecules of HMWK carry factor XI, another proenzyme, to the surface, and where it meets factor XIIa, XI will be converted to XIa.

This chain of events will be completed in contact with an activating surface even in the absence of calcium ions. Further events, no longer needing this interface, will include interactions at the surface of phospholipid micelles, briefly as follows. Factor XIa will activate factor IX, which, with its γ-carboxyglutamic acid residues, will be bound by calcium ions to the negatively charged micelle surface, where it will unite with a molecule of factor VIII (the antihemophilic factor). Factor VIII is probably bound with hydrophobic bonds to the lipid hearts of the micelles. The combination of factor IXa and its active cofactor, VIIIa, enzymatically converts another protein, factor X, into its active form, Xa. In a remarkable repetition of interaction patterns, Xa is bound to a phospholipid micelle surface via its γ-carboxyglutamic acid groups and calcium ion bridge, where it is joined by a cofactor, V, to form another complex that enzymatically converts the thrombin precursor, prothrombin, to thrombin. Thrombin is an extremely potent enzyme that not only removes two polypeptides from fibrinogen so that it becomes a monomer able to polymerize into fibrin, but also has many other activities [29]. Perhaps most significant at interfaces with whole blood, thrombin causes drastic changes in blood platelets that include liberation of the type of phospholipid micelles that catalyze clotting.

Each step, with its activation of one enzyme or another, leads to an approximately 1000-fold increase in reaction rate of the total process of coagulation. It should therefore not be surprising that the process includes not only positive but also negative feedback, so that a small wound would not cause the blood to coagulate through the entire body. The destruction of certain enzymes by thrombin itself, the complex interactions with various antithrombins, and the activation of plasminogen into the fibrin digesting plasmin are all important events that remove us further and further from the surface we introduced. A textbook should be consulted (e.g., [30]).

The entire chain of intrinsic clotting factor interactions—including events that require phospholipid micelles—may, however, remain more localized than one would expect. The following simple experiment shows this. Place a drop of citrated, somewhat diluted normal plasma that is still intact (i.e., has never before been in contact with a markedly coagulation activating surface such as glass) on a glass slide. Rinse it off with buffer after 30 s to 5 min. Cover the wet slide with citrated congenitally factor XI- or XII-deficient plasma to which, immediately before use, calcium chloride had been added. Such recalcified plasma usually clots in about 10 min rather than in the normal time of 2 to 4 min. Several minutes after having been placed onto the slide, the deficient plasma will form fibrin on the surface area where the drop of normal plasma had been. The technique is in fact useful for mapping the presence of factor XI or XII where a pattern of deposition and activation is anticipated [12].

Among the ramifications of events that may remain more local than expected, those involving the immune system are most remarkably complex.

C. Immune System; Complement Activation

A textbook (e.g., [31]) should be consulted to appreciate the many interactions that help defend us against infections and that cause rejection of foreign proteins and tissues while recognizing and not rejecting the body's own components. Limiting ourselves to a discussion of plasma proteins at interfaces, we will not discuss how the immune system is able to form antibodies; we merely want to know what, if anything, an antibody or any other immunoglobulin molecule can do at interfaces. As we described above, immunoglobulins and most notably molecules of IgG are adsorbed out of plasma onto many surfaces and will reside there for a period that depends on their concentration, on the nature of the material surface and its texture (narrow spaces), and on flow. IgG is a more-or-less Y-shaped molecule, of which the arms (Fab sections) contain the variable sections that provide specific antibody activity (sites able to combine with matching antigen sites), while the tail of the Y (the Fc section) is not variable and has at least two important functions: exposed under certain conditions, groups of Fc tails will activate the complement system and will cause adhesion of certain white blood cells (granulocytes).

By combining with a layer of antigen molecules such as one offered by an invading particle, the antibody molecules will arrange themselves with the Fab arms embracing the antigen and their Fc tails sticking out into the bulk plasma. The particle is thus "opsonized" and will cause attachment of the six-tailed complement component C1q, which will activate C1r, which will activate C1s. C1s activates C4, which will help C1s to activate C2. C2 is then split and its activated part will join C4 at the interface to become one active entity called C3 convertase, an enzyme able to convert large amounts of C3 into fragments C3a and C3b. Many cells have receptors for C3b. Also, C3b combined with C3 convertase will form a complex of C423 called C5 convertase because it will split C5 into C5a and C5b. Although the chain of events extends for another four protein molecules, leading to a complex with detergent action that can break up the cells on whose surface the chain reaction took place, again we want to limit ourselves to the foreign surface we introduced.

There are at least two ways in which the complement system may become involved: (1) where IgG is adsorbed, so that two or more Fc tails are facing the bulk plasma, C1s may attach itself and start the complement activation sequence; and (2) where a plasma protein is adsorbed, in such a way that its conformation is changed sufficiently to expose a "new" site—one that is not recognized as belonging to the "self" of the organism—antibody molecules may be available to attach themselves to this altered protein, and the antibody–antigen complex will initiate complement activation.

C3a and C5a are anaphylatoxins; they are able to cause release of granules from certain leukocytes (basophils) that liberate histamine and cause allergic reactions and edema. They are also chemotaxins; phagocytic cells will migrate from areas of low to areas of high chemotaxin concentrations. During hemodialysis, while the blood is exposed to large areas of dialyzer membrane, the complement system may be activated either via one of the pathways described above or via an alternative pathway directly through activation of C3b and several other proteins. Depending on the nature of the membrane material [32], one consequence to the patient may be a temporary loss of white cells (granulocytes), and another may be pulmonary hypertension.

III. CONCLUSIONS

When hands and feet are bound and the mouth is covered with tape, a person's response to questions is rather limited; even the person's own quite reasonable range of questions will not be richly expressed. So it is with blood removed from its body and bereft of its own blood cells. Somehow we do get answers from experiments with plasma—answers that we can at least classify in a simple enough system. It would be useless to look for example, for proteins left on material that had been exposed to whole blood, because to the scores of plasma proteins we would have to add the hundreds of

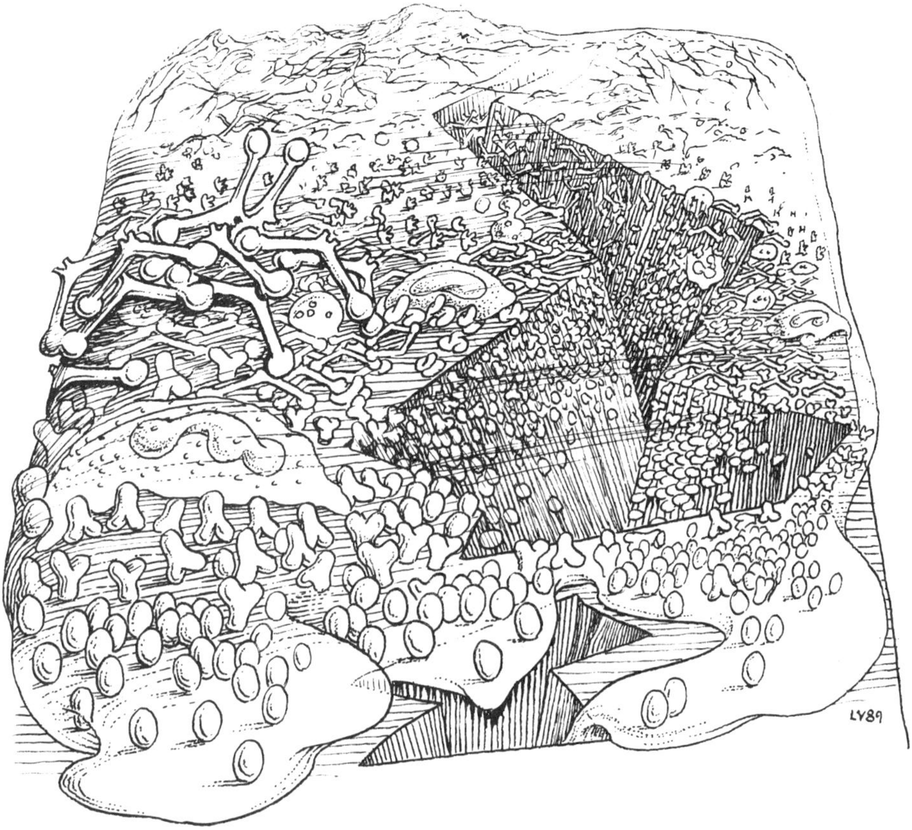

FIG. 1 Blood sketched as if flowing toward the observer, over an imperfect surface, with the leading edge representing contact that began less than 1 s ago. Here albumin (drawn egg-shaped) is being adsorbed. Slightly beyond, immunoglobulins (IgG, drawn Y-shaped) are displacing albumin. At left, a granulocyte is adhering to the Fc tails of the IgG. Beyond, fibronectin (small area of molecules drawn as twisted rods) and fibrinogen molecules (drawn as broken halters) are displacing IgG. Platelets (small round or discoid particles containing granules) are adhering to the deposited fibrinogen, and a monocyte is adhering to deposited fibronectin. In the distance, high-molecular-weight kininogen and factor XII have already displaced fibrinogen, and at the horizon, have enlisted the help of phospholipid micelles liberated by platelets to cause clotting (fibrin formation). A fragile film of proteins is also adsorbed, within a very few seconds, at the air—blood interface. It consists primarily of fibrinogen, and is insoluble. Only a fragment of it is shown cast over the scenery, at left. Note that the relationship of molecular to cellular size had been sketched about five orders of magnitude too large, for the sake of clarity.

proteins that may have been left by platelets and leukocytes (granulocytes, monocytes, and others). The reasonable clues to look for then would be provided by the microscope; it would tell us what formed elements are adhering, and where. To find out why, we would have to look under these cells and find what protein film caused them to adhere. But we can probably not remove cells and hope this film will remain intact; so we will have to identify the attractive proteins before the attracted cells get there.

If, on the other hand, we place plasma without cells on a surface, fibrin will form either overtly or secretly. The most striking macroscopic phenomenon displayed by blood is its sudden solidification 4 to 10 min after it had been taken from its own world and poured into a glass tube. Thanks to the invention of glass, this miracle happens quickly and clearly. It is, of course, quite possible that all reactions described in this chapter occur in our circulating blood all the time but on a single-molecule scale. Only by synchronizing the activation of thousands of molecules (e.g., of factor XII), bringing them all in contact with an activating surface simultaneously, can we enlarge the shackles of their chain reaction enough to hear it clang.

Perhaps one way of approaching the surface close enough to see it on a molecular scale is by greatly reducing the increment of time between observations. We are trying to do that with plasma by diluting it and by observing it in narrow spaces, slowing down protein–protein interactions enough to observe each one, and then calculating what their speed should be under physiological conditions. Once we have done that, we can try to compute the arrival times of platelets and of various white blood cells to match their schedule with arrival and departure times of the proteins for which the arriving platelets and cells have specific receptors, and end up with predictions of what will adhere where and when (see Fig. 1). Even then we should keep in mind that the ability to predict is not necessarily proof of the ability to understand.

REFERENCES

1. I. Langmuir, Protein monolayers, *Cold Spring Harbor Symp. 6*: 171–189 (1938).
2. L. Vroman, Effects of hydrophobic surfaces upon blood coagulation, *Thromb. Diath. Haemorrh. 10*:455–493 (1964).
3. R. E. Baier and R. C. Dutton, Initial events in interactions of blood with a foreign surface, *J. Biomed. Mater. Res. 3*:191–206 (1969).
4. J. L. Brash and P. ten Hove, Effect of plasma dilution on adsorption of fibrinogen to solid surfaces. *Thromb. Haemostasis 51*:326–330 (1984).

5. L. Vroman and A. L. Adams, Findings with the recording ellipsometer suggesting rapid exchange of specific plasma proteins at liquid/solid interfaces, *Surf. Sci. 16*:438–446 (1969).
6. T. A. Horbett, Mass action effects on competitive adsorption of fibrinogen from hemoglobin solutions and from plasma, *Thromb. Haemostasis, 51*:174–181 (1984).
7. L. Vroman, A. L. Adams, G. C. Fischer, and P. C. Munoz, Interaction of high molecular weight kininogen, factor XII, and fibrinogen in plasma at interfaces, *Blood 55*:156–159 (1980).
8. A. H. Schmaier, L. Silver, A. L. Adams, G. C. Fischer, P. C. Munoz, L. Vroman, and R. W. Colman, The effect of high molecular weight kininogen on surface-adsorbed fibrinogen, *Thromb. Res. 33*:51–67 (1983).
9. L. Vroman, Problems in the development of materials that are compatible with blood, *Biomater. Med. Devices Artif. Organs 12*:307–323 (1985).
10. L. Vroman, A. L. Adams, G. C. Fischer, P. C. Munoz, and M. Stanford, Proteins, plasma, and blood in narrow spaces of clot-promoting surfaces, *Adv. Chem. Ser. 199*:265–276 (1982).
11. L. Vroman and A. L. Adams, Adsorption of proteins out of plasma and solutions in narrow spaces, *J. Colloid Interface Sci. 111*:391–402 (1985).
12. M. F. Stanford, P. C. Munoz, and L. Vroman, Platelets adhere where flow has left fibrinogen on glass, *Ann. N.Y. Acad. Sci. 416*:504–512 (1983).
13. M. B. Zucker and L. Vroman, Platelet adhesion induced by fibrinogen adsorbed onto glass, *Proc. Soc. Exp. Biol. Med. 131*:318–320 (1969).
14. M. B. Zucker, S. Brownlea, and J. McPherson, Insights into the mechanism of platelet retention in glass bead columns, *Ann. N.Y. Acad. Sci. 516*:389–406 (1987).
15. H. R. Gralnick, S. B. Williams, and B. S. Coller, Fibrinogen competes with von Willebrand factor for binding to glycoprotein IIb/IIIa complex when platelets are stimulated with thrombin, *Blood 64*:797–800 (1984).
16. E. W. Salzman, J. Lindon, G. McManama, and J. A. Ware, Role of fibrinogen in activation of platelets by artificial surfaces, *Ann. N.Y. Acad. Sci. 516*:184–195 (1987).
17. B. Kudrik and L. Vroman, unpublished observations (1982).
18. L. Vroman, A. L. Adams, M. Klings, and G. C. Fischer, Fibrinogen, globulins, albumin and plasma at interfaces, *Adv. Chem. Ser. 145*:255–287 (1975).
19. P. A. Cuypers, G. M. Willems, H. C. Hemker, and W. T. Hermens, Adsorption kinetics of protein mixtures: a tentative explanation of the Vroman effect, *Ann. N.Y. Acad. Sci. 516*:244–252 (1987).

20. J. L. Bohnert and T. A. Horbett, Changes in adsorbed fibrinogen and albumin interactions with polymers indicated by decreases in detergent elutability, *J. Colloid Interface Sci. 111*: 363–377 (1986).
21. L. Vroman, A. L. Adams, M. Klings, G. C. Fischer, P. C. Munoz, and R. P. Solensky, Reactions of formed elements of blood with plasma proteins at interfaces, *Ann. N.Y. Acad. Sci. 283*:65–76 (1977).
22. S. I. Miekka and I. Gozze, Anticomplementary activity of human immunoglobulin/G. 1. Mechanism of the artefactual increase in anticomplementary activity of IgG during the assay. *Vox Sang. 29*:101 (1975).
23. F. Grinnell and T. V. Phan, Deposition of fibronectin onto artificial material surfaces from human plasma and blood under clotting and non-clotting conditions, *J. Cell. Physiol. 116*:289–296 (1983).
24. J. L. Brash and P. ten Hove, Transient adsorption of fibrinogen on foreign surfaces: similar behavior in plasma and whole blood, *J. Biomed. Mater. Res. 23*:157–169 (1989).
25. L. Vroman, A. L. Adams, and M. Brakman, Lack of exchange among plasma proteins in narrow spaces on glass, demonstrated with metal oxide coatings, *Haemostasis 15*:300–303 (1985).
26. L. Vroman and A. L. Adams, Rapid identification of proteins on flat surfaces, using antibody-coated metal oxide suspensions, *J. Immunol. Meth. 93*:213–216 (1986).
27. R. W. Colman, C. F. Scott, A. H. Schmaier, Y. T. Wachtfogel, R. A. Pixley, and L. H. Edmunds, Jr., Initiation of blood coagulation at artificial surfaces, *Ann. N.Y. Acad. Sci. 516*:253–267 (1987).
28. M. Silverberg and S. Vest Diehl, The activation of the contact system of human plasma by polysaccharide sulfates, *Ann. N.Y. Acad. Sci. 516*:268–279 (1987).
29. J. W. Fenton II, Thrombin, *Ann. N.Y. Acad. Sci. 485*:5–15 (1986).
30. R. W. Colman, J. Hirsh, J. Marter, and E. W. Salzman, *Hemostasis and Thrombosis: Basic Principles and Clinical Practice*, 2nd ed., J. B. Lippincott, Philadelphia, 1987.
31. E. Benjamini and S. Leskowitz, *Immunology: A Short Course*, Alan R. Liss, New York, 1988.
32. D. E. Chenoweth, Complement activation during hemodialysis: clinical observations, proposed mechanisms and theoretical implications, *Artif. Organs 8*:281–287 (1984).

II

Membranes

7

Contribution of Water in the Stabilization of Biological Membranes

MARKO S. MARKOV Department of Biophysics, Sofia University, Sofia, Bulgaria

I.	Water Structure	153
II.	Membrane Structure and Hydration	154
III.	Water in Living Cells	156
IV.	Water and Proteins	158
V.	Water and Lipids	161
VI.	Nucleic Bases and Water	163
VII.	Participation of Water in the Formation of Pores or Channels	165
VIII.	Freezing	168
	References	169

I. WATER STRUCTURE

The structure of membranes and their properties, including membrane transport, involve questions of basic significance in modern biophysics. Lipid organization in bilayer structure, the protein arrangement and conformational changes, and lipid–protein interactions have been the subject of numerous studies. The fluid mosaic model of Singer and Nicolson [1] is accepted as the most probable model of biological membranes. In this model the membrane is considered to be a two-dimensional solution of globular integral proteins in the lipid bilayer. The model is based on the thermodynamics of membrane components in a water medium.

Henderson [2] considers the extraordinary significance of water for living organisms: including heat capacity, heat conductivity, expansion of cooling near the freezing point, density of ice, heat of fusion, heat of vaporization, freezing point, dielectric constant, ionizing power, and surface tension.

Let us consider the appearance of the first membrane a billion years ago. It is now well known that the first membrane appeared in a water medium. At this time inorganic molecules in a solution of water began to form peptides, amino acids, and so on. From an evolutionary point of view these processes were accompanied by the formation of special structures that separated the prebiological structures from the surrounding water. It is quite probable that the prebiological membrane was formed by biopolymers whose composition and architecture were very similar to the structure of present-day lipids. One of the most important features of these structures is considered to be the existence of two parts: hydrophobic and hydrophilic. The hydrophobic portion of the lipids tries to get away from water molecules, while the hydrophilic portion is willing to stay in close contact with water molecules. In this way the formation of the double lipid layer separating the cell constituents from the outside medium can be explained.

Bernal and Fowler [3] emphasize the existence of two centers of positive charge associated with the hydrogen nuclei, and two centers of negative charge corresponding to the unshared electrons of the oxygen. Thus each molecule contains two hydrogen-bond donors and two acceptors located at the corners of a tetrahedron.

The first continuum model was proposed by Pople [4], who visualized liquid water as a structure in which most molecules are hydrogen-bonded to their nearest neighbors, but in which these bonds are often distorted from the linear O—H···O arrangement.

The high polarity of a water molecule makes it apt to both generate and respond to the local electric field. This response will somewhat limit the dynamic orientation freedom of the water molecule, originated in principle as a non-zero-order parameter. This is the basis of cooperative aspects of water structure. Recent experimental, theoretical, and computer-experimental studies concur in pointing out that bulk water is to be viewed as a random three-dimensional network of hydrogen bonds, lying above the critical percolation threshold at biologically important temperatures.

II. MEMBRANE STRUCTURE AND HYDRATION

The electrical properties of the water molecule make it particularly susceptible to the effects of electric charges, particularly those of ions. The presence of positive ions, except the large positive ions, potassium, rubidium, and cesium, polarizes the water structure in their immediate neighborhood. In the case of small positive ions a group of four or six water molecules attach themselves to the ion and move with it, forming a hydrated ion. Hydration also accurs around negative ions and not only with oxygen-covered negative ions. The small ions produce disturbances in the water structure which are strictly local. Thus, besides the normal form of pure liquid water, a great many alternative forms, important in biology, exist.

Edsall and McKenzie [5] considered the hydration of ions, comparing their hydration with the hydration of apolar molecules in aqueous solution. They concluded that when a nonpolar molecule is presented in water, the molecules of water in the immediate vicinity of solute molecules must rearrange themselves in a more ordered structure in which there is less randomness and better hydrogen bonding than in the bulk solvent at the same temperature.

A simple model of hydration involves a central ion surrounded by a spherical region of immobilized water molecules, a spherical region of water molecules in which there is a structure breaking, and a region of "normal" liquid water. A fundamental problem in the study of ion hydration is the determination of the hydration number (ligancy) of the ion. It is important to realize that the exchange rates of water of hydration vary very considerably from one ion to another.

One principal question arises when we consider the physical properties of water in close proximity to the membrane: Does this water exhibit the properties of water present in aqueous solution? In addition, what role does water play in cells and cell membranes? Since osmotic activity is an expression of a decrease in the activity of water present, the question can be formulated as follows: What component of the living cell causes lowering of the activity of the bulk or cell water? On the other hand, the hydrophilic part of the membrane lipids is mainly charged. Because of that, polar lipid head groups electrostatically interact with the surrounding membrane.

Molecular interactions between proteins and phospholipids in biomembranes have been shown to play an important role in the regulation of both the structural and dynamic parameters of biomembranes and the biological function of membrane proteins. Intrinsic proteins show nonspecific hydrophobic interactions with the hydrocarbon moiety of lipid molecules and changes in the lipid environment that reflect the protein activity.

The concept of electrical activity in cell membranes dates back to the time of Galvani and Volta and their controversy in the late eighteenth century concerning "animal electricity" [6] and has been much advanced in the intervening years by biophysicists and electrophysiologists. The electrical surface potentials are useful concepts in bilayer membranes having relatively simple and well-defined surfaces. Biological membranes have highly irregular surface topographics. Their lipid and protein components contain diverse charge groups, so that both lateral and transverse heterogeneities are to be expected in the surface charge distribution and hence the electrical potential profile. The water-activity-reducing effect is seen as a consequence of polarization and immobilization by the propagated polarization emanating from the oxygen atoms and the consequent reduction of translational and rotational *partition functions* of the water molecules in statistical mechanical terms [7].

In many studies the membrane is regarded as part of a system consisting of two homogeneous compartments separated by a planar homogeneous membrane. The two bulk phases consist of the solvent water and any number of nonelectrolytes to which the membrane may or may not be permeable. The volume of both compartments is assumed to be large enough and at the same time, the volume of the membrane is small compared with the bulk phase volume.

Regarding the hydration properties and forces in bulk phase and in close vicinity to the membrane, one should take into account that the behavior of water in both cases is quite different. This fact seems to be one of the principal reasons for the observed differences in hydration in the bulk phase and surface area as well as for the differing extent of ions and biomacromolecules. It is well known that water in disperse systems has properties that deviate from those of the usually observed "bulk" properties. Several states of water in contact with the surface have been discussed, but often a crude distinction into "free" and "bound" water has been thought to be useful.

Water is effective in dissolving inorganic electrolytes because of its high dielectric constant, and its molecules tend to interact with ions or charged molecules to form hydrated structures. Both of these properties are related to the large electric dipole moment of the water. The structural region of the polar part is characterized by lateral interaction mainly through hydrogen bonding. Sphingolipids are (because of the donor and acceptor groups) capable of forming rigid networks of hydrogen bonds.

The observed low solubility of hydrocarbons in water is due to unfavorable entropy effects, tentatively associated with a structuring of water near interfaces between water and nonpolar molecules [8]. Experimental data are in accord with the conception that nonpolar solutes introduce a water structure that has both lower energy and lower entropy than those of an ice-like water structure.

III. WATER IN LIVING CELLS

Participation of water in biochemically active systems can be considered as involving competition between bulk water and various equilibria in which the biological molecules are engaged. The physical presence of water is essential as a reaction medium; it also acts as a controlling dynamic participant, the activity of which can enforce displacement of inter- and intrastereochemical equilibrium.

Bulk water encompasses dynamic equilibria of water association and dissociation which depend on the binding forces of the individual water molecules. The presence of neutral molecules, ions, and macromolecule surfaces can be considered as affecting water structures on different scales of interaction. First are interactions on the order of 10 Å or less in which the water molecules are closely bonded to something like an ice structure. Thus a water molecule in the immediate neighborhood of protein is bound to protein and travels with it [9]. Second, in the intermediate layer ranging from 10 to 100 Å, the structure of water is dominated by the neighboring molecular surfaces. In this state of water there is no limitation on the movement of small molecules and ions in solution. Finally, there are structural effects that appear to distances on the order of 4000 Å—here there is no question of the structure being different from that of free water. All these aspects of water structure have important biological significance [10].

The concept of a "hydrophobic effect" (the tendency for nonpolar molecules or parts of molecules in aqueous solution to aggregate in order to reduce a nonpolar surface area exposed to water) has been a central idea in many attempts to come to grips with the thermodynamics of protein structure [11]. A hydrophobicity scale giving the free-energy difference for the transfer of residue originally in a helix in water to a helix in a nonpolar phase lacking hydrogen-bonding capacity was developed by von Heijne and Blomberg [12]. The presence of hydrophobic groups in aqueous solutions results in the enforcement of corresponding water structures with lower entropies.

Hydrophobic groups of biological molecules are usually present in aqueous media in excess of their individual chemically permissible solubility. By their enforced intrusion, the hydrophobic groups cause the adjoining water molecules to adopt a more organized water structure than is normal under the existing conditions of temperature, pressure, and solute concentrations. Because hydrophobic groups enforce extensive water organization around themselves, they must reduce the availability of water for interactions such as those involved in acid-base ionization and in the solubility of other substances. The characteristics of hydrophobically enforced water organization are such that the interfacial tension of the water in contact with the hydrophobic group, and the surface tension of the solution—air interface, must be lower than that of the original water medium with air. The enforced lower entropy of the neighboring water structure is the reason that the individual hydrophobic groups will tend to adhere. This adherence results in an increase in the water entropy. Adjoining more organized water structure going to bulk water will adopt a less orgasnized structure. Thus the major driving force for the adherence of hydrophobic groups is the increase in the entropy of the water as it transforms itself from the lower entropy state of the more organized structures into the higher entropy state of the normal, less extensively organized water structures [13].

Wheatley et al. [14] have pointed out several important conclusions concerning cell water: (1) free molecules in cells are eluted by osmotic withdrawal of bulk-phase water; (2) resident acid-extractable pool molecules are only partially released from cells that have had up to 90% labile cell water removed; (3) exchange reactions are predominantly similar to those occurring under isotonic conditions; and (4) although the acid-extractable pool could exist in the more firmly bound water layer (water molecules associated with proteins), the findings do not preclude other possibilities (e.g., complexing of amino acids with other substances).

Measuring cell volume by means of the hematocrit (cell number) method, Lee et al. [15] show that the density behavior of red cells is completely predictable from water content. Funder and Wieth [16] found a direct correlation beween the Na^+, K^+ content in red blood cells and that in cell water. The value of studying single red blood cells can be much improved by knowing the relationship between element content and cell water [17].

The importance of water in the formation and turnover of amino acid pools is implicit, although it is conceivable that should amino acid molecules exist as complexes inside cells, they might not behave as part of the bulk aqueous phase. Gould and Measures [18] consider that cell metabolism is arrested by salt-induced dehydration. Some protein synthesis must take place within the dehydrated cell when the measured level of total protein level is very low.

IV. WATER AND PROTEINS

Kauzmann [19] draws attention to the fact that nonpolar groups of native, globular proteins are located in interior regions of the protein molecule, withdrawn from exposure to solvent water. To elucidate the nature of the forces that tend to drive nonpolar molecules away from aqueous media, the thermodynamics of the transfer of hydrocarbons from organic solvents is considered. Available experimental data show that hydrocarbons are strongly soluble in water. The low solubility of hydrocarbons in water is due to unfavorable entropy effects tentatively associated with a structuring of water at interfaces between water and nonpolar molecules. Experimental data are in accordance with the concept that nonpolar solutes introduce water structures with lower energy and lower entropy than those of pure water (ice-like water structures). On the other hand, the low solubility of nonpolar substances in water is due to the strong attraction of water molecules for one another [8].

When the introduction of the individual hydrophobic groups into the medium causes the adjoining water molecules to adopt a lower energy, the process of breakup of these structures must be accompanied by an increase in entropy. Only if the water structures adjoining the hydrophobic groups are in a state of randomness equal to that of the surrounding bulk is the presence of water not of energetic concern.

Being concave with respect to the convexity of the hydrophobic groups, the hydrophobically enforced water structure displays toward bulk water a convex face with outwardly directed H··· donors and positive polarity characteristics. It was Kauzmann [20] who developed the concept of hydrophobic bonds (or hydrophobic interactions). When hydrocarbon groups are immersed in water, the negative change in entropy and the positive change in heat capacity could be interpreted in increased order of the water molecules around the hydrocarbon groups.

The introduction of solute molecules or ions into water may have a marked effect on this *order–disorder system*. When nonpolar molecules such as those of hydrocarbons are dissolved in water, there is a considerable loss of entropy. This entropy may be interpreted in terms of increased order in the arrangement of the water molecules in the neighborhood of the solute. Ion–water binding depends greatly on the quality of the ion. The activation energy values of many dilute electrolyte solutions are close to those of pure water, so there is a close connection between ion movements and the dipole rotation of water.

It is well known that every globular protein is biologically active only when it is hydrated. Thus the water molecules bound to a protein molecule are considered to play an essential role in its biological function. Their molecular structures as found in the crystal also involves a considerable amount of water, and these structures have been supposed to be on going from crystals to aqueous solutions [21]. For example, tightly bound water molecules (and possibly weakly bound water molecules also) are nearly intrinsic on the surface of an insulin molecule, and even in aqueous solution their sites and orientations are similar to those in crystal.

On most protein molecules there is the regular formation of ice in two areas, internal and external, associated with protein molecules. Many protein molecules are hollow inside. Inside such molecules the water molecules are arranged in definite order and are held by hydrogen bonds. What is most important, it is impossible for ions to penetrate the ice.

The degree of orientation of water molecules on extended surfaces varies considerably. The water near the surface, up to a distance of 10 to 20 Å, will be bound in this form of ice. Up to 100 Å the water will not be so tightly bound and thus able to contain ions. This is about the limit to which the structure of water itself is liable to be affected by hydrophobic surfaces [10].

A complete representation of the water structure surrounding a protein molecule would include all possible coordinate networks for each water molecule, characterized both by the probability of its occurrence at a given site per unit and by an appropriate thermal factor [22].

The entropy increase due to changes in the structure of water is the dominant factor for making total free energy negative, and hence hydrophobic interactions will stabilize the helix. Two-thirds of the free energy of denaturation is allotted to this *entropy bounding* of water. Therefore, it was assumed that the intramolecular state of order in dissolved protein molecules had been adapted to the intermolecular state of order in the solvent water and that it was adjustable by the water. Since all further steps in structure formation depend on the helical conformation of the protein molecules, it is possible that the parameters of water structure also regulate the process of "aging" [23].

According to the polarized multilayer hypothesis of cell water [7], the maximum effect on water polarization, solvency reduction, and water-activity reduction occurs when the "matrix protein" chain is fully extended and is highly oriented, as in the highly organized living cell interior. Very important is the relation between the protein content and the distances between protein chains where cell water is found. The question is: How would the content of these proteins affect the number of water molecules found between the nearest-neighboring chains?

The following process of a natural polypeptide chain in water leads from an essential ensemble of extended chain conformation to a single compact structure. In this process the complete system consisting of chain and water attains a state of minimal free energy. The stabilization energy is on the order of 50 kJ/mol and therefore rather small. The so-called hydrophobic forces play an essential role. They are produced by increasing water entropy on decreasing the contact area between water and nonpolar parts of the chain. Accordingly, the conformation of a polypeptide chain in nonwater solvent bears no resemblance to the native structure. The free energy of the peptide bond in water is rather small compared with other covalent bonds, so can be reestablished easily [24].

It has been shown by Ling [7] that 13 globular proteins, including hemoglobin, which show no or very little effect on water structure and water solvency acquired the ability to do so when these proteins were exposed to urea or guanidine HCl. It was argued that native globular proteins affect water little or not at all because their NHCO groups are locked in α-helical or other macromolecular H-bonds. Although the electrically neutral polymers lack the NH groups of extended polypeptide chains, they possess oxygen atoms at regular distances apart and like the oxygen atoms of peptide CO groups, the distances between the nearest-neighboring oxygen atoms are roughly equal to twice the diameter of water molecules. As anticipated, they cause soluble exclusion from the surrounding water.

The random coil polypeptide form is bound to be extensively hydrated in aqueous solution via its $>C{=}O$ and $>N{-}H$ groups. It is also involved with water via the hydrophobic attachment of water to amino acid side-chain groups of OH (serine, tyrosine, threonine), carboxylic (glutamic, aspartic), ammonium (lysine), iminazole (histidine), and guanidinum (arginine) and the water structures in contact with hydrophobic groups (alanine, leucine, and valine) and possibly in a hydrophobic/hydrophilic bond.

The question of whether or not proteins adsorb reversibly has been the subject of controversy. There has been a widespread belief that adsorption of protein is irreversible at the air–water interface. This is based on the observation that protein molecules change from a globular shape in solution to an unfolded state at an interface. However, it is well known from polymer chemistry that large molecules change their configuration when they move from one energy environment to another. The surface is very different from the bulk solution [25].

Raman resonance spectroscopic evidence that water molecules are located in close proximity to the SB group of the chromophore has been published. It seems that these molecules are essential to the function of a bacteriorhodopsin chromophore by stabilizing its configurational and conformational structure. After these molecules were removed by dehydration of the membrane, a new chromophore appeared. In contrast to the hydrated membrane, where the primary event is photoisomerization, in the case of a dehydrated membrane, a photo-induced proton-transfer mechanism is suggested.

V. WATER AND LIPIDS

Lipid bilayers have often been used as models of biomembranes. The orientation of lipid molecules produces an electrostatic potential, called the dipole membrane potential, which originates from the dipoles of polar head groups and ester linkages of lipid molecules. In addition to the dipole potential, the diffuse double-layer potential is induced at the membrane–solution interface in the solution side when the acidic lipids are used as the membrane molecules. Electrostatic interaction of the membrane lipids with different ions will influence the extent of sorption of these ions. Organic ions, for instance, show a complicated sorption behavior because of the balance of electrostatic and hydrophobic interaction with the lipid membrane. To elucidate the contribution of electrostatic and hydrophobic interactions to the sorption of these ions into the membrane, Matsumura and Furusawa [26] studied the difference in the sorption behavior between a lipid-soluble anion and a lipid-soluble cation. They found that the lipid-soluble anions are sorbed at the inner side of the membrane, not being so hindered by the potential, induced by the sorption of these ions themselves. The lipid-soluble cations are sorbed near the membrane surface, being hindered effectively by the potential induced by the sorbed ions.

According to the hydration model, which is based on x-ray analysis, one water molecule forms a hydrogen-bonded bridge between two peptide carbonyl oxygens on two different protein chains in the collagen threefold helix. It was suggested that another water molecule formed a hydrogen-bonded bridge between a peptide N—H bond and a carbonyl oxygen, also in two different chains [27].

To describe fully the intermolecular forces in a bilayer, one must understand the packing arrangement and constraints of the lipids and water comprising them. The molecular packing density of bilayers is revealed macroscopically by the partial specific volume of the components as a function of composition. It has been commonplace to assume that the partial specific volumes of water (ν_W) and lipids (ν_L) in bilayers are independent of the degree of hydration of the bilayer. White and King [28] pointed out that neither ν_W nor ν_L has ever been measured for any lipid lamellar phase containing less than excess water. It is known that the partial specific volume of water in extremely concentrated electrolyte solutions is considerably less than 1 ml/g. Because the polar groups of lipids must surely form a very concentrated electrolyte solution, the water is likely to behave in a very nonideal way. The water in the lipid mixture exhibits a large negative deviation from ideality, as would be expected if the water is strongly attracted to the head groups. The arrows mark the water content corresponding to 11 to 13 water molecules per lipid, which is the generally accepted range for the number of waters in the primary hydration shell of phosphatidylcholine lipids. The nonideality, which is clearly associated with the water of the polar group hydration shell, suggested that the partial specific volume of the water in bilayers at low hydrations might be lower than 1 ml/g. Measurements of changes in area per molecule (A) and spacing (d) with hydration lead immediately to the partial specific volume because the volume of a lipid molecule and the (n) water molecules associated with it is equal to Ad/2.

According to the Raman results, some rearrangement of the hydrocarbon chains occurs when the water layer thickness decreases.

The electrostatic and structural interactions within the water layer influence the molecular arrangement within the lipid bilayers [29].

On the basis of neutron diffraction data [28] it was evaluated that a bilayer is much less deformable than expected on the basis of the partial specific volume assumption. On the other hand, it was shown that the partial specific volume of the lipid might depend on the hydration [30]. The molecular volume of water suggests very strongly expressed electrostrictive effects in the polar head groups. At a hydration value above 10 water molecules per lipid were found lipid and water volumes of 1432 and 24.8 Å, respectively [28].

At a very low hydration level (one or two water molecules) the head groups are not effectively packed, perhaps due to an excluded volume effect. As additional waters enter the head groups, they bind tightly to the head groups within the existing excluded volume spaces and cause minor changes in volume. As more waters are added after the filling of the hydration shell, they behave more normally and osmotic pressure drives the head groups apart. They farther apart they are, the smaller the excluded effects will be. This model ignores packing changes in the hydration region.

The polar head group of glycosphingolipids has a major influence on their molecular behavior in biological interfaces. Most observed effects depend on the type of oligosaccharide chain present in gangliosides and other glycosphingolipids. We have obtained evidence that the type of polar head group present in a particular glycosphingolipid plays a major role in determining its phase behavior [31].

The fluid like surfaces of bilayers and biological membranes are highly mobile due to the thermal motions of the hydrated lipid head groups and to fluctuations in bilayer thickness, estimated to be on the order of a few tenths of a nanometer. Israelashvili and Pashley [32] propose that the origin of the strong, short-range repulsion hydration force seen in many different systems is the same and is due primarily to the ordered layering of water molecules bound to hydrated surface groups.

VI. NUCLEIC BASES AND WATER

In this section we discuss briefly interactions between nucleotide bases and water. Experimental data show that nucleic acid bases form stacked associates in water. Coplanar base pairs in water have been detected successfully. At the same time, there are some theoretical suggestions that coplanar base pairs in vacuum have a lower energy then stacked base pairs of the same composition. Analysis of the theoretical and experimental data leads to the conclusion that water plays an important role in the formation of stacked associates. It is not clear why stacks are more favored in water than are coplanar base pairs. Monte Carlo simulation of a pure cluster of 200 water molecules as well as of the cluster containing uracil gives evidence that a decrease in water—water interaction energy in a system with a stacked dimer is caused by an increase in water structurization [33].

For the cluster chosen, the stacked dimer is also energetically more favorable than the coplanar base pair, the water—water interaction change contributing mainly to the total value. This is caused by the formation of a greater number of H-bonds between water molecules around the stack than around the base pair. The experimental data indicate quite clearly that the structurization of water around the stack is also higher than around the base pair.

The formation of a double strand of DNA from its respective two complementary strands involves detachment of the water molecules H-bonded to the amino, imino, carbonyl, and ternary nitrogen groups into the bulk solvent. Water intercalation can be a transitional intermediate state facilitating the energy of drugs and intercalating agents into nucleic acids [34]. Model binding indicates that water can act as a "spacer" and stabilizing a DNA helix [35].

Water-bridging activities can conveniently be considered under two headings:

1. *Water bridging.* The water molecules span widely separated H··· acceptor sites and H··· donor sites by simultaneous H··· donation and acceptance via OH groups.
2. *Ion—water binding.* In this situation two widely separated similarly charged groups, both possessing H-bonding potential, are bridged by a hydrated counterion, the water molecules being bonded by both dipolarity and H-bonding—for example, water bridges supporting the Na^+ ions between the phosphate groups of the two complementary strands of DNA. When a water bridge forms part of a cooperative closed resonance circuit system, it should possess a total strength greater than that of the sum of the normal corresponding H··· bonds in the β form of the double helix of DNA that may possess this potential.

In DNA the water-bridge ends are connected to the H-donor N—H··· and to an H-acceptor such as an oxygen in a phosphate, the low pair of electrons of the nitrogen of a basic amino group, a carbonil or deoxyribose group, or a ring nitrogen [13]. It has been proved by Monte Carlo simulation that water molecules fill in both minor and major grove of β-DNA and thus stabilize the double helix of DNA [36].

VII. PARTICIPATION OF WATER IN THE FORMATION OF PORES OR CHANNELS

Plasma membranes are known as diffusion barriers to hydrophobic solutes despite their very inhomogeneous structure and composition. The permeation of ions and hydrophobic nonelectrolytes occurs with high preference via specialized and controlled pathways exhibiting the properties of carriers or channels. One of the principal functions of biological membranes is selective control of the flow of different molecules of ions. Simple inorganic cations (e.g., Na^+ or K^+) can be transported across membranes; the difference in ionic concentrations at both sides of the membrane brings about electrical potential differences of basic importance, especially in a neuron's firing membranes.

Edmonds [37] supposes that the gating charge moves within a pore which has an effective dielectric constant ($\varepsilon = 80$) as high as that of bulk water. It is, in fact, highly unlikely that all the charge is screened by water or that the water molecules within a pore diameter of only 10 Å will so effectively screen any charge within it, particularly if the water is itself electrically ordered.

One possible mechanism explaining how an ion crosses a membrane is migration through a molecular transmembrane channel. It is accepted that the peptide gramicidin A can act as a molecular channel specifically for monovalent cations. An analysis of the structure of water molecules interacting with a gramicidin A channel was carried out by Foruili et al. [38]. They discuss the number of water molecules that can be placed inside the gramicidin A channel, the interaction energy of the water molecules with the channel and among themselves, and the structure of the water molecules at the two extremes of the channel.

Two positions of gramicidin A hydration sites can be obtained rather easily by probing with one water molecule in or near the channel. Because of the electrostatic charge distributions of the water near gramicidin A molecules for a given position of the water molecule relative to the gramicidin A, there corresponds only the optimal orientation for the water's OH bonds. It is evident that the gramicidin A channel is very attractive to water molecules. Monte Carlo simulation gives evidence that water molecules are hydrogen bonded together, as clearly shown either from the bottom inset or from the statistical distribution. A hydrogen bond exists between two water molecules whenever O—O internuclear separation is not larger than 3.5 Å and the O—H bond of one water molecule points toward the oxygen atom of the second molecule with an $\widehat{OHO}$ angle between 180° (linear bond) and 150° (bent hydrogen bond).

The water—water network is also strongly connected at the two channel ends. Inside the channel, due to its narrowness, one can have an accurate count of the water molecules and of a hydrogen-bonding pattern. Comparing the two insets, one can note that the water molecules of the single configuration have a structure that corresponds very closely to the probability distribution. The existence of hydration effects outside and inside a cell medium, as well as water film covering the lipids and proteins in the membrane bilayer, is extremely significant for such membrane electrical properties as conductance, capacitance, surface charge distribution, and transmembrane potentials (Fig. 1).

Several of the present theories regarding water structure are very important for the problem being treated: the ability of the water molecules to form hydrogen bonds, the possibility of the formation of associates of water molecules, the hexagonal structure of these associates, the presence of cooperative effects, and the hydration of ions and macromolecules.

Especially in the case of membrane transport, it is very important to consider the participation of water in forming ionic channels and pores. It is quite probable that local changes in the structure of water adjoining a membrane produce energy release that might be sufficient for pore formation. On the other hand, if for any reason a pore appears (Fig. 2), its edge has to be covered by a film of water (Fig. 3). If the pore edge is covered by a film of water molecules, two important problems should be considered: (1) a decrease in pore or channel diameter because of the water film (compare Figs. 2 and 3), and (2) a decrease in the speed of transport

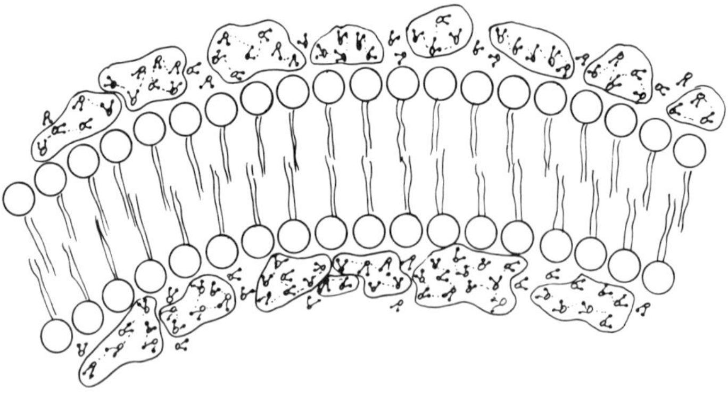

FIG. 1 Water aggregates and monomeric water molecules hydrate a membrane bilayer.

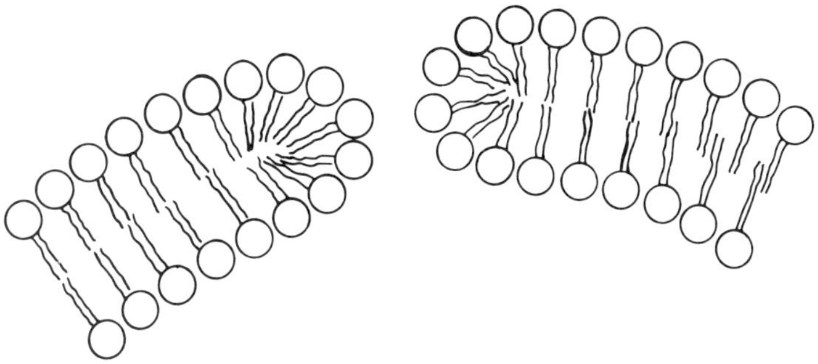

FIG. 2 Traditional illustration of a lipid pore.

of ions through the membrane. The water film covering the pore edge is important when the speed of ions through the pore is examined. Even a layer of monomeric water molecules (thickness of 2.8 Å) will decrease the effective pore radius and thus will alter the possibility of passing ions of definite size.

Weaver et al. [39] also make the assumption that a bound monolayer of water exists on the interior surface of the pore, so the effective cross-sectional area of a pore is reduced (the monolayer thickness is 2.8 Å). Then only a relatively small subpopulation of pores can participate in diffusive transport, since only a pore whose radius is higher than the sum of the water diameter and ion radius can pass a solute molecule that has an effective molecular radius r_S. Similar considerations are valid for channels, organized on the base of proteins (Fig. 4).

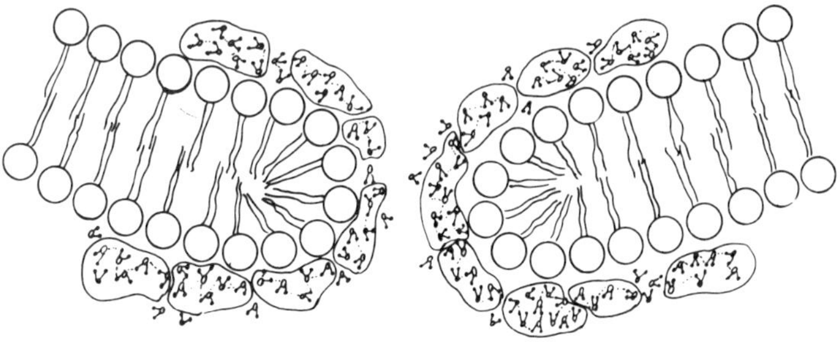

FIG. 3. Pore edges are covered by a film of water.

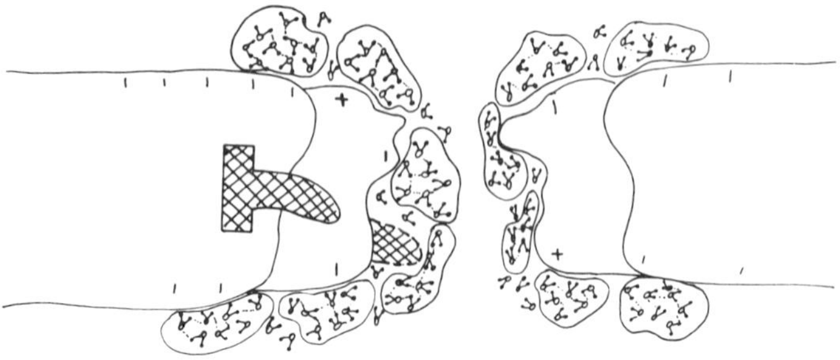

FIG. 4 Ensembles of water molecules and monomeric water molecules also cover channel edges.

VIII. FREEZING

The participation of water in stabilizing membrane structures and in regulating membrane transport properties is also reflected in the freezing behavior of water near the membrane surface and the membrane itself. The physicochemical basis of cell freezing is modified according to the following scenario: Upon nucleation at a subfreezing temperature, ice is formed initially in the extracellular solution, giving rise to a water activity difference across the plasma membrane. The ice depresses the extracellular water activity compared to the liquid intracellular water. In response to this imbalance of activities, water is expressed from the intracellular volume so as to osmotically relax toward a state of thermodynamic equilibrium. As the temperature progressively decreases, the extracellular ice mass continues to grow, resulting in a further reduction in water activity and a concomitant cell volume diminution. At relatively fast cooling rates, the extracellular activity drops more rapidly than can be compensated by cellular dehydration. Consequently, the intracellular water activity remains at an elevated level, thereby increasing the likelihood of intracellular ice formation with its associated injury. At relatively slow cooling rates the intracellular and extracellular water activity can be maintained nearly equal via continual adjustment of the intracellular solution concentration by dehydration.

In conclusion I would like to stress again that hydration of the important biopolymers, as well as the physical-chemical properties of water, play an essential role in formation of the membrane structure and in membrane function.

REFERENCES

1. S. J. Singer and G. L. Nicolson, *Science 175*:720 (1972).
2. L. J. Henderson, *The Fitness of the Environment*, Macmillan, New York, 1913, p. 317.
3. J. D. Bernal and R. H. Fowler, *J. Chem. Phys. 1*:515 (1933).
4. J. A. Pople, *Proc. R. Soc. Lond. A205*:163 (1951).
5. J. T. Edsall and H. A. McKenzie, *Adv. Biophys. 10*:137 (1978).
6. J. Tasaki, in *Biophysics and Bioengineering*, Vol. 3 (A. Noordagraff, ed.), Academic Press, New York, 1983.
7. G. N. Ling, *Physiol. Chem. Phys. Med. NMR 15*:155 (1983).
8. A. Hviot, *Physiol. Chem. Phys. Med. NMR 15*:501 (1983).
9. M. F. Perutz, *Trans. Faraday Soc. 428*:187 (1946).
10. J. O. Bernal, in *The State and Movement of Water in Living Organisms* (G. F. Fogg, ed.), University Press, Cambridge, 1965, p. 17.
11. G. von Heijne, *Curr. Top. Membr. Transp. 24*:151 (1985).
12. G. von Heijne and C. Blomberg, *Eur. J. Biochem. 97*:175 (1979).
13. S. Lewin, *Displacement of Water and Its Control of Biochemical Reactions*, Academic Press, London, 1974, p. 367.
14. D. N. Wheatley, M. S. Inglis, and J. S. Clegg, *IRCS Med. Sci. 11*:594 (1983).
15. P. Lee, R. G. Kirk, and J. F. Hoffman, *J. Membr. Biol. 79*: 119 (1984).
16. J. Funder and O. O. Wieth, *Scand. J. Clin. Lab. Invest. 18*: 167 (1966).
17. D. C. Tosteson and J. F. Hoffman, *J. Gen. Physiol. 44*:169 (1960).
18. G. W. Gould and J. C. Measures, *Phil. Trans. R. Soc. Lond. B278*:151 (1977).
19. W. Kauzmann, *Adv. Protein Chem. 14*:1 (1959).
20. W. Kauzmann, in *The Mechanism of Enzyme Action* (W. D. McElroy and B. Glass, eds.), John Hopkins Press, Baltimore, 1954, p. 71.
21. N. Nagashima, E. Suzuki, and M. Tsuboi, *Bull. Chem. Soc. Jpn. 57*:1409 (1984).
22. A. C. T. North and J. C. Smith, *Int. J. Biol. Macromol. 7*: 223 (1985).
23. R. Huttenrauch and S. Fricke, *Naturwissenchaften 71*:426 (1983).
24. G. E. Schulz, in *Biophysics* (W. Hoppe, ed.), Springer-Verlag, Berlin, 1983.
25. F. MacRitchie, *J. Colloid Interface Sci. 105*:119 (1985).
26. H. Matsumura and K. Furusawa, *J. Colloid Interface Sci. 110*: 137 (1986).

27. G. N. Ramachandran and K. Chandrasekharan, *Biopolymers* *6*:1649 (1968).
28. S. H. White and G. J. King, *Proc. Natl. Acad. Sci. USA 82*: 6532 (1985).
29. J. Lindstrom, *Structure of Biological Membranes* (S. Abrahamsson and J. Pascher, eds.), Plenum Press, New York, 1977.
30. M. J. Janiak, D. M. Small, and G. G. Shirpley, *J. Biol. Chem.* *254*:6068 (1979).
31. B. Maggio, T. Ariga, J. M. Startevan, and R. K. Yu, *Biochemistry 24*:1081 (1985).
32. J. N. Israelashvili and R. M. Pashley, *Nature 306*:249 (1983).
33. V. I. Danilov, I. S. Tolkh, V. I. Poltev, and G. G. Malenkov, *FEBS Lett. 167*:245 (1984).
34. A. Banerfee and H. M. Sobell, *J. Biomol. Structure Dynam.* *1*:253 (1983).
35. T. Srikrishnan and R. Parthasarathy, *Arch. Biochem. Biophys.* *239*:38 (1985).
36. E. Clementi and G. Corongin, *Int. J. Quantum Chem. 22*:585 (1982).
37. D. T. Edmonds, *Biological Membranes*, Vol. 5, Academic Press, New York, 1984.
38. S. L. Fornili, D. P. Vercautern, and E. Clementi, *Biochim. Biophys. Acta 771*:151 (1984).
39. J. C. Weaver, K. T. Powell, R. A. Mintzer, S. R. Sloan, and H. Ling, *Bioelectrochem. Bioenerg. 12*:405 (1984).

8

Electromagnetic Field Influence on Membranes

MARKO S. MARKOV Department of Biophysics, Sofia University, Sofia, Bulgaria

I. Primary Aspects of Matter in Electromagnetic Fields 172
A. Biologically significant electromagnetic fields 172

II. Macromolecular Organization in Electromagnetic Fields 174
A. Frequency-varying fields 175
B. Static fields 175
C. High-voltage pulses 176
D. Electromanipulation 176

III. Window Effects 177

IV. Electromagnetic Fields and Cellular Function 178
A. DNA in an electromagnetic field 180
B. Proteins in an electromagnetic field 180

V. Molecular Interactions in Membranes Exposed to Electromagnetic Fields 181
A. Ions in electromagnetic fields 184

VI. Mechanisms of Electromagnetic Field Action on Membranes 184

References 188

I. PRIMARY ASPECTS OF MATTER IN ELECTROMAGNETIC FIELDS

It is now generally accepted that life on Earth originated and evolved in the presence of electromagnetic fields that differed in their frequency and amplitude. From the physiological point of view it is known that electrical potentials originate in biological systems as a result of concentration gradients of ions across the membrane. Because of that, the role of the ionic distribution across the membrane in influencing the regulatory pathways of cell growth control is a question of considerable interest.

An externally applied electromagnetic field is considered as a vectorial perturbation for chemical or orientational distribution involving interacting molecules or molecular organization. Major structural–chemical changes under electromagnetic fields require the presence of ions, ionized groups, and permanent or induced dipolar charge configurations, preferably in macromolecular structures.

A. Biologically Significant Electromagnetic Fields

The biological effects of time-varying electromagnetic fields (EMFs) are not yet studied sufficiently because of the multiplicity of electrical parameters. The interest in EMF effects on macromolecules was revived appreciably when it was found that electromagnetic fields are capable of producing structural–conformational changes in biopolymers and membranes. Dielectric measurements have shown that polypeptides in viscous organic solvent may undergo an intramolecular helix coil of DNA transitions in the presence of electric fields [1].

In living organisms electromagnetic fields of sufficiently high intensity and variations large enough to affect chemical processes are encountered not only within membrane phases, but also near the surface of membranes and protein organization. The observed membrane potential difference $\Delta\Psi_m$ of about 100 mV may correspond to average values of the electric field strength

$$E = \frac{\Delta\Psi_m}{d}$$

of about 100 $kV \cdot cm^{-1}$ when the thickness d of the dielectric membrane is about 10 nm.

The field intensities, which are experimentally accessible, are limited by dielectric breakdown. In aqueous solutions, fields up to 150 $kV \cdot cm^{-1}$ may be controlled over distances in the millimeter and centimeter range. In ionic solutions electric fields cannot be maintained for a long time. The maximum homogeneous fields that can be achieved experimentally are comparable to the maximum values of electric fields encountered in biomembranes.

In the past decade an increasing number of investigations have demonstrated the magnetosensitivity of different biological systems. Behavioral or physiological responses to weak magnetic fields of 10 to 100 μT are known to be present in fish, amphibians, reptiles, mice, and humans [2]. Recent studies have shown that electrical activity in guinea pigs, pigeons, and rats can be altered by Earth-strength magnetic field. Various theoretical proposals point to possible magnetoreception by retinal photoreceptors.

Studies on the interaction of magnetic fields and biological systems are centered on the influence of the applied magnetic fields on the physiology and behavior of organisms, and a search for magnetic sources within the organisms. Evidence continues to accumulate that a wide range of organisms can detect ambient magnetic fields or use them for orientation and navigation.

It has been suggested [3] that to fine-tune our knowledge we define at least three groups of electric variables: (1) the induction waveform, (2) the topography of electromagnetic fields, and (3) the dielectric characteristics of living tissues. It seems to us that the most important biotropic parameters of electromagnetic fields to consider [4] are the following: (1) type of field, (2) induction, (3) gradient, (4) vector, (5) frequency, (6) shape of the pulse, (7) localization of the influence, and (8) exposure time. One or another of these parameters may be responsible for a particular result, but for better reproduction of experiments, as many of these parameters as possible have to be reported when the method of experiment is explained.

The existence of a frequency dependence of the conductance, impedance, and dielectric properties of biological systems is established. The dispersion of the above-mentioned parameters is closely connected with polarization phenomena. Schwan [5] has categorized the dispersion of the dielectric constant ε_r and specific resistance ρ of the tissues as the α, β, and γ dispersion regions. Each dispersion is defined fairly well by either a single relaxation time or a small spectrum of relaxation times. In a low-frequency region the cell membrane is considered to be a capacitor with a low-loss dielectric of enormous dielectric constant. α-Dispersion is a result of the variability of the apparent outer cell membrane capacitance with the frequency, β-dispersion is a Maxwell–Wagner relaxation factor resulting from the charging of cell membranes, and γ-dispersion is characterized by the relaxation time of free water molecules.

II. MACROMOLECULAR ORGANIZATION IN ELECTROMAGNETIC FIELDS

Thermodynamics and kinetics provide the formalism to describe the observed dependencies of classical–conformational reactions on external physical state variables: temperature, electric field, and magnetic field. In the present account the theoretical foundations for the analysis of electrical–chemical processes are developed only on an elementary level. In most treatments of electric field effects on chemical processes the theoretical expressions are based on the *homogeneous field approximation* of the continuum relationship between the total polarization and the electric field strength [6].

Basic information about macromolecular dimensions, size, and shape have been derived from a relaxation of field-induced changes in optical properties and electrical parameters of the electrically and optically anisotropic systems. It has been established that the extremely large dipolar moments that the electric field produces by displacement of the counterion atmosphere parallel to the long axis of the polyions are responsible for their orientation in the direction of the external field.

A practically instructive example for effective interaction of macromolecular organization with electric field is the membrane-bound bacteriorhodopsin. In this macromolecular system electric fields cause major structural transitions that involve orientation changes of the chromophores' retinal tyrosine and/or tryptophan residues and pK changes of at least two types of H^+-binding sites. The conformational changes are based on a saturable dipole mechanism associated with an extremely large anisotropic (electric) polarizability. On a molecular level the induced polarization appears to involve a restricted electric displacement of ionic groups (ion pairs) within the protein in a highly cooperative manner. The electric field effects observed in bacteriorhodopsin membrane fragments are of functional importance for this light-driven H^+-pumping system [7,8].

Recent electrooptical data on linear polyelectrolytes such as the K^+ salt of polyriboadenylate demonstrate that the dissociation of counterions from the inner atmosphere, and coupled to it, the destacking of the adenine bases in high electric fields, is strongly anisotropic [9]. The anisotropy of counterion movement along polyionic surfaces suggests that counterion exchange as well as influx and efflux of counterion substrates or hormones occur preferably at the borderlines of the ionic atmospheres that cover polyionic region on macromolecular enzyme and receptor proteins and membranes. Once part of the ion cloud, such substrates and activator substances may reach the active sites via surface diffusion. In this model the border regions of the counterion atmosphere serve as a preferable cross section for trapping counterionic substrates [6]. The origin of the membrane electric polarizability induced by the electric field is considered to be connected with the volume, Maxwell—Wagner, or surface polarizability of the membrane fragments [10].

A. Frequency-Varying Fields

It has been found that the membrane resistance in many living systems changes after exposure to electromagnetic field action. A decrease in this parameter was reported for snail neurons [11], muscle [12], and plants [13]. Changes in the permeability of erythrocyte membrane irradiated with microwaves 1 to 3 GHz were established [14]. These changes are developing very slowly and have long-lasting after-effects. Their existence was explained as an influence of microwaves on the conformational state of lipid—protein complexes as well as influence on the transport properties of the membrane without any general disturbances of the membrane.

It has been demonstrated [15] that microcalorimetery could be useful in investigation of the interaction of electromagnetic fields with biological membranes. It was found that the intensity of all thermally induced transitions decrases considerably after irradiation of erythrocyte ghosts by a 330-MHz electromagnetic field. One can assume that EMF affects the water bound to outer membrane components of the erythrocyte ghosts, and this leads to a disturbance of hydrogen-bond formation near the membrane surface. Because the intensity of the applied EMF was not high (92 $mW \cdot g^{-1}$), to get enough energy from the field for molecular rearrangements, it is necessary to assume concentrating the field because of the membrane heterogeneity. If this idea is shown to be correct, the EMF could affect the protein channel portion, leading to disturbance of intact hydrogen-bond formations of structural channel water that is reflected on the infrared spectrum. Thus the protein structure is slightly destabilized.

B. Static Fields

The static magnetic fields effects are often discussed in terms of three factors: (1) geomagnetic background, (2) hygienic standards, and (3) NMR diagnostics and magnetotherapy. Specific behavior in Earth's magnetic field was observed in microorganisms and plants. Magnetotropism manifests itself in the orientation of seed roots, which germinate toward power lines in Earth's magnetic field. A variety of reported orientational properties of plants, animals, and birds to Earth's magnetic field are closely connected with the sensitivity to acting fields. The hygienic aspects include an estimation of the amount of magnetic field energy that reaches human organisms during professional activities and everyday life. NMR diagnostics have become widely utilized in recent years. By using strong magnetic fields (0.5 to 5 T) for short periods, NMR equipment is not as dangerous for patients, but complete monitoring of the magnetic field to which operators are exposed is obligatory. The application of different magnetic or electromagnetic fields in magnetotherapy makes the problem of the influence of these fields on living tissues a question of principal importance.

C. High-Voltage Pulses

The effects of high-voltage pulses on erythrocytes have been studied [16–18]. Hemolysis of human erythrocytes was observed for a field strength of 2 to 4 $kV \cdot m^{-1}$ and pulses of 20 to 40 μs. High-voltage pulses have also been shown to have a lethal effect on bacterial and yeast cells [16,19]. Permeability changes induced by electric pulses in vesicular membranes were also reported [20]. The external field induces a transmembrane potential and it is assumed that when this reaches a critical value of about 1 V, a dielectric breakdown occurs. This voltage is proportional to both field strength and cell radius; hence a large cell cannot withstand the same field strength as a small cell. It was shown [21] that human leukocyte cell membrane damage occurs at a critical peak electric field in the sample of about 2.6 $kV \cdot cm^{-1}$.

Electric fields connected with an electric discharge are accompanied by a current flow through suspension and a temperature rise. On the cellular level, defects of the cell membrane were detected by changes of cell functions. The new pores, channels, or lesions in the cell surface induced by the electric treatment are subject to repair processes that restore the original state.

D. Electromanipulation

Applying a series of short but high-voltage electric pulses on the cell suspension, one can observe permeability changes in membranes. It is generally accepted that this reversible or irreversible increase in membrane permeability or membrane conductivity occurs upon applying electric field pulses of critical strength. The following events are observed in experimental studies: (1) electroportation, (2) electrofusion, (3) electroorientation, (4) electroseparation, (5) electrorotation, (6) electrotransformation, (7) electrotransfection, (8) electroincorporation, (9) electroinjection, and (10) electroporative gene transfer. All these effects of an electric field will not be discussed because they are the subjects of other chapters in this volume.

III. WINDOW EFFECTS

The sensitivity of biological tissues to small variations in electromagnetic fields may explain the window effects described elsewhere. In a number of experiments designed to study a large range of inductions in a constant magnetic field (CMF) applied to different levels of organization biosystems (microorganisms, plants, animals) [22, 23], a specific dependence on the extent of magnetic field effects on the value of magnetic induction has been observed—the optimum of the effect at a magnetic induction of 45 mT. It was indicated as a "resonance" mechanism of the magnetic field influence. The existence of "permitted" levels that biosystems could attain under constant magnetic field action is assumed (Fig. 1). When the induction of CMF is equal to that necessary for the transition, the system achieves a "stationary" state at which it can remain for a certain

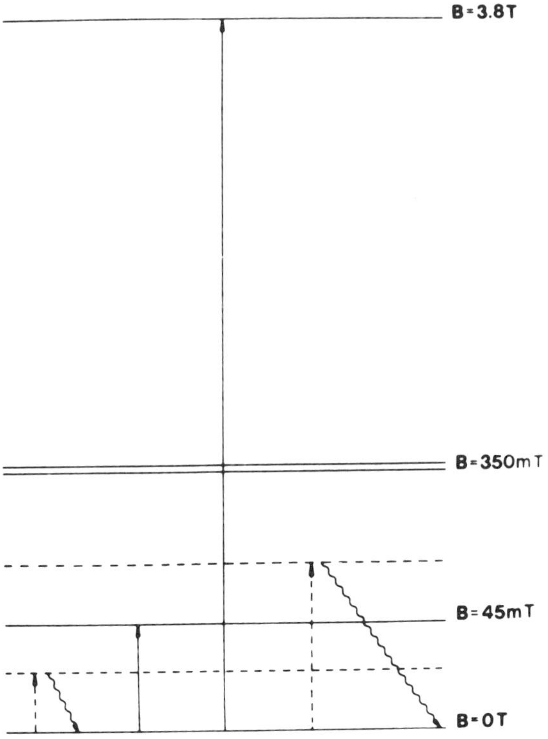

FIG. 1 Resonance levels of constant magnetic field action.

period of time. Any other energy will bring the system to a state other than the stationary state. Because of the fact that the system exchanges energy intensively with the environment, the observed effects are lower and at the same time they quickly disappear at these "nonpermitted" states [4]. In many respects this idea is similar to the idea of "window" action in electromagnetic fields.

A resonance hypothesis has been introduced to explain the frequency-dependent response in brain tissue Ca^{2+} efflux for both low-frequency fields and for radiofrequency fields modulated with low frequencies [24,25]. The cyclotron resonance hypothesis [26] maintains that ion selective (Ca^{2+}, K^+, Mg^{2+}...) resonance conditions in ion channels are produced for specific combinations of dc and ac ELF magnetic fields, following the cyclotron resonance conditions specified by the ionic charge to mass ratio. Not only does this hypothesis explain many of the results observed in the Ca^{2+} experiments, but it has also been used to design a series of predictive experiments, some of which have been successful [27]. The finding of a transient sensitivity of cAMP-independent protein kinase activity to microwaves exposure suggest a "time-window." There is evidence that these field-induced changes involve the cell membrane. At the same time, evidence for amplitude "windows" have been discussed elsewhere.

One particular example of window levels in the case of a constant magnetic field is shown on Fig. 2. It seems to us quite probable that different types of windows exist in biological systems. These particular windows are separated in their reaction to field strength, frequency, time of exposure, and so on.

IV. ELECTROMAGNETIC FIELDS AND CELLULAR FUNCTION

Exogeneously applied weak electromagnetic fields are known to modify or alter normal cellular functions [28,29]. The potentially large number of cellular processes could be affected by EMF. To date, no obvious linear relationship between the shape of signal and the effectiveness of a given signal was found [29].

Theoretical considerations of Frohlich [30,31] have suggested that living objects might contain systems exhibiting large-amplitude oscillations at high frequencies, a conjecture with quite far-reaching consequences for our understanding of biological processes. In particular, Frohlich predicted that there would be a very high frequency oscillations in the frequency range of 100 GHz. Raman spectroscopy study [32] indicated that when metabolically active, living cells produce oscillations in the very high frequency range.

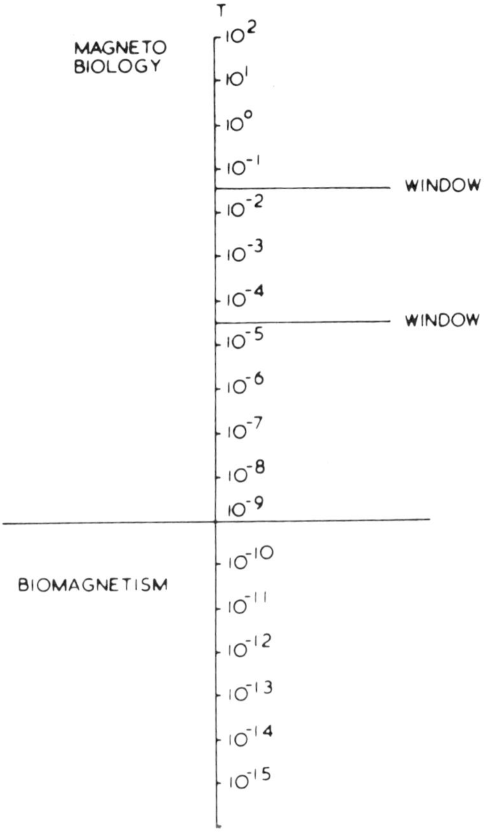

FIG. 2 Biological windows in the intervals of magnetobiology and biomagnetism.

EMF effects are known to play an important role in many biological processes at the cellular level. Phenomena as different as nerve excitation, electrogenic ion transport, neurostimulated secretion of hormones and transmitter substances, and photosynthesis of ATP involve cell functions in which biochemical reactions are inseparately coupled to electromagnetic field forces [6]. A study of the orientation of erythrocytes of a patient suffering from sickle cell anemia was reported [33]. Differences in orientation between normal and pathologic erythrocytes observed would help distinguish the cell in clinical diagnostics. Orientation of sickle cells was also observed in homogeneous magnetic fields due to magnetic anisotropy in the heme molecule in deoxyhemoglobin, whereas normal erythrocytes do not show any orientation in a magnetic field. The effects observed in a study of sick erythrocytes probably are a result of the different shape of the cells and different dielectric properties of membrane which lead to the generation of membrane voltage.

Recently, electromagnetic heating of cancer cells has been used therapeutically. Different kinds of electromagnetic fields are distinguished by the waveform and wavelength of frequency ranges involving microwaves, ultrahigh frequency, and radio frequency [34].

A. DNA in an Electromagnetic Field

Electrooptical methods have been widely used for the analysis of biological macromolecules. Electrooptical data proved to be particularly useful for the characterization of structures with some symmetry, such as nucleic acids. Application of electric field pulses to biological macromolecules requires special caution, since the field pulses may induce conformational changes. In most of the cases investigated, the reaction is driven by the dissociation of ion complexes accompanied by polarization of the ion atmosphere around the polymer molecules. The ion displacement may result in dissociation of the ordered structures. It seems that the action of an electric field upon biopolymers is always "destructive" [35]. Evidence that electric fields may have a "constructive" action in interactions between nucleic acid helices has been published. Changes in the light scattering of DNA induced by dielectric fields have been attributed to the orientation of DNA helices.

According to many studies, electric fields induce dissociation of macromolecular structures. Thus a field-induced association reaction of DNA has really not been expected, since the high negative charge density of DNA helices leads to repulsion, especially at low ionic strength. It is generally accepted that the DNA dipole moment results from polarization of its ion atmosphere. Because of that one can imagine one of the possible mechanisms of electromagnetic field action on DNA structure—disturbances in the original polarization of a DNA ion atmosphere.

B. Proteins in an Electromagnetic Field

The mechanism by which the cell senses a change in electromagnetic field, ultimately translating this alteration into a change in its behavior, is not yet clear. The delayed initial response of some cells to the application of an electric field has prompted a number of investigators to suggest that time is needed for the redistribution of membrane receptors [36].

It has been shown that electrostatic forces may influence the exact disposition of protein inserted into the membrane [37]. The membrane potential is considered connected with the changes in protein disposition in artificial membranes, and this leads to specific predictions about the placement of charged amino acids in biosynthetically inserted proteins.

The proteins that mediate ion translocation are themselves more often than not sensitive to an electric field. The carbohydrate portion of membrane glycoproteins offer a pathway by which alterations in cell behavior could be achieved. An increase in [^{3}H] glycosamine incorporation into plasma membrane glycoproteins in response to an electric field, suggesting some alterations in cell behavior has been reported [38].

Changes in protein kinase activity in cultured lymphocytes exposed to modulated microwave fields have been observed. Protein kinases are known as enzymes with general properties of phosphorylating other enzymes or proteins. Protein kinases have been implicated in the regulation of many intracellular processes, including glycose and lipid metabolism, protein synthesis, and membrane permeability. They have observed a large reduction in histone kinase activity in lymphocytes during the exposure of cells to relatively low levels of modulated microwave radiation [26].

V. MOLECULAR INTERACTIONS IN MEMBRANES EXPOSED TO ELECTROMAGNETIC FIELDS

Perturbation of surface topography by application of electric fields is likely to alter molecular interactions within the plane of the cell membrane. Changes arising in biological membranes under the action of electromagnetic fields have still not been studied completely. One of the mechanisms for alteration of the functional properties of membranes can be their structural rearrangement.

All biological membranes seem to be associated with an electric membrane potential; all membrane constituents and membrane processes are exposed to electric forces [39]. Pohl [40] believes that there is a real probability that cells produce oscillating electric fields. Living cells of many types give evidence of the presence of naturally oscillating electromagnetic fields.

It has been recognized that macromolecules and macromolecular organizations such as biological membranes are particularly effective media for the coupling of high electric fields with biochemical reactivity. Indeed, all biomembranes appear to be associated with electric membrane potentials. The elucidation of the possible connection between the membrane electric field and the orientational and conformational properties of membrane proteins could not only be essential for further refinements in excitable membrane theories, but could also have a more general impact on an understanding of the connection between membrane structure and function [41]. Electrical chemical membrane processes are most evident in the rapid electric communication system of living entities.

Many membrane molecules are capable of transition within the plane of the membrane. Since many of these molecules bear a net electric charge and the charged moieties will probably lie in the aqueous surface of the membrane, it is reasonable to expect that tangential application of an electric field to the cell membrane surface will cause directed migration of the molecules relative to the uncharged portions of the membrane matrix.

The purple membrane of *Halobacterium halobium* is a distinct example of the very important role of the membrane electric field in the membrane protein function. It has been shown that the drastic change in the purple membrane color upon deionization, low pH, and so on, is accompanied by a drastic change in the purple membrane function (i.e., the proton translocation through the bacteriorhodopsin channel is blocked). The electrooptic investigation of the blue form of purple membrane [42], has also shown a drastic drop in the purple membrane permanent dipole moment which is obviously connected with the membrane function. There are several indications that lipids and proteins can move laterally within biomembranes. This movement is strongly dependent on the physical parameters of EMF applied.

It is probable that the ubiquitious ionic double layers at the numerous interfaces within the cells play a role in the natural oscillations of cells. In this connection it may be noted that mechanically induced electromagnetic radiation from vibrated systems having ionic double layers may play a role in cellular responses [43]. Another mechanism for producing resonant oscillations is that which could arise from local fluctuations in the charge density at the surface of the cell. It could affect the charge distribution around the periphery of the outer ionic double layer of the cell—medium interface.

The distribution of cell surface macromolecules can be modulated by various factors, such as extracellular ligands, cell—cell or cell—substrate contacts, and extracellular electric fields [44]. In a number of cell types, static electric fields cause the migration and accumulation of plasma membrane macromolecules toward one side of the cell. Steady-state accumulation of charged membrane components in the presence of a dc field depends on at least three molecular processes on the cell surface: lateral electrophoresis, Brownian diffusion, and solvent drag due to the electroosmotic flow of fluid near the cell surface [45].

In addition to the use of an electric field as a tool in studying cell surfaces, there is a separate issue concerning the biological implication of electromigration; that is, is there an endogeneous electric field within tissue effective in modulation membrane topography? In particular, are pulsed repetitive electric fields, which are characteristic of endogeneous fields associated with neural activities, capable of inducing the migration of plasma membrane components? If so, what are the intensity and frequency requirements of the effective field [44]?

Electrooptic data of aqueous suspensions of purple membranes demonstrate that high electric field pulses (1 to 30 $\times$ 10^5 $V \cdot m^{-1}$, 1 to 100 μs) cause structural transitions in two kinetically different phases. The rapid mode (1 μs) represents a concerted change in the orientations of both retinal and some tyrosine and/or tryptophane side chains. The slower mode (100 μs) only involves changes in the microenvironment of aromatic amino acids [46]. The kinetics of the electrooptic signals on the electric field strength suggests a saturable induced dipole mechanism which is indicative for appreciable cooperativity in the probably unidirectional transversal displacement of ionic groups on the surfaces.

In situ separation of membrane proteins when the membrane is exposed to an electric field has been discussed [47]. At a field strength level beneath the threshold where protein will migrate without prior complexation, specific ligand binding will induce the labeled species to move.

It has been shown [48,49] that the lipid and protein phases of erythrocyte membranes under the action of radio-frequency radiation undergo structural changes that can be explained by elevated temperature produced by radiation. Pickard and Barsoum [50] have revealed at the membrane level a contribution of the nonthermal component to radio-frequency-induced changes. Using fluorescent probes in erythrocyte membranes to study the lipid–protein interactions under radio-frequency radiation as well as changes occurring in the membrane lipid phase, the following were found [51]: (1) an increase in fluidity of the lipid phase at the site of probe localization, (2) changes in the state of the lipid–protein contact region, and (3) a decrease in shielding of the lipid bilayer by proteins.

It has been shown that electric pulses on the order of 3 kV cm^{-1} (which generate transmembrane potential of 1 V) and of duration 10 to 100 μs induce hemolysis of erythrocytes in an isotonic medium. The effect was attributed to dielectric breakdown of red cell membranes [20,52,53]. It was found that the voltage pulse does not lead to a general breakdown of the cell membrane. Pores that allow passage of Na^+ and K^+ ions are implanted in the red cell membranes. These pores are small enough to prevent the permeation of sucrose, and leakage of hemoglobin is unlikely. Subsequent hemolysis is due to a secondary effect [18,54]. Once the membrane is rendered permeable to Na^+ and K^+ ions, these ions approach the Donnan equilibrium within seconds. The excess osmolality of hemoglobin inside the cytoplasm drives a water influx and the red cells swell. At 25°C this process continues for several hours; then the cell membranes are ruptured and hemoglobin is released from the treated cells.

The size of electric pulse-induced pores can be varied by changing experimental conditions: pulse duration, field strength, and the ionic strength of the suspending medium. The conductance of a dense suspension of erythrocytes increases when the cell membranes are perforated with an electric pulse. The time course of the pore formation can be followed by the evolution of membrane conductance [55].

A. Ions in Electromagnetic Fields

Ion transport in aqueous solutions is known to be a result of individual ion flows driven by electrochemical potential gradients. The bioelectrochemical quantities in a cell exposed to an electric field have been computed [56]. The net ion exchange between the external medium and the cytoplasm can be obtained; then modifications of the ion concentration inside the cell are computed and the time of beginning the possible biological effects is discussed.

The presence of electromagnetic fields increases the dissociation of weak acids and bases and promotes the separation of ion pairs into the corresponding free ions. The free ions or ionized structures may then move in the direction of the electric field, and a field-dependent stationary state in the ion distribution may be established. It is generally accepted that electromagnetic fields may interact with cell membranes, resulting in physiologically significant changes in ion transport and/or binding.

Electrostatic forces significantly affect the turnover capacity of ion pores. In addition, they exert a major influence on an ion's ability to enter a pore. The ion pore former/membrane water system is considered to be a problem of applied electrostatics. The electrical shielding by the pore former may greatly reduce the image potential of an ion traversing a pore [57]. A substantial fraction of the voltage drop due to the application of a transmembrane potential may occur in the aqueous regions exterior to the pore [58].

VI. MECHANISMS OF ELECTROMAGNETIC FIELD ACTION ON MEMBRANES

Electromagnetic field effects play an important role in many biological cell processes. The absence of general interaction mechanisms between electromagnetic fields and biological systems is a major problem in the study of the biological effects related to electromagnetic fields.

It has been recognized that macromolecules and macromolecular organizations such as biological membranes are particularly effective media for the coupling of electromagnetic fields with biological reactivity. All biomembranes appear to be associated with the electric membrane potential.

One of the main aims of investigating electric field effects on chemical and biological transformations is to determine the reaction mechanisms. It is known that the majority of biological processes involve ionic species in aqueous environments. The stability of many biopolymers requires a finite ionic strength. Electric field effects in these systems have to be studied in conducting solutions and suspensions.

The mechanism by which the cell senses the application of an electromagnetic field, ultimately translating this stimulus into biochemical and behavioral alterations, is presently unknown. At least one component of this information transfer probably involves changes in the concentration of cations at the cell surface and/or within the cell. One possible mechanism may be that changes in the concentration of protein-bound ions may induce conformational changes in associated molecules [59], ultimately affecting both the extracellular and intracellular processes. With regard to molecules that operate in the extracellular environment, alterations that produce conformational changes in plasma membrane components may induce concomitant changes in either the activity of accessibility of these components. With regard to intracellular processes, the glycosylation apparatus of the cell appears to be a candidate target for changes by alterations in net ion flux. The conformation of protein substrates for various polycosidic enzymes may be changed, signaling the transfer of additional sugar residues. In addition, changes in the conformation of the enzymes themselves could affect transferase activity [60].

To provide a rational theory of EMF influence, it seems reasonable to explain certain principles governing electrostatic interactions of the important biomolecules. Proteins and their conjugates (lipoproteins and glycoproteins) maintain their solubility due to their hydrophobic nature. These macromolecules maintain optimal hydration at slightly alkaline pH, due to the presence of electronegative charge.

The molecular or ionic events that couple with the electromagnetic signals are still unknown. There is some evidence that the plasma membrane ATPases are involved in signal coupling. Two possible couplings have been emphasized; one is the ionic microenvironment across the plasma membrane and other biological structures;

the second is the plasma membrane itself. Colacicco and Pilla [61] are in agreement with the possibility that the low-energy electromagnetic signals perturb metastable equilibria and thus trigger a cascade of cooperative phenomena whose energetics could be paid by metabolic processes; coupling of the signal with macromolecular cooperative processes cannot be excluded.

Energetically, amplification can be visualized in terms of a casade of metabolic processes. In contrast, a model of direct electrical amplification of the pulses of induced current defies description by classical thermodynamics. The possible participation of the cell membrane in signal coupling is consistent with several observations and hypotheses:

1. The plasma membrane has been attributed a role in the spinning of suspended live cells under the action of high-frequency electromagnetic fields [62].
2. According to the electrical information transfer hypothesis, the electromagnetically induced currents must couple nonfaradaically with ion transport across the membrane [63].
3. The same very low frequency fields that stimulated Ca uptake by the chick tibia also promoted Na^+ flux and ATPase activity across red blood cell membranes.
4. Since the same fields stimulate DNA synthesis and cell proliferation as are regulated by the plasma membrane Na,K-ATPase, the role of membrane ATPase in coupling the signal with the biological processes becomes inevitable.
5. If a bicarbonate-dependent Ca^{2+} ATPase or Ca^{2+}-activated ATPase regulates Ca^{2+} transport across cell membranes, the involvement of this ATPase may explain the effect of bicarbonate on the stimulation of Ca mobilization in very low frequency fields.

Electric chemical membrane processes are most evident in the rapid electric communication system of living entities. For example, the generation and rapid transmission of electric signals such as nerve excitation are based on interactions between electric field and biological membrane organization [65].

Adey [66] has proposed a means by which the extended membrane region surrounding the neutral membrane may be sensitive to electric charge movement or to electric fields. This region, perhaps 2000 Å thick, is referred to as the glycocalyx. Cooperative conformational changes is considered to occur in the glycocalyx.

A fundamental question is whether stimulation of the biological processes effected by the EMF is electric or magnetic. According to the electrochemical information transfer hypothesis [63], stimulation and response are electrical; thereby, loops of electromagnetically induced current couple with transmembrane ionic currents. Two widely different (by frequency) fields may be coupling with totally different or only quantitatively different processes.

The biological response induced by exposure to electromagnetic radiation at microwave and radio frequencies and at high intensities is generally considered to involve temperature elevation caused by deposition and subsequent thermalization of electromagnetic energy in the sample [67]. It is accepted that one possible mechanism of microwave influence on biological systems is connected with heat effects. Akoev et al. [68] suppose that a thermal mechanism of microwave influence on the ionic channel is present. At the same time it is quite probable that there are nonthermal effects as well as heating of particular parts of the membrane.

The quantum energy associated with an EMF is directly proportional to the frequency. At a power frequency (extremely low frequency range) the corresponding quantum energy is on the order of 10^{-34} J. Obviously, the quantum energy associated with extremely low frequency EMF is unlikely to produce significant bioeffects. Classical concepts of the possible interaction between EMF and living organisms are of little value in the extremely low frequency region, since the wavelength precludes any resonance phenomena and the energy delivered is far less than the kT value of the living systems.

It is known that magnetic fields can induce currents that will flow in closed circuits within the body. It can be calculated that a 1-G, 60-Hz magnetic field induces a $2\text{-mV}\cdot\text{m}^{-1}$ electric field in a circuit of radius 0.1 m composed of material having the same resistivity as body tissues. This internal field produces a current density that is comparable to that produced by a $2.5\text{-kV}\cdot\text{m}^{-1}$ external electric field. However, the current pathways are not the same and the magnetic field-induced current is proportional to the area enclosed by the circuit. The direct interaction of a magnetic field with biological matter involves energies that are very much smaller than the thermal energies [69].

Becker [70] proposes another concept for the possible mechanism of EMF influence. He supposes a specific sensing mechanism evolutionarily developed to derive information of biological value from Earth's normal magnetic field. Earth's normal electromagnetic spectrum is a result of electromagnetic radiation from the sun and space interacting with the ionosphere and Earth's magnetic field. This results in an electromagnetic spectrum consisting of three parts: (1) the dc magnetic field and associated micropulsation frequency ranging from less than 1 Hz to approximately 100 Hz with major power density centers about 10 Hz, (2) a band of frequencies centered in the 10-kHz region due to lightning strikes, and (3) the narrow zone of visible light associated with nonvisible ultraviolet and infrared spectra derived from sunlight.

Of particular interest in a discussion of electromagnetic field effects in biological structures is the observation of threshold phenomena. It has been found that electric impulses above a certain threshold intensity are capable of triggering conformational transitions in metastable polynucleotide structures. A similar threshold effect is associated with electric-field-induced permeability changes in vesicle membranes [71,72] as well as cellular systems [73,74,76]. Nonlinear field dependencies of base stacking in single-stranded polynucleotides have been discussed as a threshold effect [75].

Concluding this chapter, we would like to note that because of the difference in their physical parameters, every particular EMF is capable of influencing different membrane or cellular components as well as the cell environment. This is probably the reason that a general mechanism of EMF action on biological membranes has not yet been proposed. It seems to us that a complex approach including cell interior, cell membrane, and cell environment (hydration properties mianly) is necessary for a study of the effect of electromagnetic fields on biomembranes.

REFERENCES

1. G. Schwarz and J. Seeling, *Biopolymers 6*:1263 (1968).
2. S. Reuss and J. Olcese, *Neurosci. Lett. 64*:97 (1986).
3. M. Hinsenkamp, M. Rooze, N. Noorbergen, B. Tuerlinckx, and C. Nicolini, in *Interactions Between Electromagnetic Fields and Cells* (A. Chiabrera, C. Nicolini, and H. P. Schwan, eds.), Plenum Press, New York, 1985, p. 557.
4. M. S. Markov, in *Charge and Field Effects in Biosystems* (M. J. Allen and P. N. R. Usherwood, eds.), Abacus Press, London, 1984, p. 319.
5. H. P. Schwan, *Adv. Biol. Med. Phys. 5*:147 (1957).
6. E. Newmann, in *Modern Bioelectrochemistry* (F. Gutmann and H. Keyzer, eds.), Plenum Press, New York, 1986, p. 97.
7. K. Tsuji and E. Neumann, *Int. J. Biol. Macromol. 3*:231 (1981).
8. K. Tsuji and E. Neumann, *FEBS Lett. 128*:265 (1961).
9. E. Neumann, K. Tsuji, and D. Schallrenter, in *Biological Structures and Coupled Flows* (A. Oplatka and M. Balaban, eds.), Academic Press, New York, 1983, p. 135.
10. G. Todorov and S. P. Stoylov, *Chem. Phys. Lett. 133*:78 (1987).
11. S. L. Arber, *J. Microwave Power 16*:15 (1981).
12. A. Portela, J. G. Voccari, S. M. Michaelson, O. Liobera, M. Brennan, A. E. Gostonyi, J. C. Perez, and H. Jenerick, *Stud. Biophys. 53*:197 (1975).

13. J. G. Barsoum and W. P. Pickard, *Bioelectromagnetics 3*:393 (1982).
14. E. Sh. Ismailov, *Sci. Rep. High School Biol. N3*:58 (1971).
15. V. L. Shnyrov, G. G. Shaden, and I. G. Akoev, *Bioelectromagnetics 5*:411 (1984).
16. A. J. H. Sale and W. A. Hamilton, *Biochim. Biophys. Acta 163*:37 (1968).
17. U. Zimmermann, G. Pilwat, G. Holzapfel, and K. Rosenheck, *J. Membr. Biol. 30*:135 (1976).
18. K. Kinosita and T. Y. Tsong, *Biochim. Biophys. Acta. 471*: 227 (1977).
19. E. G. Niemann, *Radiat. Environ. Biophys. 18*:221 (1980).
20. E. Neumann and K. Rosenheck, *J. Membr. Biol. 14*:194 (1973).
21. K. H. Mild, L. Lovedahl, K. G. Lovstrand, and S. Lovtrup, *Bioelectromagnetics 3*:213 (1982).
22. M. S. Markov, *Proc. 6th International Conference on Magnet Technology*, Bratislava, 1977, p. 384.
23. M. S. Markov, *Trans. IEEE* 2334 (1981).
24. S. M. Bawin and W. R. Adey, *Proc. Natl. Acad. Sci. USA 73*:1999 (1976).
25. S. M. Bawin, L. K. Kaczmarek, and W. R. Adey, *Ann. N.Y. Acad. Sci. 247*:74 (1975).
26. A. R. Liboff, *J. Biol. Phys. 13*:99 (1985).
27. A. R. Liboff, S. D. Smith, and B. R. McLeod, in Mechanical Approaches to Interactions of Electric and Electromagnetic Fields with Living Systems (M. Blank and F. Findl, eds.), Plenum Press, New York (1987).
28. R. Goodman, C. A. L. Bassett, and A. Henderson, *Science 22*:1283 (1983).
29. R. Goodman and A. S. Henderson, *Bioelectromagnetics 7*:23 (1986).
30. H. Frohlich, *Int. J. Quantum Chem. 2*:641 (1968).
31. H. Frohlich, *Nature 228*:1093 (1970).
32. S. J. Webb, *Phys. Rep. 60*:201 (1980).
33. J. Vienken and U. Zimmermann, *Naturwissenschaften 71*:158 (1984).
34. W. S. Yamanashi, M. R. Fesen, D. W. Anderson, J. Valentine, A. M. Sy, C. Crandall, and P. D. Lested, *Physiol. Chem. Phys. Med. NMR 16*:491 (1984).
35. D. Porschke, H. J. Meier, and J. Ronnenberg, *Biophys. Chem. 20*:225 (1984).
36. R. Nuecitelli and C. A. Erickson, *Exp. Cell Res. 147*:195 (1983).
37. R. Blumenthal, C. Kemp, J. van Renswonde, J. N. Weinstein, and R. D. Klausner, *J. Cell. Biochem. 22*:55 (1983).
38. M. S. Markov and M. Blank (eds.), Electromagnetic Fields and Biomembranes, Plenum Press, New York (1988).

39. K. Tsuji and E. Neumann, *Biophys. Chem. 17*:153 (1983).
40. H. Pohl, in *Current Excitation in Biological Systems* (H. Frohlich and F. Kramer, eds.), Springer-Verlag, Berlin, 1983, p. 199.
41. S. P. Stoylov, G. Todorov, and A. Zhivkov, *Bioelectrochem. Bioenerg. 12*:49 (1984).
42. S. G. Taneva, G. Todorov, I. B. Petkanchin, and S. P. Stoylov, *Eur. Biophys. J. 14*:415 (1987).
43. H. Pohl, in *Bioelectrochemistry* (H. Keyzer and F. Gutmann, eds.), Plenum Press, New York, 1980, p. 273.
44. S. Lin Liu, W. R. Adey, and M. M. Poo, *Biophys. J. 45*:1211 (1984).
45. S. McLaughlin and M. M. Poo, *Biophys. J. 34*:85 (1981).
46. E. Neumann, *Stud. Biophys. 94*:107 (1983).
47. S. H. Young, M. McCloskey, and M. M. Poo, *Receptors 1*:511 (1987).
48. Y. W. Allis and B.L. Sinha, *Bioelectromagnetics 2*:13 (1981).
49. M. J. Ortner, M. J. Galvin, D. I. McRee, and C. F. Chiguell, *J. Biochem. Biophys. Meth. 5*:157 (1981).
50. W. F. Pickard and Y. H. Barsoum, *J. Membr. Biol. 61*:39 (1981).
51. Y. A. Kim, S. S. Fomenko, T. A. Agafonova, and I. G. Akoev, *Bioelectromagnetics 6*:305 (1985).
52. A. J. H. Sale and W. A. Hamilton, *Biochim. Biophys. Acta 163* (1968).
53. F. Rieman, U. Zimmermann, and C. Pilwat, *Biochim. Biophys. Acta 394*:449 (1975).
54. K. Kimosita and T. Y. Tsong, *Proc. Natl. Acad. Sci. USA 74*: 1923 (1977).
55. K. Kinosita and T. Y. Taong, *Biochim. Biophys. Acta 554*:479 (1979).
56. A. Casaleggio, L. Marconi, G. Morgavi, S. Ridella, and S. Rolando, *J. Bioelec. 3*:305 (1984).
57. P. C. Jordan, *Biophys. Chem. 13*:201 (1981).
58. P. C. Jordan, *Biophys. J. 39*:157 (1982).
59. A. A. Pilla, in *Bioelectrochemistry* (H. Keyzer and F. Gutman, eds.), Plenum Press, New York, 1980, p. 353.
60. S. J. Fischer, J. Dulling, and S. D. Smith, *J. Bioelectr. 5*: 253 (1986).
61. G. Colacicco and A. A. Pilla, *Bioelectrochem. Bioenerg. 12*:259 (1984).
62. U. Zimmermann, J. Vienken, and G. Pilwet, *Z. Naturforsch. 36c*:173 (1981).
63. A. A. Pilla, *Ann. N.Y. Acad. Sci. 238*:149 (1974).
64. G. Colacicco and A. A. Pilla, *Z. Naturforsch. 38c*:465 (1983).

65. A. Katachalski and E. Neumann, *Int. J. Neurosci.* *3*:175 (1972).
66. W. R. Adey, in *Behavior and Brain Electrical Activity* (H. Altsmer and B. Burch, eds.), Plenum Press, New York, 1974.
67. C. F. Blackman, S. G. Benate, D. E. House, and W. T. Joines, *Bioelectromagnetics* *6*:1 (1985).
68. I. G. Akoev, V. V. Tjagelov, O. V. Kolomitkin, S. I. Alekseev, and P. A. Grigoriev, *Proc. Acad. Sci. USSR Biol N1*:41 (1985).
69. A. R. Sheppard and M. Eisenbud, *Biological Effects of Electric and Magnetic Fields of Extremely Low Frequency*, New York University Press, New York, 1974.
70. R. Becker, *J. Bioelectr.* *4*:133 (1985).
71. E. Neumann and K. Rosenheck, *J. Membr. Biol.* *10*:279 (1972).
72. E. Neuman, *Angew. Chem. Int. Ed.* *12*:356 (1973).
73. U. Zimmermann, P. Scheurisch, G. Pilwat, and R. Benz, *Angew. Chem.* *93*:332 (1981).
74. J. Teissie and T. Y. Tsong, *Biochemistry* *20*:1548 (1981).
75. D. Porschke, *Biopolymers* *15*:1917 (1975).
76. M. S. Markov and J. P. Guttler, *Studia Biophysica* *130*:211 (1989).

9

A Critical Reevaluation of the "Nonspecific" Adsorption of Plasma Proteins and Dextrans to Erythrocytes and the Role of These in Rouleaux Formation

JOHAN JANZEN Department of Pathology, University of British Columbia, Vancouver, British Columbia, Canada

DONALD E. BROOKS Department of Pathology and Chemistry, University of British Columbia, Vancouver, British Columbia, Canada

I.	Introduction	193
II.	Solution Analyses	198
	A. Mathematical description	198
	B. Results	204
	C. Discussion	216
III.	Pellet Analyses	220
	A. Mathematical description	220
	B. Results	221
	C. Discussion	235
	Symbols	244
	References	245

I. INTRODUCTION

Erythrocytes of many species can be reversibly aggregated by various protein macromolecules and large polymers [1–4]. It has been proposed that this aggregation is due to adsorption and reversible *cross-bridging* of the membrane components of adjacent cells by the macromolecules [5–7]. This mechanism presumes that significant nonspecific interactions occur between a wide range of macromolecules and the membrane components of cells. However, direct measurements confirming such interactions are infrequent.

The cross-bridging mechanism of cellular aggregations is well supported where relatively specific or high-affinity molecular interactions occur [8]. Evidence for weaker, apparently nonspecific molecular interactions has been more difficult to obtain. Dextran adsorption to human erythrocytes has been studied [9,10] and correlated with qualitative measures of aggregation [11,12] but similar correlations are not available for reversible aggregation by other polymers or by any plasma proteins. Indeed, the occurrence of significant nonspecific adsorption of macromolecules to red cells is disputed [13–16]. Thus it has been proposed that nonspecific red cell aggregation [15,17], embryonic retinal cell aggregation [16], and differential cell adhesion in embryonic tissue in general [18] are due to mechanisms that do not involve macromolecular adsorption.

The cross-bridging mechanism is appealing, as it is readily evident that the free-energy reduction associated with adsorption is available to do the work required to aggregate the cells [12]. The free energy available must be related to the nature of the individual interactions, the number of interactions per macromolecule, the number of macromolecules adsorbed, and the degree to which the interactions act independently. A quantitative test of the cross-bridging mechanism has not been reported.

Aggregation of cells by a mechanism that does not require adsorption or requires nonadsorption has received much less attention. One quantitative explanation for particle aggregation without surface adsorption appears to originate with Asakura and Oosawa [19]. They proposed that aggregation could be effected by the restrictions imposed on the configurations of molecules in the region adjacent to a surface. Thus the configurational entropy of macromolecules near the surface is reduced and it is possible, in the absence of sufficient adsorption energy, that the concentration of the molecules will be similarly reduced [20]. When two surfaces are close enough to allow the regions of depletion to overlap, the local osmotic pressure is reduced relative to the bulk solution and a force acting to reduce the separation results. This mechanism has subsequently been developed by Joanny et al. [21], Fleer et al. [22], and Evans et al. [23]. The mechanism proposed has been described in the context of the colligative properties of polymer solutions, but it may also apply to solutions of protein macromolecules, as it has been successfully invoked to explain the aggregation of lipid vesicles in polymer and protein solutions [23].

Our objective here is to review the direct evidence for nonspecific adsorption to cells so that limits may be placed on the cell surface concentrations of adsorbed molecules. This will help place in context the relevance of the nonspecific surface interactions of various macromolecules and may clarify the conditions under which the proposed aggregation mechanisms are effective. It may be that different macromolecules cause aggregation by different mechanisms. We have restricted our consideration to reports where the amounts of interacting species were estimated from measurements of species concentration and volume and which deal with the interaction of red cells with plasma proteins such as albumin and fibrinogen or polymer molecules such as dextran. We have found that the surface concentrations reported for plasma proteins range widely even though experimental conditions appear similar. Thus the surface concentrations of weakly adsorbed macromolecules have in some cases been overestimated and the perception that cells such as erythrocytes are in general "coated" with adsorbed proteins when suspended in plasma may be incorrect [24,25].

Our interest is restricted to reversible red cell aggregation because a quantitative measure of the free-energy change that characterizes the phenomenon is available (i.e., the surface affinity for erythrocyte aggregation) [26,27]. Macromolecular adsorption and surface affinity measurements will be compared elsewhere in a separate paper. The requirement that aggregation be reversible limits the species of molecules that we will consider. Aggregation of red cells by immunoglobulins can result in bonding between cells, which is usually irreversible in the face of washing or moderate shear forces and can be strong enough to rupture the cells if sufficient stress is applied [28]. In these cases (hemagglutination) the interactions are generally agreed to be specific, involving the binding of antibody(s) with cell surface antigen [29]. On the other hand, normal red cells in plasma form aggregates which are easily disaggregated. These are described by the term *rouleaux*: "a stack of coins" [30,31]. After initially random encounters, rouleaux form because the biconcave, disk-shaped cells realign to increase their area of contact [32,33]. Relatively weak fluid shear stress disaggregates rouleaux [34—36] and a single wash with isotonic saline normally removes enough of the aggregating agent(s) to abolish reaggregation [7]. In cases where the aggregation behavior of red cells is reversible, the interactions responsible are also likely to be reversible.

The term *nonspecific* is applied to an interaction by default rather than on the basis of a demonstrated mechanism. Specific interactions are associated with the concept of a receptor and are partially defined for experimental purposes by demonstration of (1) a correlation between quantity of material bound and effect, (2) inhibition of the effect by structurally similar molecules, and (3) identification of the species of molecule or molecular substructure which represents the receptor [37]. Thus dextran adsorption is considered nonspecific because it does not meet the second or third conditions, although it appears to meet the first condition. More important from our point of view, red cell aggregation by dextran is reversible under most conditions [38]. As noted earlier, no correlations between cell surface adsorption of protein and rouleaux formation have been reported, so that the surface interaction is also nonspecific in these cases. In addition, there are as yet no known specific inhibitors of rouleaux occurring in plasma. Although the term *nonspecific* is problematic in that it depends on a lack of specificity, it is of use in that it seems to correlate with those cases where aggregation is reversible, evidently because the interactions are relatively weak.

The present discussion concerns direct measurements of adsorption (i.e., measurement of adsorbate quantity). Adsorption inferred from secondary effects such as aggregation electrophoretic mobility [39], or heats of reaction [24] cannot yet be interpreted unequivocally in terms of the amount of material associated with the surface. Two types of direct measures exist and will be referred to as the "solution" and "pellet" analysis methods. Solution analyses determine the change in the amount of the adsorbing molecule in solution as a result of adding cells and suspending medium. Thus it is necessary to measure the concentration of the molecule and the volume of the solution before and after the addition. The main limitation of this analysis is that the volumes of the solution and cells are difficult to determine to the required accuracy [9]. Pellet analysis methods measure the amount of a molecule associated with the cells and requires that the cells be separated from the solution in which they are suspended. Separations achieved by centrifugation alone are not complete [i.e., the cell pellets contain entrained (trapped) intercellular solution]. In general, correction for solution entrained with the cells involves the use of a volume-marking molecule that ideally is similar in size to the adsorbed molecule, yet does not adsorb to the cells. A more complete separation can be achieved by washing the cells or passing them through a phase devoid of the adsorbed molecule. Then, however, the time required for the separation must be significantly less than the time required for the molecule to desorb from the surface [40]. This condition does not appear to be met in the case of rouleaux-like aggregation. Thus molecules retained by cells after a wash with buffer are probably not those directly responsible for normal aggregation in plasma. Techniques employing centrifugation and rapid washing have been used where high-affinity (specific) binding or slow off-rate nonspecific adsorption occurs, however [40].

Both the solution and pellet analysis methods aim to describe the distribution of a soluble molecule between the solution and cell surface compartments of a cell suspension. Molecules associated with the cells are said to be adsorbed, but in fact little information is available regarding the actual location of this material. At equilibrium the distribution may be described by the ratio of the surface concentration of adsorbed molecules to their concentration in solution (i.e., a distribution ratio). This is particularly useful in the cases of interest here because for weakly associated material the amount bound is usually directly proportional to the equilibrium concentration over a considerable range, making the distribution ratio a constant over that range. We will at times refer to the adsorbing species as a ligand but do not imply by this usage any specificity of binding. In what follows we discuss all adsorption data in terms of the amount adsorbed per unit volume or mass of cells and, where necessary, use the mean cellular area-to-volume or area-to-mass ratio to calculate surface concentration. Thus we define a practical distribution ration P_j for the jth molecular species by

$$P_j = \frac{C_{jc}}{C_{js}} \tag{1}$$

where C_{jc} is the amount of species j adsorbed per unit volume or mass of cells and C_{js} is the equilibrium concentration of j, per unit volume or mass as appropriate, in solution.

The determination of distribution ratios is made more complex when more than one species is interacting with the surface or when more than one detectable species is present and these distribute differently between the compartments. Either of these cases is best dealt with by doing experiments with isolated species. However, when residual impurities can be identified, it is also possible to correct an apparent distribution ratio for the presence of the impurities. This approach requires separate experiments to determine the distribution ratios of the impurities and is of increasing importance when very low levels of adsorption are to be measured or where molecules of interest can cross the external cell membrane. This applies in particular to experiments that employ radiotracers such as ^{125}I or ^{131}I since free iodide can cross cell membranes via an anion-exchange channel [41,42]. Thus in addition to intercellular trapping, intracellular trapping must be considered.

In adsorption experiments where the concentrations of two species may be added linearily (i.e., detection sensitivites are equal), the apparent distribution ratio P due to two species is given by

$$P = r_{1s}P_1 + r_{2s}P_2 \tag{2}$$

where P_1 and P_2 are the distribution ratios of the isolated species and r_{1s} and r_{2s} are the respective fractional concentrations of species 1 and 2 relative to the total apparent concentration of the speces C_s (i.e., $r_{1s} + r_{2s} = 1$).

II. SOLUTION ANALYSES

A. Mathematical Description

Solution analyses are based on experiments in which cell suspensions and cell-free solutions are combined. The analysis requires knowledge of the concentration of the adsorbing species in solution, the cell volume (or mass) fraction, and the volumes of the cell suspensions both before and after mixing. A general expression for the distribution ratio in terms of these parameters can be derived from conservation of the adsorbing species and total volume (or mass). It is assumed that the changes in the partial molar volumes of the various solution components upon mixing are insignificant; thus

$$V_s'' + V_s' + V_c' = V_s + V_c \tag{3}$$

where V is a volume (or mass), the nonprimed volumes are those after mixing, the single primed volumes are associated with the cell suspension before mixing, the doubly primed volume is that of the cell-free solution, and the species subscript have been omitted for clarity. Conservation of the adsorbed species is given by

$$C_s''V_s'' + C_s'V_s' + C_c'V_c' = C_sV_s + C_cV_c \tag{4}$$

where the notation is analogous to that described above. Since C_c' is a measure of the adsorbing species present on the cells before they are exposed to the experimental solution, this approach describes the general case by which either increments or decrements in adsorption may be treated. The cell suspension volumes before and after mixing are defined respectively by

$$V_i \equiv V_s' + V_c' \tag{5}$$

and

$$V_e \equiv V_s + V_c \tag{6}$$

The cell volume fractions before and after mixing are given respectively by

$$H_i \equiv \frac{V_c'}{V_i} \tag{7}$$

$$H_e \equiv \frac{V_c}{V_e} \tag{8}$$

Thus the volumes V_s, V_s', and V_s'' may be expressed as

$$V_s = \left(H_e^{-1} - 1\right) V_c \tag{9}$$

$$V_s' = \left(H_i^{-1} - 1\right) V_c' \tag{10}$$

$$V_s'' = V_e - V_i$$

$$= \left(H_e^{-1} - H_i^{-1} \frac{V_c'}{V_c}\right) V_c \tag{11}$$

The change in the amount of cell associated material is given by

$$C_c V_c - C_c' V_c' = C_s'' V_s'' + C_s' V_s' - C_s V_s \tag{12}$$

Division of both sides of Eq. 12 by C_s and V_s and substitution for V_s'', V_s', and V_s using Eqs. (9), (10), and (11) leads to

$$\frac{C_c - C_c' V_c' / V_c}{C_s} = \frac{C_s''}{C_s}\left(H_e^{-1} - H_i^{-1} \frac{V_c'}{V_c}\right) +$$

$$\frac{C_s'}{C_s}\left(H_i^{-1} - 1\right)\frac{V_c'}{V_c} - \left(H_e^{-1} - 1\right) \tag{13}$$

Provided that the cells do not change volume during the experiment, the right-hand side (RHS) of Eq. (13) is equal to the difference in the distribution ratio before and after exposure of the cells to the experimental solution and hence is proportional to the change in the surface concentration of the adsorbed species. Thus to determine absolute surface concentrations and distribution ratios, the changes have to be summed from zero surface concentration. Desorption occurs when $C_s' > C_s''$; adsorption occurs if $C_s'' > C_s'$.

Two classes of the general solution analysis experiment was generally carried out. In the first, desorption or the wash-off behavior of the adsorbed species is studied. Hence the concentration in the cell-free solution is zero ($C_s'' = 0$), while the initial cell adsorbed and solution concentrations are nonzero ($C_s', C_c' > 0$). In this case Eq. (13) reduces to

$$\frac{C_c - C_c' V_c' / V_c}{C_s} = \frac{C_s'}{C_s}\left(H_i^{-1} - 1\right)\frac{V_c'}{V_c} - \left(H_e^{-1} - 1\right) \tag{14}$$

In these experiments the solution concentration of the ligand in equilibrium with the surface decreases. When binding to the surface is reversible, desorption is expected; these will be referred to as desorption experiments. A distribution ratio may be determined directly for conditions where the amount adsorbed is proportional to the solution concentration (i.e., when the distribution ratio is independent of concentration). The left-hand side (LHS) of Eq. (14) may be rewritten as

$$P = P' \frac{C_s'V_c'}{C_sV_c} = \frac{C_s'}{C_s}\left(H_i^{-1} - 1\right)\frac{V_c'}{V_c} - \left(H_e^{-1} - 1\right) \tag{15}$$

and if the distribution ratio is independent of ligand concentration (i.e., $P = P'$),

$$P = \left[\frac{C_s'}{C_s}\left(H_i^{-1} - 1\right)\frac{V_c'}{V_c} - \left(H_e^{-1} - 1\right)\right]\frac{C_sV_c}{C_sV_c - C_s'V_c'} \tag{16}$$

In the second case of general interest the surface adsorbed and solution concentrations of the adsorbing species are initially zero ($C_s' = C_c' = 0$), while the concentration of ligand in the cell-free solution is nonzero ($C_s'' > 0$). These are the usual conditions for studying adsorption. In this case Eq. (13) reduces to

$$P = \frac{C_c}{C_s} = \frac{C_s''}{C_s}\left(H_e^{-1} - H_i^{-1}\frac{V_c'}{V_c}\right) - \left(H_e^{-1} - 1\right) \tag{17}$$

The RHS of Eq. (17) determines the surface concentration and distribution ratio directly. This advantage may explain why the second variant of the solution analysis method has been used more frequently.

The factor V_c'/V_c is the ratio of the initial and final cell volumes. Red cell volume is known to change in response to changes in the osmotic strength of the suspending medium [43]. The possibility of volume changes in response to hydrostatic pressure associated with centrifugation has also been discussed [44–46]. Equations (7) and (8) give the ratio of the initial to final cell volumes as

$$\frac{V_c'}{V_c} = \frac{H_iV_i}{H_eV_e} \tag{18}$$

In some of the reports we have reviewed, it appears that the importance of this factor has not been appreciated. However, where the volume changes are associated with changes in osmotic strength, we have recalculated results using a semiempirical relation reported by Ponder [43]. It describes red cell volume as a function of the tonicity and volume of the media added to a suspension of cells in plasma or isotonic media:

$$\frac{V_c}{V_c'} = Rw \frac{p - pT}{pT + 1} + 1 \tag{19}$$

Here w is the fractional water volume of a red cell, p is the ratio of the volume of media added to the volume of cell water, the tonicity T is the ratio of the freezing-point depression (or osmolarity) of the added media to that of plasma, and R is an empirical parameter. Variations in R have been noted with changes in anticoagulant, red cell crenation, temperature adaptation of rat red cells, and extreme tonicity [43]. Ponder concludes that a simple osmotic theory (i.e., one where the osmotic properties of the red cell interior are constant) is generally not possible. Thus we view our reconstruction of results using this information only as an indicator of the possible range of adsorption or trapping estimates.

Where the volume of the added media is large compared to that of the initial cell suspension, the equilibrium osmolarity of the suspension approaches that of the added media and the tonicity factor in relation (19) reduces to $(1 - T)/T$. Figure 1 shows the ratio of the final to the initial cell volumes under these conditions as a function of the equilibrium osmolarity where the initial osmolarity is taken to be 289 mOsm (isotonic). The cell water volume fraction w is 0.7, and the empirical factor R is 0.50, 0.83, or 0.95. The extreme values of R are those reported by Ponder to encompass the normal range of red cell behavior. The intermediate value is that implied from data reported by Evans and Fung [47].

When the volume of the suspending medium is underestimated, the changes in solution concentration described above may also be interpreted as being due to pure dilution into the volume surrounding the cells in the initial cell suspension rather than due to surface adsorption. The equivalence of these explanations for protein concentration changes is evident when an expression for the trapped volume is written. If the ligand distribution ratio is known, Eqs. (16) and (17) may be rewritten so that the initial cell volume fraction H_i may be calculated. In desorption experiments, if the distribution ratio is independent of ligand concentration, Eqs. (16) and (18) lead to the following:

$$\left(1 - \frac{C_s'V_c'}{C_sV_c}\right)P = \frac{C_s'}{C_s}\frac{V_c'}{V_c}\left(H_i^{-1} - 1\right) - \left[H_i^{-1}\left(\frac{V_c'}{V_c}\frac{V_e}{V_i}\right) - 1\right]$$

$$= H_i^{-1}\frac{C_s'V_c'}{C_sV_c} - H_i^{-1}\left(\frac{V_c'}{V_c}\frac{V_e}{V_i}\right) - \frac{C_s'V_c'}{C_sV_c} + 1$$

$$= H_i^{-1}\left(\frac{C_s'}{C_s} - \frac{V_e}{V_i}\right)\frac{V_c'}{V_c} - \frac{C_s'V_c'}{C_sV_c} + 1$$

$$H_i^{-1}\left(\frac{C_s'}{C_s} - \frac{V_e}{V_i}\right)\frac{V_c'}{V_c} = P\left(1 - \frac{C_s'V_c'}{C_sV_c}\right) - \left(1 - \frac{C_s'V_c'}{C_sV_c}\right)$$

$$= (P - 1)\left(1 - \frac{C_s'V_c'}{C_sV_c}\right)$$

$$H_i = \left(\frac{C_s'}{C_s} - \frac{V_e}{V_i}\right)\frac{V_c'}{V_c}\,\frac{C_sV_c}{C_sV_c - C_s'V_c'}\Big/(P - 1)$$

$$= \left(\frac{C_s'}{C_s} - \frac{V_e}{V_i}\right)\frac{V_c'}{V_c}\,\frac{C_sV_c}{C_s'V_c' - C_sV_c}\,\frac{1}{1 - P} \tag{20}$$

Similarly, in adsorption experiments Eqs. (17) and (18) lead to

$$P = \frac{C_s''}{C_s}\left(H_i^{-1}\frac{V_c'}{V_c}\frac{V_e}{V_i} - H_i^{-1}\frac{V_c'}{V_c}\right) - \left[H_i^{-1}\left(\frac{V_c'}{V_c}\frac{V_e}{V_i}\right) - 1\right]$$

$$= H_i^{-1}\left[\frac{C_s''}{C_s}\frac{V_c'}{V_c}\left(\frac{V_e}{V_i} - 1\right)\right] - H_i^{-1}\left(\frac{V_c'}{V_c}\frac{V_e}{V_i}\right) + 1$$

$$P - 1 = H_i^{-1}\left(\frac{C_s''}{C_s}\frac{V_c'V_e}{V_cV_i} - \frac{C_s''}{C_s}\frac{V_c'}{V_c} - \frac{V_c'V_e}{V_cV_i}\right)$$

$$= H_i^{-1}\frac{V_c'}{V_c}\left(\frac{C_s''}{C_s}\frac{V_e}{V_i} - \frac{C_s''}{C_s} - \frac{V_e}{V_i}\right)$$

$$H_i = \frac{V_c'}{V_c}\left[\frac{C_s''}{C_s}\left(\frac{V_e}{V_i} - 1\right) - \frac{V_e}{V_i}\right]\frac{1}{P - 1}$$

$$= \left[\frac{C_s''}{C_s}\left(1 - \frac{V_e}{V_i}\right) + \frac{V_e}{V_i}\right]\frac{V_c'}{V_c}\,\frac{1}{1 - P} \tag{21}$$

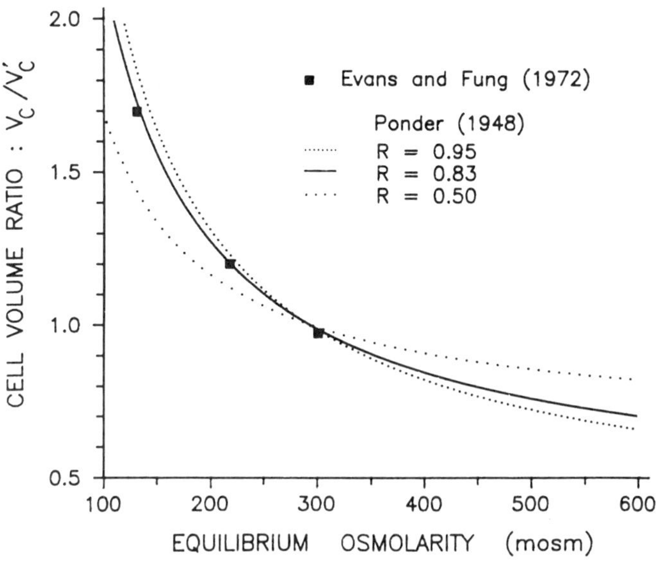

FIG. 1 Human erythrocyte volume expressed relative to that at 289 mOsm as a function of the equilibrium solution osmolarity. The isosmotic volume (97 μm^3) was determined by fitting results reported by Evans and Fung [47] to Eq. (19). With the fractional water volume w of the cell equal to 0.7, the empirical parameter R was found to be 0.83. The volume ratio for the extreme values of R reported by Ponder [43] are shown by the dashed lines.

Thus the true cell volume fraction can be measured if the surface adsorption is known. If $P < 0$ is excluded, the calculated value of H_i is a minimum, $\underline{H}_i$, when the distribution ratio is zero (zero adsorption). Thus, if H_i is measured and is greater than $\underline{H}_i$, a maximum trapped volume fraction, $\underline{H}_t$, may be calculated:

$$\underline{H}_t \equiv H_i - \underline{H}_i \tag{22}$$

From either of Eq. (20) or (21) it also follows that

$$\underline{H}_t = H_i P \tag{23}$$

This equation expresses the equivalence of two explanations for the changes in ligand concentration associated with the mixing of a cell suspension with a cell-free solution. On the one hand, the change may be attributed to overestimation of the initial cell volume fraction by hematocrit measurements, while on the other hand, it may be explained by ligand adsorption to the cells. In hematocrit measurements the trapped volume decreases with increased centrifugation rates and times [48,49] and may also be affected by changing the deformability of the cells. However, in normal erythrocyte suspensions trapping need not exceed a few percent of the actual cell volume fraction so that distribution ratios greater than a few percent would be readily accessible.

B. Results

Since 1935 six papers on the nonspecific adsorption of plasma proteins to erythrocytes, determined by solution analysis, have been published. Three report upon adsorption from plasma [13,50,51] two discuss adsorption from purified albumin solutions [14,52], and one deals with adsorption from solutions of purified fibrinogen [53]. The papers of Monaghan and White [13] and Brown [14] were written in response to those of Bellis and Scott [50] and Furchgott and Ponder [52], respectively, and oppose the view that proteins adsorb nonspecifically to erythrocytes. The following summarizes these published results and those of experiments carried out by ourselves [54].

Bellis and Scott [50] reported that the total protein present in bovine serum increased when defibrinated bovine blood was diluted with isosmotic solutions of dextrose, NaCl, $CaCl_2$, or $BaCl_2$. This was interpreted as indicating that measurable amounts of protein exchanged between the surface membranes of the cells and the solution due to changes in the solution concentration of the serum proteins. On the other hand, dilution with hyperosmotic solutions (approximately nine times isosmotic) was reported to result in a decrease in the total protein in solution. This was interpreted to mean that adsorption increased and that the amount of protein adsorbed depended more strongly on solution osmolarity than on the concentration of protein in solution. Protein was determined by tungstate precipitation and Kjeldahl nitrogen analysis. Cell volume fractions were determined by measuring hematocrits.

Monaghan and White [13] reported that they had repeated the Bellis and Scott experiment in a modified form and concluded that the earlier results should not be interpreted as exchangeable surface adsorption. The results of Bellis and Scott implied that no protein would be found in solution after hypertonic dilution of packed red cells. Thus, in their experiments, Monaghan and White diluted packed cells from defibrinated beef blood with either isotonic or hypertonic saline. Protein was found in the supernatant fluid of the resuspended cell pellets implying that protein adsorption, if any, was substantially less than predicted from the earlier results. Also at odds with the results of Bellis and Scott was the observation that hypertonic and isotonic dilution both gave similar results. Monaghan and White interpreted the protein in solution as being due to intercellular serum in the packed cell pellets.

Table 1 summarizes the results reported by Bellis and Scott and Monaghan and White for isosmotic saline dilution of bovine blood. In addition, their results have been expressed in terms of the distribution ratio P and the maximum trapping fraction $\underline{H}_t$. Bellis and Scott did not report equilibrium protein concentrations or cell volume fractions; however, these may be calculated where the ratio of the initial and final cell volumes is known. Thus distribution ratios have been calculated for the isosmotic dilution experiments, where it may be assumed that the cell volume remained constant. Equations (16) and (22) were used to calculate P and $\underline{H}_t$, respectively; however, Eq. (16) was valid only where P is constant. Analysis of the results reported by Bellis and Scott for three series of isosmotic dilutions from constant initial conditions indicate that this condition was probably met. The changes in the adsorbed protein concentration ΔC_c and the equilibrium protein concentration C_s were calculated and were found to be linearly related (Table 1). Thus the amount adsorbed is also linearly related to the equilibrium concentration. If the protein adsorption was reversible, the amount adsorbed would be proportional to the equilibrium concentration and thus P would be constant. It was not possible to demonstrate from these experiments that the adsorption went to zero with the equilibrium protein concentration. Similar results were reported for dilution by isosmotic dextrose and $CaCl_2$. There were insufficient data to determine whether or not dilution with isosmotic $BaCl_2$ gave a similar linear relation.

If the distribution ratios determined from the results reported by Bellis and Scott are taken to represent protein adsorption, these correspond to approximately 30 g of protein per liter of bovine erythrocytes when the solution concentration is that of normal bovine plasma. This is equivalent to a surface concentration of approximately 2 μg/cm² for bovine erythrocytes. Bovine erythrocytes have a volume of 50 μm³ [55] and their area was estimated to be about 90 μm² by assuming similarity with human erythrocytes [i.e., $A/A_h = (V/V_h)^{2/3}$, where $A_h = 135\ \mu m^2$ and $V_h = 97\ \mu m^3$]. If it is assumed that the mean partial specific volume for protein is 0.7 cm³/g, this surface concentration corresponds to a 120-Å layer of protein at the cell surface. When compared with the dimensions estimated for bovine serum albumin (38 × 150 Å) [56], this thickness suggests that the cells would be covered with at least a monolayer of protein molecules.

TABLE 1 Serum Distribution Ratios Versus Trapping in Erythrocyte Pellets Estimated from the Dilution of Bovine Blood by Isosmotic Saline. Data from Bellis and Scott [50] and Monaghan and White [13]

		Equilibrium				
Dilution (diluent: blood)	Change in solution protein[a] (g/L)	Solution protein conc., C_s (g/l)	Cell volume fraction, H_e	Change in adsorbed protein conc., ΔC_c (g/L)	Protein distribution ratio[b] P	Maximum trapping fraction[c] $\underline{H}_t$
		Bellis and Scott: $C_s' = 81$ g/L, $H_i = 0.35$				
10:90	+2.39	71.2	0.315	-4.4	0.46	0.16
20:80	+5.22	62.3	0.280	-9.7	0.52	0.18
50:50	+9.46	35.6	0.175	-17.6	0.39	0.13
70:30	+12.7	20.4	0.105	-23.6	0.39	0.14
		Monaghan and White: $C_s' = 80$ g/L, $H_i = 1.0^d$, 0.9^e				
1:1	—	6.4	0.50^d	-6.4^d	0.087^d	0.087^d
1:1	—	5.7	0.50^d	-5.7^d	0.077^d	0.077^d
1:1	—	6.4	0.45^e	$+1.1^e$	-0.015^e	-0.013^e
1:1	—	5.7	0.45^e	$+1.9^e$	-0.026^e	-0.023^e

[a]The quantity reported by Bellis and Scott is the total protein in solution at equilibrium divided by the initial serum volume minus the initial serum protein concentration.
[b]Calculated using Eq. 16.
[c]Calculated using Eq. 22.
[d,e]Respectively, the upper and lower values assumed for the initial cell volume fraction and the calculated values of H_e, ΔC_c, P, and H_t that follow.

Monaghan and White reported equilibrium protein concentrations and assumed an initial cell volume fraction of 1. Thus the maximum adsorption corresponds to a distribution ratio of about 0.08 and would be less if the initial cell volume fraction were in fact less than 1. For their experiments, zero adsorption corresponds to an initial cell volume fraction of between 0.91 and 0.92. As shown in Table 1, assuming an even smaller initial cell volume fraction implies negative trapped volume fractions and distribution ratios. No data were reported concerning changes in trapped volume fraction with changes in equilibrium protein concentration. However, trapped volume fractions could be expected to be relatively independent of protein concentration, and this would be equivalent to apparent distribution ratios, which are also independent of protein concentration.

The results reported by Monaghan and White for hyperosmotic dilution of blood are in disagreement with those of Bellis and Scott. Monaghan and White reported that equilibrium protein concentrations after isosmotic and ninefold hyperosmotic dilution were not significantly different, thus implying that either trapped volumes before dilution or desorption due to dilution were unaltered by solution osmolarity. On the other hand, Bellis and Scott reported that hypertonic dilution consistently resulted in adsorption (i.e., C_c increased as C_s decreased). A calculation reported by Bellis and Scott indicated that dilution of 90 cc of blood with 10 cc of 8.1% NaCl resulted in an apparent reduction in cell volume of 0.37 cc. The ratio V_c'/V_c was thus equal to 1.012. The osmolarity of the equilibrium solution would have been about 460 mOsm if a volume of water equivalent to 70% of the cell volume had been transferred to the solution and about 620 mOsm if only 0.37 cc had been transferred. However, these results are at odds with the osmotic response of red cells reported elsewhere [43,47]. The volume ratio indicated can be achieved in hypertonic solutions of less than 300 mOsm [47]. Thus it appears that Bellis and Scott seriously underestimated the decrease in the cell volume they effected, and thus overestimated the adsorption in their hypertonic dilution experiments. Can the reported "adsorption" be explained by changes in cell volume? If adsorption were zero, the change in adsorption would also be zero and Eqs. (14) and (18) may be used to calculate H_e and the V_c'/V_c values for the experiment. The volume ratio associated with zero adsorption is 1.48 and is in the range of volume ratios (1.33 to 1.90) predicted using the semiempirical relation describing red cell osmotic volume changes [43]. Thus the adsorption reported by Bellis and Scott under hyperosmotic conditions may be entirely explained by an undetected volume change.

Table 2 summarizes the mean distribution ratios for the isosmotic dilution experiments reported by Bellis and Scott. These include

TABLE 2 Distribution Ratios for Protein Components of Bovine Plasma and Serum Determined by Isosmotic Dilution[a]

Protein	C_s' (g/L)	Isosmotic diluent NaCl	Dextrose	$CaCl_2$	$BaCl_2$
Serum	81	0.4 ± 0.1 (n = 4)	0.3 ± 0.1 (n = 4)	0.4 ± 0.1 (n = 4)	0.6 ± 0.0 (n = 2)
Fibrinogen	4	0.5 ± 0.3 (n = 4)	36 ± 56 (n = 4)	—	—
Albumin	40	0.8 ± 0.1 (n = 4)	0.7 ± 0.3 (n = 4)	—	—
Albumin	40	1.2 ± 0.3 (n = 4)	1.5 ± 0.9 (n = 4)	—	—

[a]Mean and standard deviation calculated from results reported by Bellis and Scott [50] using Eq. 16. The initial albumin and fibrinogen concentrations C_s' are mean values reported for cattle [55].

experiments where the serum or plasma in equilibrium with the cells after dilution were fractionated to determine whether a particular protein was responsible for the effects observed. On a mass basis, albumin was found to be the major component. Distribution ratios, on the other hand, compare adsorption relative to the solution concentration of a component. The distribution ratios were calculated using Eq. (16). The initial protein concentrations are those given by Doxey [55], as these were not reported by Bellis and Scott, while the final concentrations were calculated from the initial concentrations and the reported protein changes [50]. It is evident from Eq. 16 and Table 2 that this calculation is very sensitive to the assumed initial protein concentrations. If taken at face value, the distribution ratios calculated for albumin are somewhat higher than those of fibrinogen and serum. Presumably, serum contains proteins with distribution ratios lower than that of albumin. However, given the sensitivity of the calculation to the assumed concentration, it is not clear that the differences are significant.

Furchgott and Ponder [52] reported that crystalbumin (carbohydrate-poor albumin) inhibited the disk-sphere transformation of mammalian red cells induced by proximity to a glass surface. They proposed that in the presence of glass, an unidentified antisphering factor initially adsorbed to the cells desorbed and bound to the glass, thus allowing the cells to sphere. Addition of crystalbumin to a suspension of sphered cells restored the discoid shape of the cells and they reported that these cells took up crystalbumin from solution (0.6 to 1.0 g per 100 cm^3 of cells). The amount of crystalbumin adsorbed was determined from adsorption-type solution analysis experiments. Protein concentrations were determined by Kjeldahl nitrogen analyses, while cell volumes were determined from cell counts and the assumption that 5×10^6 cells/mm^3 correspond to a cell volume fraction of 0.42. A 1% saline solution was used to prepare the crystalbumin solutions and was (apparently) also used in the pretreatment of the red cells.

Brown [14] disputed the results reported by Furchgott and Ponder and attributed the antisphering activity of crystalbumin to its buffering action with respect to glass-adsorbed HCO_3^-. A similar adsorption-type solution analysis experiment was performed and was reported to show that the amount of protein bound to the cells was less than the sensitivity of the experiment (−0.2 ± 0.4 g per 100 cm^3 of cells). Protein concentrations were also determined by Kjeldahl nitrogen analyses, but Brown noted that studies subsequent to those made by Furchgott and Ponder had identified "many sources of error not appreciated before" in this methodology.

However, Brown's work was not without its difficulties. In these experiments the cells were pretreated in 0.111 M NaCl and the crystalbumin was prepared in 0.166 M NaCl. The final suspension salt concentration would have been about 0.161 M. The osmolarity of the initial and final cell suspensions would thus have been about 204 and 296 mOsm and the ratio of the initial and final cell volumes calculated from Ponder's relation would have been between 1.13 and 1.29 for R values between 0.50 and 0.95. The distribution ratios for the results reported by Furchgott and Ponder [52] and Brown [14] are summarized in Table 3. Distribution ratios for Brown's results were calculated assuming either no change in cell volume or a decrease in cell volume of at most 22% due to osmotic effects. Equations 17 and 18 were used in these calculations and it can be shown that for $C_S''/C_S > 50/45$, the distribution ratio increases with increasing cell volume ratio, whereas for $C_S''/C_S < 50/45$, the distribution ratio decreases with increasing cell volume ratio. Brown's results suggest that the experiment as carried out was not capable of detecting changes in the amount of protein in solution.

The results reported by Furchgott and Ponder [52] imply adsorption of about 9 g of albumin per liter of human erythrocytes at an equilibrium protein concentration of 0.7 g/L. The corresponding surface concentration is about 0.65 $\mu g/cm^2$ or about 8×10^6 albumin molecules per cell (mpc). This amount is about 30% of that reported by Bellis and Scott [50]; however, the equilibrium protein concentrations

TABLE 3 Distribution Ratios for Crystalbumin (Carbohydrate-Poor Albumin) Calculated from Adsorption Solution Analyses Reported by Furchgott and Ponder [52] and Brown [14].

Initial cell volume (cm^3)	Protein conc. ratio, C_s''/C_s	Initial cell volume fraction, H_i	Equilibrium: Cell volume fraction, H_e	Equilibrium: Protein conc., C_s (g/L)	Distribution ratio, P
Furchgott and Ponder: $V_c' = V_c$					
0.24	1.164	0.048	0.0048	0.60	10.9
0.25	1.195	0.050	0.0050	0.65	16.1
0.18	1.153	0.036	0.0036	0.62	11.5
0.17	1.159	0.034	0.0034	0.73	13.7
					13 ± 2
Brown: $V_c' = V_c$[a]					
0.40	1.0734	0.080	0.0080	2.01	-3.2
0.37	1.1071	0.074	0.0074	0.74	0.5
0.28	1.0858	0.056	0.0056	0.66	-3.1
0.39	1.1223	0.078	0.0078	0.86	2.3
0.35	1.0817	0.070	0.0070	1.87	-2.8
					-1.3 ± 2.5
Brown, $V_c' = 1.29(V_c)$[b]					
0.40	1.0734	0.080	0.0063	2.01	-4.5
0.37	1.1071	0.074	0.0058	0.74	0.4
0.28	1.0858	0.056	0.0044	0.66	-4.2
0.39	1.1223	0.078	0.0061	0.86	2.7
0.35	1.0817	0.070	0.0055	1.87	-3.9
					-1.9 ± 3.2

[a]Distribution ratio calculated assuming a constant cell volume.
[b]Distribution ratio calculated assuming a 22% decrease in cell volume due to increased solution osmolarity.

in these experiments were very different If the adsorption reported by Bellis and Scott was proportional to the equilibrium protein concentration, the binding at a protein concentration of 0.7

g/L would be about 0.3 g/L erythrocytes. Thus at equivalent protein concentrations the adsorption reported by Furchgott and Ponder is about 30 times greater than that reported by Bellis and Scott. Brown [14] used radiotracer techniques ([^{131}I] crystalbumin) to show that an insignificant amount of crystalbumin was retained by washed cells. Although this result repudiated the interpretation that albumin remained bound to buffer-washed cells, it did not eliminate the possibility that the protein bound reversibly to the erythrocytes.

Traber and Kolmen [53] reported binding of canine fibrinogen to canine erythrocytes at 37°C. Their experiments were adsorption-type solution analyses. Fibrinogen protein was determined as fibrin and it was assumed that centrifugation at 3500 × g for 15 minutes reduced the fluid entrained with the cells to insignificant levels. However, Chien et al. [57] reported that the cell volume fraction determined from radioiodine-labeled albumin retention in pellets of canine erythrocytes was less than that determined from hematocrit measurements (i.e., 97% with centrifugation at 15,000 × g for 30 min and 95% at 1500 × g for 5 min). If these results were not due to adsorption of albumin to the erythrocytes and albumin were an accurate volume marker with respect to fibrinogen, the binding reported by Traber and Kolmen would be an overestimate. Table 4 summarizes the results of Traber and Kolmen and expresses them in terms of the distribution ratio calculated with initial cell volume fractions of both 1.0 and 0.96. The amount of fibrinogen adsorbed was reported to increase in proportion to the solution concentration of the protein. Thus the distribution ratio is constant with changing protein concentration (Table 4). The upper estimate for the distribution ratio at the highest solution protein concentration is equivalent to 1.1 × 10^5 fibrinogen mpc or a surface concentration of about 55 ng/cm^2 for cells of 71 μm^3 volume [58] with an estimated area of 110 μm^2. The calculated apparent thickness of an 0.7-cm^3/g layer of protein is about 4 Å and suggests that the adsorbed fibrinogen molecules (50 × 450 Å) [59] are widely dispersed on the cell surface.

A similar analysis with similar assumptions was reported by Kikuchi and Koyama [51] for plasma adsorption to human erythrocytes at 37°C. These authors commented on the entrainment of saline with the cells added to the plasma sample; however, they did not correct for this effect in the results reported. Protein concentrations were determined by refractometry and equilibrium cell volume fractions were determined from hematocrit measurements. Distribution ratios calculated from the results reported and assuming that H_i = 1 are summarized in Table 5. The distribution ratio is as low as the maximum value reported by Monaghan and White [13]. However, Kikuchi and Koyama's experiments were carried out at an equilibrium serum concentration about 10 times greater.

TABLE 4 Distribution Ratios for Canine Fibrinogen Calculated from Adsorption Solution Analyses Reported by Traber and Kolmen [53][a]

Equilibrium: Adsorbed protein conc., C_c (g/L)	Equilibrium: Solution protein conc., C_s (g/L)	Initial solution protein conc., C_s'' (g/L)	Protein distribution ratio, P: $H_i = 1.0$	Protein distribution ratio, P: $H_i = 0.96$
1.058	4.75	5.10	0.222	0.196
0.584	3.72	3.92	0.156	0.121
0.389	2.04	2.17	0.192	0.158
0.157	1.03	1.08	0.150	0.115
			0.18 ± 0.03	0.15 ± 0.04

[a] $V_c' = V_c$, $V_i = 1$ mL, $V_e = 4$ mL.

As Bellis and Scott had earlier, Kikuchi and Koyama reported that increasing the osmolarity of the cell suspension resulted in further adsorption to the cells. They reported that adsorption doubled as a result of increasing the osmolarity of the solution from 290 mOsm to 400 mOsm. The dilution effect due to osmotically induced water transfer from the cells was taken into account. From the results they reported and estimates of the erythrocyte volume at 290 and 400 mOsm [43], we estimate that the protein concentration changes due to dilution by cell water were between 30 and 70% of the total observed concentration changes. The hematocrit changes observed by these authors were not reported, but we estimate these to be between 1 and 6% in this experiment and thus near the limits of experimental reliability.

Janzen [54] determined the adsorption of isolated, radioiodinated human fibrinogen and albumin to washed human erythrocytes under isosmotic conditions at 21°C. Cell volume fractions were determined from microhematocrit measurements and were converted to a cell mass fraction by either (1) a calculation based on values for the densities of the red cell (1.084 g/mL) [60] and the suspending medium with the assumption that no suspending medium was entrained in the cell pellets, or (2) an empirical relation between the hematocrit of cell

TABLE 5 Distribution Ratios for Human Serum Calculated from Adsorption Solution Analyses Reported by Kikuchi and Koyama [51][a]

Change in solution protein conc., $C_s - C_s''$	Equilibrium: Cell volume fraction, H_e	Equilibrium: Solution protein conc., C_s (g/L)	Protein distribution ratio, P (H_i = 1.0)
3.00	0.435	67.00	0.058
1.98	0.285	68.02	0.073
1.98	0.290	68.02	0.071
0.98	0.186	69.02	0.062
0.98	0.112	69.02	0.113
			0.075 ± 0.022

[a] $V_c' = V_c$, $C_s'' \simeq 70$ g/L, isosmotic.

suspensions determined using a standardized centrifugation procedure and the cell mass fractions of the same suspensions. The cell pellet used to prepare the suspensions was itself prepared by ultracentrifugation. The latter method has the advantage that measured volume fractions are transformed to mass fractions using data for pellets in which the entrained mass fraction is constant (i.e., that of the ultracentrifuged material), whereas it has been demonstrated for normal hematocrit determinations that the apparent cell volume fraction increases due to increase in the entrained volume fraction as the distance from the axis of the centrifuge decreases [48]. Distribution ratios calculated from cell mass fractions determined by method 1 were consistently 0.01 unit higher than those calculated from cell mass fractions determined by method 2. In a number of experiments the ratio of the initial and final cell masses were determined according to Eq. (18), and these were found to be within 1% of unity (Table 6).

Preparations of radioiodinated proteins contain variable amounts of radiolabel not bound to protein which increase with time [61,62]. In preliminary adsorption experiments, we found that free radioiodide was responsible for changes in the apparent binding of radiolabel by cells. Specifically, the apparent distribution ratio decreased

TABLE 6 Cell Mass Ratio (W_c'/W_c) in Isosmotic Adsorption Solution Analyses Reported by Janzen [54]

Protein	Initial cell volume fraction,[a] H_i	Final cell volume fraction, H_e (range)	Cell mass ratio,[a] W_c'/W_c	
			Analysis 1[b]	Analysis 2[c]
Albumin				
1	0.997 ± 0.001 (n = 2)	0.42 - 0.79 (n = 8)	1.004 ± 0.005	1.001 ± 0.005
2	0.993 ± 0.001 (n = 2)	0.48 - 0.80 (n = 12)	0.998 ± 0.004	0.995 ± 0.004
Fibrinogen				
1	0.948 ± 0.002 (n = 4)	0.48 - 0.76 (n = 9)	1.005 ± 0.008	1.000 ± 0.006
2	0.917 ± 0.002 (n = 6)	0.45 - 0.70 (n = 12)	0.992 ± 0.009	0.990 ± 0.009

[a]Mean and standard deviation.
[b]Cell mass fractions calculated from cell volume fractions and density.
[c]Cell mass fractions determined from cell volume fractions and a calibration curve determined from resuspended, ultracentrifuged red cells.

with increasing equilibrium cell concentration (volume fraction), and this effect was related to the amount of free radioiodide present in protein preparations, the iodide distribution ratio, and the equilibrium cell concentration [54]. The expression derived to describe this dependence was used to determine the amount of free radioiodide in protein preparations, and these agreed with values determined by gel filtration. Generally, the fraction of the radiolabel as iodide was less than 2%. The mass distribution ratio for iodide in erythrocyte suspensions was found to be 0.80 ± 0.02, in approximate agreement with the volume distribution ratios reported by Dalmark and Wieth [41]. From Eq. (2) the correction to the apparent mass distribution ratio due to 1% free iodide would be approximately 0.008. Although this would be a small correction if distribution ratios were no smaller than those reported by Kikuchi and Koyama [51], we found that distribution ratios were significantly smaller and that corrections for free iodide were significant.

Apparent distribution ratios for radioiodinated albumin and fibrinogen were determined over a range of equilibrium protein and cell concentrations and corrections to these were determined to account for radioiodide contamination. Figures 2 and 3 show the distribution ratios determined for albumin and fibrinogen, respectively, as functions of the equilibrium concentrations of these proteins. The data in each figure are derived from two separate preparations of radioiodinated protein. The albumin data are derived from a total of 18 determinations, while the data for fibrinogen are from 21 determinations. From each determination two estimates of the distribution ratio were computed. The first estimate is based on cell mass fractions calculated by method 1 above and is not corrected for radioiodide contamination. This estimate is our upper estimate for distribution ratios. This estimate was made because it is uncertain whether or not the cells were compressed by the ultracentrifugation used in determining cell mass fractions by method 2. Dellimore and Gosling [46], however, reported that centrifugation at accelerations between 40 and 6000 × *g* did not alter the ionic concentration of the suspending medium. A similar result was observed under our centrifugation conditions (202,000 × *g*) [54]. Apparently, if compression occurs it is reversible or involves both water and ions. In addition, the radioiodide correction was omitted from the upper estimate so that this estimate would be entirely free from systematic errors that might occur as a result of determining radioiodide from the change in apparent adsorption with cell concentration. The second estimate is based on cell mass fractions determined by method 2 and is corrected for radioiodide contamination. This is our lower estimate for distribution ratios.

Regression analyses were carried out to determine whether either of the estimates for P were linear functions of the equilibrium ligand and cell concentrations (i.e., is $P = a_0 + a_1C_s + a_2H_e$?). The coefficient a_2 was not significantly different from zero ($p = 0.05$) in any of the experiments. For the C_s dependence there were no significant correlations in half of the experiments, while both positive and negative coefficients occurred in the remainder, suggesting that the distribution ratio estimates are independent of protein concentration in the ranges investigated, but may be subject to systematic errors. Thus we have chosen to report the mean upper and lower estimates for the distribution ratio for each of the experiments: albumin experiment 1: 0.010 ± 0.005, to -0.004 ± 0.006 (n = 6); experiment 2: 0.022 ± 0.008, to 0.008 ± 0.008 (n = 12); fibrinogen experiment 1: 0.016 ± 0.006, to 0.002 ± 0.006 (n = 9), experiment 2: 0.000 ± 0.022, to -0.012 ± 0.021 (n = 12). To convert these mass distribution ratios

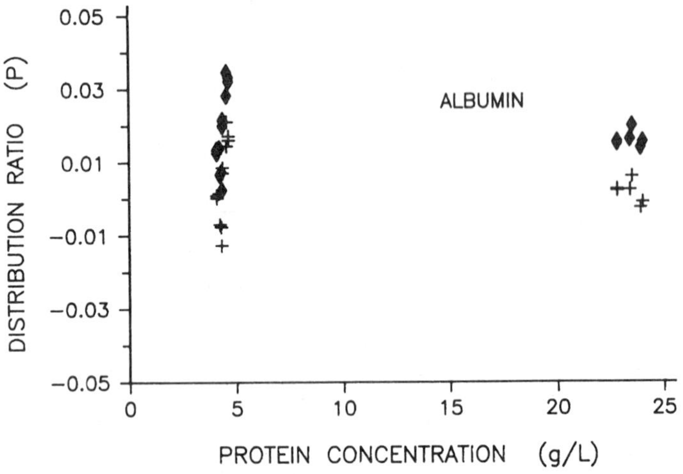

FIG. 2 Upper and lower estimates of albumin adsorption to human erythrocytes as a function of albumin concentration reported by Janzen [54] and expressed as a distribution ratio. ♦ Upper estimate; + lower estimate.

to volume distribution ratios, the mass distribution ratios must be multiplied by the ratio of the cell and the suspending medium densities. Thus the volume distribution ratios are between 8 and 9% greater than the mass distribution ratios. In terms of molecules per cell, the highest and lowest estimates for the mean albumin adsorption at the highest equilibrium albumin concentration investigated (about 25 g/L) are $+5.3 \times 10^5$ mpc and -1.0×10^5 mpc. The corresponding estimates for fibrinogen at 12 g/L are $+3.5 \times 10^4$ mpc and -2.6×10^4 mpc.

C. Discussion

The results of the solution analyses are summarized in Table 7; it is clear that most of the estimates for the protein distribution ratios do not agree. Those for albumin range from 13 to zero and are contradictory when the quoted estimates of precision are considered. We noted that overestimation of the initial cell volume fraction in solution analysis experiments results in overestimation of the distribution ratio. Monaghan and White [13] assumed that a trapped volume fraction of about 0.08 was responsible for the protein observed in the equilibrium solution of their experiments. However, such large

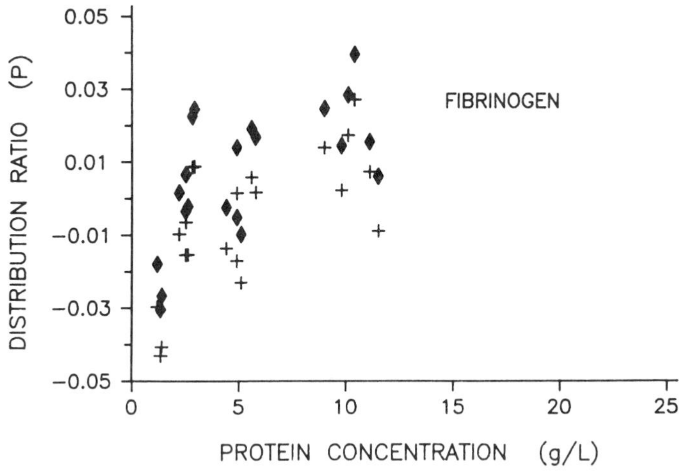

FIG. 3 Upper and lower estimates of fibrinogen adsorption to human erythrocytes as a function of fibrinogen concentration reported by Janzen [54] and expressed as a distribution ratio. ♦ Upper estimate; + lower estimate.

trapped volume fractions occur only at very low rates of centrifugation. Today, by the use of standard microhematocrit technique the trapped volume in red cell pellets can readily be reduced to 1 or 2%, so that it is unlikely that the disagreement in Table 7 can be attributed to trapping. Other variables in the experiments reviewed here are the temperature at which the experiments were carried out, the animal species from which the cells and proteins were derived, and whether or not the adsorption estimate was for an isolated protein or a mixture of proteins. On the basis of these reports, a complete discussion of the significance of each of these factors separately is not possible. However, the significantly different estimates reported for albumin, where these factors are constant, implies other causes. Finally, no saturation of the putative adsorption was evident, so that the differences can not be due to differences in ligand concentration.

It is evident that the apparent distribution ratios in Table 7 are approximately inversely related to the equilibrium cell concentration, although we reported [54] that a correlation between P and H_e was not observed. A potential explanation for this observation is based on the prediction that for polymer adsorption to solids a decrease in the average amount adsorbed with increasing adsorption area can

TABLE 7 Summary of Solution Analyses

Report	Ligand	Cell volume fraction, H_e (range)	Distribution ratio,[a] P	$\frac{P}{H_e^{-1} - 1}$	Surface concentration[b] (μm^{-2})
Furchgott [52]	Albumin	0.003	13 ± 2	0.04	85,000
		0.005		0.07	
Brown [14]	Albumin	0.004	−(2 ± 3)	−0.01	−13,000
		0.008		−0.02	
Bellis [50]	Serum	0.1	0.3 to 0.6	0.03 to 0.07	1600 to 3100
		0.3		0.08 to 0.26	
	Albumin	0.1	0.7 to 1.5	0.08 to 0.17	3600 to 7800
		0.3		0.30 to 0.64	
	Fibrinogen	0.1	0.5	0.06	510
		0.3		0.21	
Traber [53]	Fibrinogen	0.25	0.18 ± 0.04	0.06	205
Monaghan [13]	Serum	0.50	0.08	0.08	420
Kikuchi [51]	Serum	0.1	0.08 ± 0.02	0.01	520
		0.4		0.05	
Janzen [54]	Albumin	0.4	−0.004 to 0.024	0.02	−26 to 160
		0.8		0.10	
	Fibrinogen	0.4	−0.013 to 0.017	0.01	−16 to 22
		0.8		0.07	

[a]Mean and standard deviation or range.
[b]Molecules per μm^2 at 1 g/L solution concentration based on the mean or range of the distribution ratio, which is assumed to be concentration independent. Serum is assumed to be primarily albumin.

occur when the polymer is polydisperse with respect to molecular weight [66]. This effect has been observed for albumin adsorption to polystyrene latex beads and correlates with the presence of albumin multimers in solution (Olal and Brooks, unpublished observation). It is proposed that this effect depends on saturation of the adsorption and displacement of lower (and energetically less favored)-molecular-weight species by those of higher molecular weight. A decrease in the average mass adsorbed per unit area occurs upon addition of adsorbate surface area if lower-molecular-weight species are then able to bind. However, it was evident from the present solution analyses that saturation of the putative binding did not occur. Thus the polydispersity explanation cannot apply.

In her criticism of the report by Furchgott and Ponder [52], Brown [14] suggested that their adsorption estimate was affected by systematic errors associated with protein concentration measurements. Certainly, their result is in disagreement with all of the subsequent adsorption estimates in Table 7. However, the results due to Furchgott and Ponder, as well as those due to Brown, are of interest as they emphasize the apparent inverse relation between the distribution ratio and the cell volume fractions in solution analyses. If, as we reported [54], there is no correlation between P and H_e^{-1}, the apparent correlation must be an artifact. It may be due to the inverse relations evident in Eqs. (16) and (17) between the distribution ratio P and the cell concentrations H_i and H_e. In both of these equations P is the difference of two terms involving concentration and volume fraction factors. However, in the adsorption-type analyses the values of H_i and H_e are such that the distribution estimates may be approximated by a single term composed of a concentration factor $(C_s''/C_s - 1)$ multiplied by $(H_e^{-1} - 1)$ [i.e., in (17) $(H_e^{-1} - H_i^{-1})V_c'/V_c$ is no less than 79% of $(H_e^{-1} - 1)$]. Thus where the protein concentration factor is held at some positive value, reduction of the equilibrium cell concentration will result in an apparent increase in the distribution ratio similar to that in Table 7. However, our own results [54] imply that for albumin and fibrinogen the protein concentration factors must vary approximately inversely with the equilibrium cell concentration. If this result is accepted, it follows that a constant protein concentration factor with changing cell concentration must be artifactual.

Artifactual values for the protein concentration factors could occur where the changes in protein concentration are too small to be measured with accuracy and a systematic error occurs. In Table 7 the quantity $P/(H_e^{-1} - 1)$ estimates the protein concentration factor, which except for those reported by Bellis and Scott [50] are between 0.01 and 0.10. Since the protein concentration changes are thus less than 10% it is possible that systematic errors are significant in these results. On the other hand, where systematic errors in protein determinations are absent and concentration changes are too small to be resolved, one would expect that the mean binding estimates would not be significantly different from zero when trapping, as discussed earlier, was not a factor. The reports due to Brown [14] and Janzen [54] allow for the possibility that protein adsorption to the cells may be zero.

The distribution estimates reviewed thus far appear to us to be at odds. Collectively, they leave us with the choice of concluding that the nonspecific adsorption of proteins like albumin and fibrinogen to red cells is given by distribution ratios of about 0.02 or less, or that an inverse correlation between equilibrium cell concentration and the distribution ratio does exist. The latter choice implies that cell–cell interactions play a role in the measurements and suggests that protein macromolecules do adsorb nonspecifically to the red cell surface.

III. PELLET ANALYSES

A. Mathematical Description

Pellet analyses begin with the determination of the amount of ligand associated with a concentrated cell suspension (i.e., a cell pellet). It is assumed that the suspension consists of two compartments, cells and solution. In this model of a cell pellet conservation of volume and the amount of ligand is described by

$$V_p = V_s + V_c \tag{24}$$

$$C_pV_p = C_sV_s + C_cV_c \tag{25}$$

The notation used here is similar to that described earlier; C and V are ligand concentration and compartment volume respectively, while the subscripts p, s, and c identify the pellet, solution, and cell associated quantities, respectively. The analysis then aims to distinguish the amount of ligand associated with the cells from the amount associated with the solution. This requires the determination of either the volume fraction of the entrained (trapped) solution

$$H_t \equiv \frac{V_s}{V_p} \tag{26}$$

of the distribution ratio of the ligand

$$P \equiv \frac{C_c}{C_s} \tag{1}$$

In the first case the distribution ratio of the ligand may be calculated from

$$P = \frac{(C_p/C_s) - H_t}{1 - H_t} \tag{27}$$

while in the second case the entrained volume fraction can be obtained:

$$H_t = \frac{(C_p/C_s) - P}{1 - P} \tag{28}$$

If P or H_t cannot be determined independently, the ratio C_p/C_s may be interpreted as the maximum value of either the distribution ratio or the trapped volume fraction. The situation is thus in part analogous to that encountered with uncertainty in the initial cell volume fractions determined for solution analyses of binding.

The use of radiolabeling to trace the redistribution of molecules makes their quantitation in mixed systems practical. In particular, pellet analyses require estimation of the ligand concentration per unit of cell pellet C_p. However, among the liabilities of radiotracer techniques is the probability that some fraction of the radiolabel will not be attached to the ligand of interest and will distribute differently. As pointed out earlier, the use of radioisotopes of iodine to trace proteins in cell pellets introduces the problem of distinguishing free radioiodide within a cell from radioiodine covalently linked to a protein, which is in turn adsorbed to the cell surface. If the relative amount of contaminating radioiodide cannot be reduced to an acceptable level, yet can be measured, it is possible to correct an apparent distribution ratio P for the effect of nonligand radiolabel by use of Eq. (2). Equation (2) is particularly applicable in the case of radioiodine-traced biological materials because quenching of the emitted radiation is generally not a problem. Thus the detection sensitivity is the same for various radiolabeled species in different compartments. In pellet analyses where the amount of ligand adsorbed is small, as little as 1% free radioiodide may contribute significantly to apparent binding C_p because the intracellular volume in which the iodide distributes is relatively large [54].

B. Results

The red cell volume fraction of blood is a long-standing hematological indicator. As such the problems associated with determining hematocrits have been reported on extensively. In particular, there are numerous studies dealing with the measurement of entrained (trapped) fluid in columns of centrifuged cells. Here we introduce only a few studies in order to demonstrate three aspects of this methodology: (1) the trapped volume fraction is overestimated when radiotracers can enter the cells; (2) a larger trapped volume is estimated when a smaller molecule is used as a volume marker; and (3) a higher rate of acceleration during centrifugation results in a lower estimate for the trapped volume. These effects are evident in Table 8, which summarizes the reports described below.

Vazquez et al. [63] have reviewed a number of earlier reports where dyes and radioactive isotopes of sodium and iron were used to relate hematocrit measurements to "true" relative cell volumes. These differed by from -5 to +9%. Much of the variation was attributed to differences in the centrifugation conditions employed. In their own work they used [^{131}I]albumin and showed that trapping decreased with increasing centrifugation rate. However, they also argued that the dilution method of determining trapping, which was used in many of the cases reviewed and is analogous to solution adsorption analysis, is not generally suitable; measurement errors greater than 1 or 2% made it unlikely that differences less than 5 to 10% could be resolved. They also noted that prior dialysis of radioiodinated albumin resulted in lower estimates of the cell volume fraction. In a later paper, this effect and a similar effect associated with prior "adsorption" of radioiodinated albumin with red cells were suggested to be due to entry of radiolabel into the cells [67].

Maizels and Remington [64] determined trapping by extraction of volume markers from cell pellets and by use of [^{131}I]albumin. These authors suggested that different-sized molecules detected different trapped volumes. At centrifugation rates of 2000 × g trapped volumes were estimated to be between 1.8 and 2.1% by hemoglobin and albumin (labeled with either iodine or Evans Blue), while for lactose and inulin trapped-volume fractions were about 3%. They were aware that reversible adsorption of dyes could occur and lead to overestimation of the trapped volumes.

Most subsequent studies have relied on various radiolabeled molecules to determine the volume entrained in cell pellets. The rates of centrifugation have increased with the availability of more sophisticated centrifuges and the apparent trapped volumes have decreased correspondingly. Chien et al. [57] and England et al. [49] used [^{131}I]albumin and Funder and Wieth [65] used ^{22}Na, [^{14}C]-sucrose, and [^{131}I]albumin. In some cases [49,57], trapping estimates appear high; however, these authors made no corrections for radioiodide entry into red cells. In reports where the significance of free radiolabel and tracer decomposition products were recognized, very small trapped volume fractions have been estimated [65].

TABLE 8 Pellet Analysis: Trapping Studies

Report	Ligand traced	Pellet centrifugation[a] (× *g*)	Trapping fraction[b] (%)		Surface concentration[c] (μm^{-2})
Vazquez [63]	Albumin	500	8.6		560
	Albumin	1,000	4.6		
	Albumin	1,500	2.6		
	Albumin	2,080	2.2		
	Albumin	2,080	2.1 ± 0.3	n = 5	
	Albumin[d]	2,080	3.1 ± 0.2	n = 15	
	Na	2,080	3.9 ± 0.3	n = 8	140
Maizels [64]	Evans Blue	2,000	1.4–2.0		
	Lactose	2,000	2.9		
	Inulin	2,000	3.2		
	Hemoglobin	2,000	2.1, 2.3		
	Albumin	2,000	2.1		
	Albumin[e]	2,000	6.1		140
Chien [57][f]	Albumin	1,500	4.0 ± 3.1	n = 6	
	Albumin	15,000	1.0 ± 2.6	n = 6	65
Funder [65]	Na	75,000	2.27 ± 0.06	n = 42	
	Sucrose	75,000	2.08 ± 0.17	n = 19	
	Albumin	4,200	2.37		
	Albumin	75,000	0.47 ± 0.08	n = 16	31
England [49][f]	Albumin	15,000	3.3–4.7		210–310

[a]Centrifugation was for 30 min with the exception of Funder (20 min) and England (5 min).
[b]Individual measurement or mean and standard deviation with number of determinations.
[c]Molecules per μm^2 at 1 g/L solution concentration based on the minimum trapping fraction, which is assumed to be concentration independent. Trapping fractions were not corrected for adsorption, so that the inferred adsorption estimates are not corrected for intercellular trapping.
[d]The radioiodinated albumin was not dialyzed prior to the experiment.
[e]Radioiodinated albumin was not adsorbed with red cells prior to the experiment.
[f]Entry of radioiodide into the red cells was not considered.

Some appreciation of the physical significance of the trapping fractions reported can be achieved by calculating the thickness of a corresponding uniform layer of fluid at the cell surface. Twice this thickness would represent the mean separation between the cells. Thus the smallest trapped volume estimate in Table 8 [65] corresponds to a mean separation between erythrocytes of about 70 Å. Looking forward to the binding estimates summarized in Table 9 and recognizing that these may be interpreted as trapped volumes where trapped volumes were not determined independently, distribution ratios less than 0.002 correspond to mean cellular separations of less than 29 Å. Clearly, the uniform film interpretation is too simple under these conditions, as the minimum dimensions of the marker molecules are greater than the average film thickness. If the surface region of the red cell was an interface similar to that between a lipid bilayer and water, these results would suggest that the trapped volume is not uniformly distributed under these conditions.

One might ask whether the glycocalyx of the red cell is a factor in the observed behavior. Certainly, the glycocalyx, with an estimated thickness of 70 to 100 Å [71,72] would appear to have the volumetric capacity to absorb macromolecules. However, Fig. 4 shows that the trend of decreasing trapped volume with increasing acceleration continues up to centrifugation levels of at least 202,000 × *g*. Evidently, the equilibrium packing configuration is still sensitive to pressure changes under these conditions, whatever its determinants.

In addition to studies where tracers present in cell pellets have been interpreted as representing entrained solution, a number of studies have used some or all of the techniques described in the Introduction to estimate adsorption to the cells (Tables 9 and 10). Rothman et al. [73], Hughes-Jones and Gardner [68], and Müller and Gramlich [69] determined binding after washing the cells. Brooks [9] used ultracentrifugation to reduce intercellular trapping. Chien et al. [10] determined the amount of material eluted from cell pellets and corrected this for intercellular trapping using a radiolabeled "volume" marker. Brooks et al. [74] used a wash analysis to assess the homogeneity of apparent adsorption after ultracentrifugation. Rampling [70] determined the amount of material in cell pellets and corrected this for intercellular trapping, while Janzen [54] used ultracentrifugation to reduce intercellular trapping and corrected the results for intracellular accumulation of radioiodide.

The report due to Hughes-Jones and Gardner [68] is a comprehensive study dealing with the exchange of lipid and protein between red cells and serum. They found that red cells incubated with isolated plasma proteins retained up to a few thousand protein

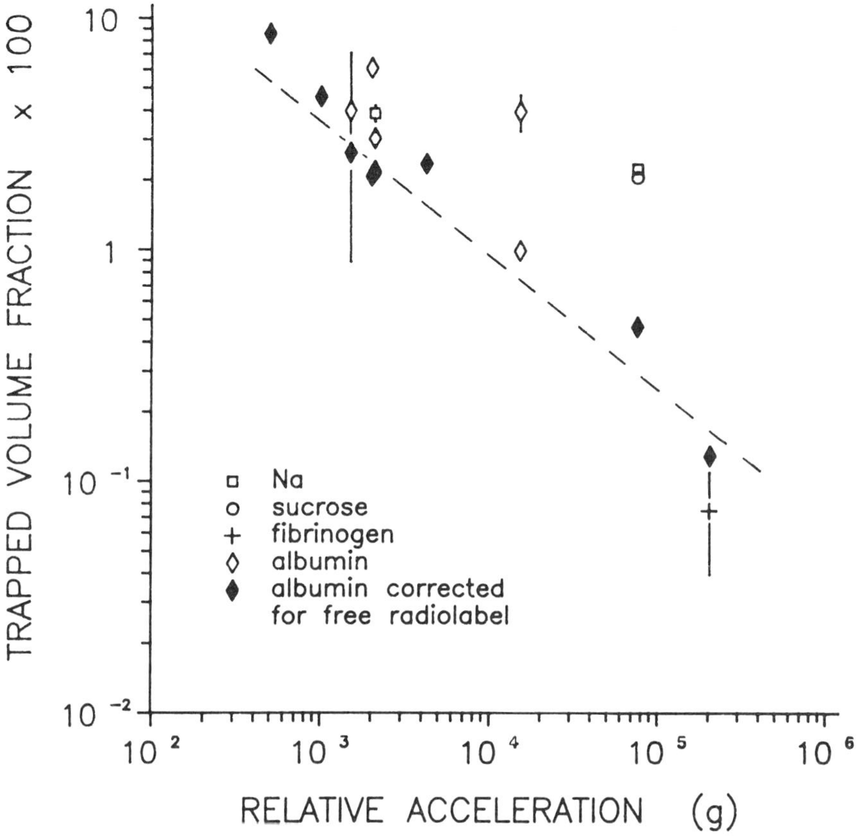

FIG. 4 Volume fraction of the entrained (trapped) solution found in pellets of human erythrocytes as a function of centrifugation rate. The volume fractions were determined using radiolabeled molecules as volume markers and the results assume that no surface adsorption occurred.

molecules per cell after exhaustive washing to saline buffer. The major portion of these adsorbed molecules was eluted by further washing of the cells with serum. Similar results were reported by Janzen [54]. The amounts bound increased to approximate saturation at solution protein concentrations of between 15 and 25 g/L. The saturating surface concentrations reported for albumin, fibrinogen, and γ-globulin were, respectively, 4400, 860, and 2900 molecules per cell. Therefore, the saturation behavior cannot be explained

TABLE 9 Pellet Analyses: Protein Adsorption Estimates

Report	Ligand traced	Pellet treatment[a]	Distribution ratio, P (× 10^3)	Surface concentration[b] (μm^{-2})	
Hughes-Jones[c] [68]	Albumin	Washed	≤ 1	7	(s)
	Fibrinogen	Washed	≤ 1	1	(s)
Müller [69]	Albumin	Washed	3.5 ± 0.4	20 ± 3	(s)
	Fibrinogen	Washed	8 ± 2	10 ± 2	(s)
Rampling [70]	Fibrinogen	Centrifuged less trapped	38	48	
Janzen[d] [54]	Albumin	Ultracentrifuged less free label	1.3—1.5	8—10	
	Fibrinogen		0.4—1.2	0.6—1.5	
Janzen [54]	Albumin	Ultracentrifuged plus wash analysis	1.0 ± 0.1	6.5 ± 0.7	(i)
			0.4 ± 0.1	2.6 ± 0.6	(s)
	Fibrinogen		0.2 ± 0.4	0.3 ± 0.5	(i)
			0.7 ± 0.1	0.8 ± 0.1	(s)

[a]Treatment used in determining the distribution ratio.
[b]Molecules per μm^2 at 1 g/L solution concentration based on the distribution ratio. These estimates are classified as intercellular (i) or slow (s) if a wash analysis is used.
[c]Binding is reported to saturate. The distribution ratios are those of the initial slopes of the adsorption isotherms.
[d]Range of distribution ratios.

by nonspecific competition for space at the cell surface as monolayer surface coverage by these proteins requires surface concentrations more than three orders of magnitude larger. This, and the serum-associated desorption, suggests a relatively specific interaction—for which a mechanism has yet to be found. The distribution ratios reported in Table 9 correspond to maximum values associated with the initial slopes of the binding isotherms.

Müller and Gramlich [69,75] determined the binding of isolated plasma proteins to red cells in the presence of anticoagulated plasma throughout a repetitive wash procedure. The amount of label associated with cell pellets decreased rapidly with the first few washes and then more slowly with subsequent washes. The second phase

TABLE 10 Pellet Analyses: Dextran Adsorption Estimates

Report	Ligand traced	Pellet treatment[a]	Distribution ratio, P (× 10^3)	Surface concentration[b] (μm^{-2})	
Rothman [73]	Dextran 90.6	Washed	0.4–1.4	1.8–6.6	(s)
Brooks [9]	Dextran 77.6	Ultracentrifuged	11	60	
Chien[c] [10]	Dextran 450	Eluted less trapped	9	8.6	
			13 (n)	12	
	Dextran 74		10	56	
			11 (n)	60	
	Dextran 40		15	160	
			12 (n)	120	
Brooks[d] [74]	Dextran 36.5	Ultracentrifuged plus wash analysis	—	7	(i)
			5.3	60	(f)
			0.7	8	(s)
	Dextran 150		0.4	1.2	(i)
			2.7	7.6	(f)
			0.8	2.4	(s)

[a]Treatment used in determining the distribution ratio.
[b]Molecules per μm^2 at 1 g/L solution based on the distribution ratio. These estimates are classified as intercellular (i), fast (f), or slow (s) if a wash analysis was used.
[c]Adsorption is reported to saturate. The distribution ratios are those of the initial slopes of the adsorption isotherms assuming a zero intercept. Ratios determined for neuraminidase treated cells are indicated by (n).
[d]Distribution ratios are for dextran concentrations less than 50 g/L.

of the radiolabel reduction was interpreted as representing a gradual removal of the protein adsorbed to the cells. Müller and Gramlich estimated the amount bound before the washing process began by back-extrapolation. A similar analysis by Janzen [54] produced similar two-component washouts. The amount and rate of elution of the rapidly eluted component implied that this component represented intracellular radioiodide [54]. The binding data reported by Müller and Gramlich are mean values determined for blood samples from a number of different individuals and are higher than those reported by Hughes-Jones and Gardner [68] (Table 9). In the Müller and Gramlich experiments the mean concentrations reported for the blood proteins were albumin 40.3 g/L, fibrinogen 2.2 g/L, γ-globulin 12.7 g/L, and α_2-macogobulin 2.0 g/L. Hence the difference between the binding of albumin and γ-globulin reported in this and the previous report is greater than is immediately evident from the distribution ratios. Hughes-Jones and Gardner [68] reported saturation of albumin and γ-globulin adsorption at approximately 25 and 15 g/L, respectively, so that the distribution ratios at these concentrations are substantially less than the maximum values given in Table 9. Although the amount of cell-bound material decreases with washing and the binding estimates reported by Müller and Gramlich [69] correspond to the amount bound before washing, whereas those reported by Hughes-Jones and Gardner [68] are for washed cells, it appears unlikely that the entire difference can be attributed to this cause. Their experimental conditions were different in that they employed tracer amounts of labeled proteins added to anticoagulated plasma rather than labeled protein fractions [68]. However, one would suppose that competition at a surface would reduce rather than increase adsorption.

Rampling [70] reported distribution ratios for fibrinogen and fibrinogen degradation products in suspensions of human erythrocytes and anticoagulated plasma or isotonic buffer. The distribution ratio for fibrinogen (Table 9) was independent of fibrinogen concentration up to 10 g/L, and of changes in anticoagulant and changes in temperature between 4 and 37°C. The distribution ratio increased with decreasing pH and increasing ionic strength. Relative to fibrinogen the distribution ratios of the fibrinogen degradation products X, Y, D, and E were reported to be respectively 0.9 ± 0.2, 4.1 ± 0.4, 0.6 ± 0.1, and 2.9 ± 0.7.

The distribution ratios were determined by use of radioiodinated proteins and calculation analogous to that described by Eq. 27. However, the contribution of free radioiodide to the total radiolabel associated with the cells was not considered. Thus the distribution ratios reported are apparent distribution ratios and could vary if the amount of free radioiodide varied, even though the adsorption remained constant. Therefore, it is not clear that the changes in the distribution ratios reported for the various fibrinogen products are significant. In fact, the distribution ratio reported for fibrinogen could be explained without reference to surface adsorption by having 5% of the radiolabel in the equilibrium suspending medium as radioiodide.

The changes in apparent binding with ionic strength and pH are also unclear. The change in the ionic strength of the suspending media was effected by changing the concentration of NaCl. However, since osmotic strength was also adjusted with NaCl, it would seem that the osmotic strength of the buffers was not held constant. Increase in the osmotic strength would result in reduction of the mean cell volume and possibly an increase in the volume of entrained suspending medium in the cell pellets rather than an increase in adsorption to the cells. Changes in pH also result in changes in red cell volume [76]; however, cell volume increases with decreasing pH so that a similar trapping explanation is not possible. It is still possible that some changes in the distribution ratio for fibrinogen may occur as the environment is changed, but the problems outlined above leave the stated results in doubt.

The volume fractions of the cell suspensions were given by hematocrit measurements corrected for intercellular trapping. The trapping was determined as described by England et al. [49], that is, by the use of radioiodinated albumin as a volume marker. However, there are no unambiguous determinations as to how much albumin is adsorbed to red cells. Thus it would appear that the distribution ratio reported for fibrinogen by Rampling [70] should be considered to be a lower estimate, as albumin adsorption would result in overestimation of trapped volumes. In addition, on the basis of the results reported in studies of trapping in cell pellets (Table 8), one would expect that the trapped volume associated with fibrinogen would be less than that associated with albumin. This is by virtue of different molecules having been used as intercellular volume markers, smaller molecules reporting larger intercellular volumes. This suggests that the glycocalyx of the cell may be acting as a region of the intercellular volume which excludes larger molecules. Some sense of the significance of the cell volume fraction in these experiments may be gained from the fact that variations of 1% in this quantity result in variations of about 25% in the calculated distribution ratio. Rampling did not report the hematocrits or correction terms that applied to his experiments.

Janzen [54] reported distribution ratios for fibrinogen and albumin in isotonic phosphate buffered (pH 7.4) saline suspensions of erythrocytes at 21°C. Ultracentrifugation (202,000 × g for 30 min) was employed to reduce the intercellular volume within cell pellets and thus to reduce this contribution to the apparent distribution ratios. The results reported assume that $H_t = 0$. The proteins were radioiodinated (^{125}I) and it was assumed that the distribution of these proteins was not affected by centrifugation. After this centrifugation treatment red cells were resuspended for wash analysis with little or no lysis and with no apparent changes in either physical appearance or in the membrane proteins detected by sodium dodecylsulfate polyacrylamide gel electrophoresis.

The apparent distribution ratios determined were dependent on the level of intracellular radioiodide. Apparent mass distribution ratios for albumin and fibrinogen varied from 0.002 to 0.011. These changes did not correlate with changes in protein concentration, but apparent distribution ratios did decrease with increase in cell volume fraction of the cell suspension before centrifugation. This correlation was due to dilution of free radioiodide by its entry into the red cells, an observation also reported by Vazquez et al. [67]. The apparent distribution ratio decreased with increasing cell concentration because the amount of free radioiodide per cell decreased.

The presence of radioiodide within the cells was corroborated in two ways. First, the amount of free radioiodide in the protein preparations, determined by gel filtration, was the same as that calculated from the change in the apparent distribution ratio with cell volume fraction. This calculation is based on the independent conservation of radiolabeled protein and radioiodide in the system and assumes that the distribution ratios for these species are independent of cell concentration. The apparent distribution ratios corrected for intracellular radioiodide determined by this method are given as ranges of values in Table 9. The ranges result from the lack of a precise value for the intercellular volume in ultracentrifuged cell pellets. The intercellular volume is a factor in estimating intracellular radioiodide. The lower and upper estimates correspond to assuming that the intercellular volume was either zero or 0.5%, respectively.

The second confirmation of the presence of free radiolabel was obtained as follows. Repetitive quantitative washing of the cells, analogous to that described by Müller and Gramlich [69], allowed determination of the amount of radiolabel eluted from the cells with a specified distribution ratio. As noted earlier, analyses of the wash behavior resolved two rates of elution for the radiolabel retained in cell pellets. The radioiodide distribution ratio can be determined because the rate of iodide transport across the cell membrane ($t_{1/2} \simeq 50$ s) [41] is such that iodide is able to reequilibrate between successive washings. The distribution ratio calculated for the rapidly decreasing component of the radiolabel washout was equal to that determined for free radioiodide. In addition, the amount of radioiodide calculated from the rapidly eluting component of the washouts was the same as both that determined by gel filtration and analysis of the dependence of apparent distribution ratios on cell concentration. This interpretation of radiolabel washout from red cells is different from the interpretation suggested by us previously [74]. The previous interpretation was that the two-component washout of fibrinogen was due to differential desorption of monomer and multimer fibrinogen.

From the wash analyses the surface concentration of the slowly eluting component in cell pellets before washing increased in proportion to the equilibrium concentration of the protein in the pellet supernatant. Expressed as a volume distribution ratio relative to the initial supernatant protein concentration, the albumin adsorption was $(4.0 \pm 0.9) \times 10^{-4}$ for albumin concentrations below 7 g/L. Similarly, the fibrinogen adsorption was $(6.6 \pm 0.7) \times 10^{-4}$ for fibrinogen concentrations of less than 12 g/L. Distribution ratios for the slowly eluting component were not calculated using the concentration of the free protein in the washings because this component did not come to equilibrium after each wash. At 1 g/L the surface concentrations of albumin and fibrinogen were 350 ± 80 and 110 ± 10 molecules per cell, respectively. These estimates are somewhat lower than those reported by Hughes-Jones and Gardner [68] (Table 9). However, at the maximum fibrinogen concentration examined, the surface concentration of this protein is greater than the saturation adsorption reported by these authors [68]. The difference between the total radiolabel present in cell pellets before wash analysis and the sum of the rapidly and slowly eluting components was found to be a positive value in most cases which could be taken to represent either intercellular volume or reversibly adsorbed protein. Interpreted as volume fractions, values of $(1.0 \pm 0.1) \times 10^{-3}$ and $(2.4 \pm 4.2) \times 10^{-4}$ were calculated from the albumin and fibrinogen data, respectively. If these counts represented surface concentrations, at a protein concentration of 1 g/L, they are equivalent to 880 ± 90 and 40 ± 70 molecules per cell, respectively.

Rothman et al. [73] reported the interaction of [^{14}C]dextran (M_n = 90,600) with human erythrocytes at room temperature (Table 10). Labeled dextran was mixed with blood to a final concentration of 1% and the red cells were isolated by differential centrifugation. The cells were washed with plasma plus 1% nonlabeled dextran until the activity in the washings was insignificant relative to that retained by the cells. Red cell binding ranged from 2.4×10^3 to 8.9×10^3 molecules per cell. Assuming that the 1% dextran concentration is the relevant solution concentration, these adsorption values correspond to distribution ratios of 0.0004 and 0.0014, respectively. The experiment is similar to that of Hughes-Jones and Gardner [68] with labeled proteins and similar distribution ratios resulted. In Table 10 dextran fractions are identified by use of the mean weight or number-average molecular weight divided by 1000 (e.g., dextran 90.6).

Brooks [7,9] determined [^{3}H]dextran (M_w = 77,600, M_n = 65,500) adsorption to saline-washed erythrocytes at 25°C. The cells were isolated from radiolabeled solution by ultracentrifugation ($368{,}000 \times g_{max}$) and use of an organic phase of density intermediate to that of the polymer solutions and cells. The adsorption reported is based on the assumption that intercellular trapping in the cell pellets was negligible. With repetitive saline washing of the cells, the activity retained in cell pellets was shown to decrease by somewhat less than one order of magnitude, to a level similar to that reported by Rothman et al. [73].

The adsorption isotherm reported is linear between 0.007 and 70 g/L dextran concentration and corresponds to an approximately constant distribution ratio of 0.011. Above about 70 g/L the slope of the isotherm increases significantly. The apparent adsorption rises to about 0.0029 g/g of cells (1.7×10^4 molecules/μm^2) at a dextran concentration of 152 g/L. The reported dextran adsorption isotherms are shown in Fig. 5. The increased slope of the isotherm at higher dextran concentrations was of particular interest as this occurred in the range of dextran concentrations that causes disaggregation of red cells aggregated at lower dextran concentrations. It was proposed that this behavior was supportive of the cross-bridging hypothesis of red cell aggregation, as desorption of bridging molecules would allow an increase in the amount bound per cell.

Chien et al. [10] reported binding isotherms for dextran fractions of mean molecular weights 40,000, 74,000, and 450,000 (denoted by the authors as dextrans 40, 70, and 500) to both normal and neuraminidase-treated human erythrocytes. The adsorption was estimated by chemically determining the amount of dextran extracted from cell pellets by repetitive washing and correcting this for dextran present in the intercellular solution. The cell pellets were produced by ultracentrifugation and the trapped volume in a pellet was estimated from the amount of [^{14}C]carboxyldextran (M = 16,000) eluted and its concentration in the supernatant of the cell suspension before centrifugation.

For adsorption to normal cells, the initial slopes of the isotherms increased from distribution ratios of about 0.009 for dextran 450 and 0.010 for dextran 74 to 0.015 for dextran 40. The initial slopes for adsorption to neuraminidase-treated cells fall between 0.011 and 0.013 and a systematic variation with dextran molecular weight cannot be discerned. The isotherms reported increased to apparent saturation at dextran concentrations between 100 and 200 g/L. The surface concentrations at saturation are between 0.13 and 0.15 pg/cell for normal erythrocytes and between 0.08 and 0.10 pg/cell for neuraminidase-treated cells. The saturation levels are significantly lower than the apparent adsorption reported by Brooks [7,9] at 152 g/L. In line with the proposal made by Brooks [7], dextran adsorption was related to aggregation by the cross-bridging mechanism.

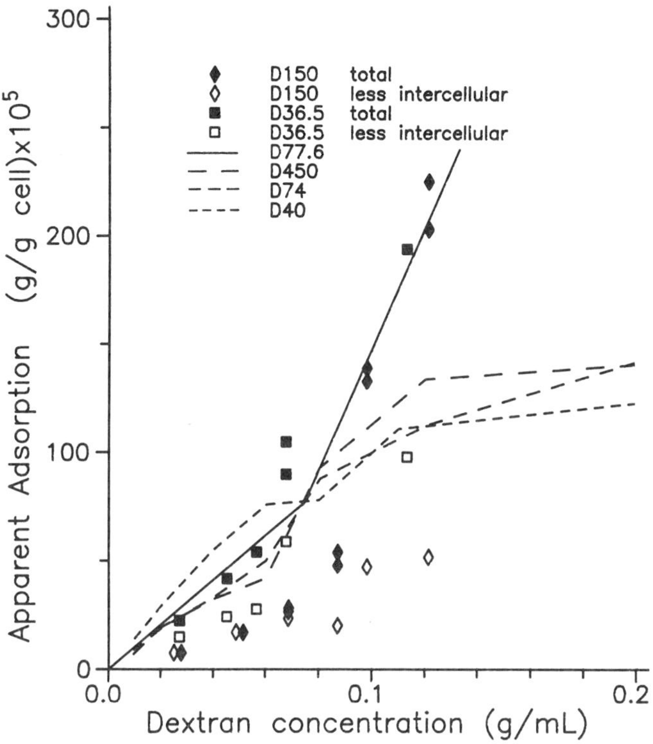

FIG. 5 Comparison of apparent dextran adsorption to human erythrocytes reported by Brooks [7,9], Chien et al. [10], and Brooks [74]. The isotherms were determined from analyses of red cell pellets after isolation from cell suspensions at the indicated bulk dextran concentrations.

The methodology of Chien et al. [10] assumes that the trapping correction determined with 16,000-molecular-weight [^{14}C]carboxyl-dextran is appropriate for substantially higher-molecular-weight dextrans. Trapping studies suggest that larger intererythrocyte volumes are measured with volume markers of lower molecular weight (Table 8). If this applies to dextran, the reported adsorption underestimates the actual adsorption. In addition, it is assumed that the volume marker does not adsorb to the cell surface. Adsorption of the volume marker to the cell would also mean that the actual amounts of dextran adsorbed were understimated. From a comparison of electron microscope observations and trapping estimates, Chien [12] claims that this negatively charged low-molecular-weight dextran does not adsorb significantly to erythrocytes. However, heparin, a negatively charged low-molecular-weight mucopolysaccharide, apparently does adsorb to red cells and causes aggregation at concentrations above 1 g % [77].

Brooks et al. [74] report the apparent adsorption of two dextran molecular-weight fractions (36,500 and 150,000) to maximum concentrations of approximately 120 g/L. The methodology was as described previously [7,9], except that the cells were pelleted at a lower centrifugation rate: 165,000 × *g*. The apparent adsorption reported for the dextran 36.5 fraction is similar to that reported for dextran 77.6 [7,9]. The dextran 150 adsorption is, however, significantly lower at concentrations below 70 g/L (equivalent to a distribution ratio of approximately 0.004) and above 70 g/L increases rapidly to values similar to those reported for dextrans 36.5 and 77.6 above 100 g/L.

The apparent adsorption was analyzed using a serial wash procedure similar to that described by Müller and Gramlich [69]. The analysis decomposed the apparent adsorption of each dextran fraction into three components. Two components were identified from the wash data by the constant fractional rates at which label was removed from the pellets. As there was no evidence of impurity in the dextran fractions, the behavior was taken to reflect characteristics of the cell surface. The amount of dextran associated with these two components before the cells were washed was determined from the wash data. The third component is the difference between the apparent adsorption and the sum of the components identified by the wash-out and is only associated with the cell pellet before washing. This component was interpreted as dextran in an intercellular volume.

The amounts associated with the slow desorption site are proportional to the dextran concentration throughout the reported range and correspond to distribution ratios of 0.7×10^{-3} and 0.8×10^{-3} for dextrans 36.5 and 150, respectively (Table 10). The amounts associated with the fast desorption site below about 50 g/L correspond to distribution ratios of 5×10^{-3} and 3×10^{-3} for dextrans 36.5 and 150, respectively. Above 60 g/L the amounts associated with the fast site are more variable but suggest that the slope of the dextran 150 isotherm increases while that of dextran 36.5 does not. For both dextran fractions, the component interpreted as intercellular is dependent on dextran concentration. For dextran 36.5 the intercellular volume fraction is approximately proportional to the dextran concentration and equals 0.010 at 100 g/L. In contrast, the intercellular volume fraction for dextran 150 is less than 0.001 at concentrations below 70 g/L and then rises to levels similar to those observed for dextran 36.5.

These results were taken to support the cross-bridging mechanism because the changes in the adsorption and the trapped solution fraction for both dextran fractions as functions of dextran concentration can be related to this mechanism. At approximately the dextran 150 concentration that results in disaggregation, both the slope of the adsorption isotherm from the rapidly eluted component and the trapped solution volume, which was constant at lower concentrations, increase. Physically, the increase in the slope of the isotherm is interpreted as being associated with the increase in surface area available for adsorption, while the increase in the trapped solution volume is seen to result from intercellular repulsion enhanced by dextran adsorption. In contrast, the adsorption isotherm and the trapped solution volume increase linearily with dextran 36.5 concentration, as cell aggregation does not occur in this dextran fraction. It was argued that adsorption of the slowly eluted component is not associated with reversible aggregation. After equilibration at high dextran 150 concentrations plus one wash with saline, the slowly eluting component is the major one retained. After this one saline wash no aggregation is observed. The amount of slowly eluting dextran bound under these conditions is higher, however, than the amount of this component bound to unwashed cells in equilibrium with lower dextran 150 concentrations where the cells are aggregated. Hence it seems unlikely that this component is responsible for reversible erythrocyte aggregation.

C. Discussion

1. Solution Versus Pellet Analyses

Comparison of the surface concentration estimates determined by solution analyses (Table 7) and pellet analyses (Tables 9 and 10) show that only the lowest estimates derived from solution analyses are in the same range as those based on pellet analyses (i.e., fewer than 200 molecules/μm^2 at a solution concentration of 1 g/L). The surface concentrations implied by the lowest trapped fraction given in each of the reports in Table 8 also fall in this range. In addition, wash analyses reveal that the material found in cell pellets generally consists of at least two components, one of which may be bound to the surface in a manner inconsistent with a role in reversible red cell aggregation. Thus barring the possibility that centrifugation has changed the equilibrium between macromolecules in solution and those adsorbed to cells, we conclude that most of the solution analyses reported overestimate the nonspecific adsorption of protein to red cells. To our knowledge, a solution analysis of dextran adsorption to red cells has not been reported.

The role of centrifugation conditions in determining weak adsorption to cells is unknown, but the pellet analyses in Tables 8, 9, and 10 appear to be the closest available approach to answering this question. The trapping studies in Table 8 do not consider adsorption. Thus, because of the symmetry of Eqs. (27) and (28), it is possible to view these results as estimates of adsorption without correction for intercellular trapping. From these data it is clear that the sum of the intercellular trapping plus adsorption depends primarily on the rate and duration of centrifugation when intracellular trapping is not a factor. If adsorption were independent of centrifugation, the lowest estimates of "intercellular volume" would most closely approximate the surface concentrations of adsorbed molecules and nonzero adsorption would result in a decrease in the slope of trapped volume versus centrifugation rate at higher rates (Fig. 4). However, such an effect is not evident. Thus it appears that adsorption cannot be determined by direct methods because intercellular trapping cannot be independently determined at very high centrifugation rates. However, we feel that a correction for intercellular trapping is not physically justifiable in view of the small mean separation between cells. For human erythrocytes an intercellular volume fraction of 0.007 corresponds to a mean separation of about 100 Å when a uniform distribution of the intercellular volume is assumed. Thus, unless the volume is very unevenly distributed, it appears that the macromolecules used as volume markers are in such close proximity to the surface that distinguishing which are adsorbed and which are free in solution is not possible by the techniques discussed here.

The maximum albumin surface concentration implied by an apparent distribution ratio of 0.007 at a solution concentration of 1 g/L is 91 molecules/μm^2 and these molecules are distributed in a region with an average thickness of less than 100 Å. Apparent distribution ratios continue to decrease with increased rates of centrifugation and it seems likely that from a molecular point of view, the environment between the cells is rather unlike that of the solution under these conditions. Thus it may be that the equilibrium between molecules in solution and those adsorbed at the cell surface has been altered by formation of a dense cell pellet. A reduction in the adsorbed surface concentration due to centrifugation would be confirmed by independent measurements that indicate a higher surface concentration before centrifugation. Solution analysis might provide this information, but unfortunately they are in disagreement. However, the lowest of these estimates (Table 7) imposes an upper limit on the amount bound and is a factor of about 20 greater than the pellet values.

We are aware of only one report that claims significant adsorption determined by an independent method. Unfortunately, we believe the result is incorrectly interpreted. Rehfeld et al. [24] reported an estimate of albumin adsorption to human erythrocytes based on a calorimetric technique. The titration calorimetry method used has been applied successfully to proton ionization and metal complex formation [78]. However, we have not found a report which indicates that this method may be used reliably to estimate adsorption of macromolecules to surfaces. Using this technique Rehfeld et al. [24] reported adsorption amounting to $(6.8 \pm 2.2) \times 10^5$ albumin molecules per cell. In their experiments they added 0.3 mL of a stock albumin solution to 2.7 mL of a 30% (v/v) suspension of washed red cells. The equilibrium albumin concentration assuming no adsorption would have been about 0.55 g/L, and assuming the adsorption reported would have been about 0.26 g/L. The adsorption could therefore have been readily verified by measuring this concentration change, but the authors did not report the equilibrium albumin concentrations of their experiments. The distribution ratio calculated from these results is 3.0 ± 1.0. This is much higher than any of the distribution ratios reported except for those of Furchgott and Ponder [52]. Since distribution ratios greater than 1 would be so readily detected yet have been reported so infrequently, we have discarded these estimates as incorrect and assume that they are due to inappropriate interpretation of the measured heat exchange.

2. Protein Adsorption and Aggregation

Since the solution analyses have not provided consistent estimates, we must rely on pellet analyses to estimate adsorption. If it is assumed that centrifugation does not displace adsorbed molecules from the cell pellet, a possibility regarding which we have no information, then as outlined in Sec. I, one can expect pellet analyses to overestimate surface adsorption. This is due to the presence of adsorbate in the solution entrained in the cell pellet (intercellular trapping) and possibly in the cellular compartment (intracellular trapping). The latter problem is unlikely to occur with macromolecules, but has been demonstrated for radioiodide [41,54], the most commonly used protein radiotracer. The adsorption corrected for these effects should then be checked for homogeneity with respect to binding affinity by analysis of the binding isotherm or by a physical analysis of the sample (e.g., a Scatchard analysis or a serial wash analysis, respectively). Only a few of the reports assembled here have considered each of these aspects of adsorption analysis.

Over the range of concentrations investigated, the protein adsorption determined from cell pellets where trapping artifacts have been considered was in each case proportional to the solution protein concentration (i.e., a constant distribution ratio was observed;

Table 9). Unfortunately, the distribution ratios associated with both intercellular and intracellular trapping are also constant with increases in the solution protein concentration. Thus, unless the adsorption isotherm is nonlinear, as reported for dextran adsorption [10], the form of the isotherm does not distinguish apparent adsorption from trapping.

The results reported by Rampling [70] are based on apparent pellet adsorption at relatively low centrifugation rates with a correction for intercellular trapping, while those due to Janzen [54] follow from ultracentrifugation and a correction for intracellular trapping. As we noted in our own results [54], apparent distribution ratios of about 0.01 were observed in experiments where some care had been taken to reduce free radioiodide in adsorbates. Rampling [70] did not consider this contribution to apparent adsorption and on this basis the adsorption he reported is probably an overestimate. The intercellular trapping correction applied by Rampling [70] was probably also overestimated, as it was based on the retention of labeled albumin in red cell pellets [49] where adsorption was *not* considered. It is thus unclear whether the reported fibrinogen adsorption is over- or underestimated. We have not been able to find evidence indicating that the surface concentrations of proteins are reduced by ultracentrifugation. However, this possibility suggests that the adsorption reported by Janzen [54] should be viewed as a lower estimate.

The protein adsorption estimates in Table 9 cannot be used to determine whether or not a homogeneous set of interactions is represented, as the apparent isotherms showed no indication of saturation. The adsorption estimates given in Table 10, on the other hand, are derived from either washed cell pellets or an analysis of radiolabel eluted during repeated washing of cell pellets. The wash analysis all indicate that part of the apparent adsorption, as determined in Table 9, is retained after washing (slow elution). Where elution was analyzed, two additional components were identified. In the case of protein adsorption, these components were interpreted as elution of intracellular radioiodide and intercellular protein [54]. However, since the free radiolabel component is completely unrelated to protein adsorption at surfaces, there are only two components of interest associated with fibrinogen and albumin adsorption.

The component slowly eluted from cells equilibrated at 1 g/L protein concentration is estimated to be between 0.8 and 10 molecules/μm^2 for fibrinogen and between 2.6 and 20 molecules/μm^2 for albumin (Table 9). Hughes-Jones and Gardner [68] reported saturation of this component of the adsorption at solution concentrations greater than 15 g/L. Other reports [54,75] imply that saturation, if it occurs, must occur at higher protein concentrations, however. At present we have no explanation for this disagreement.

This component of protein adsorption, although retained by cells when washed with saline solution, is largely eluted when cells are washed with protein-containing solutions [54,68]. This desorption therefore appears to depend on an exchange or displacement of adsorbed molecules by molecules in solution. This type of behavior has also been described for albumin adsorbed to polyethylene [79]. Since exchange does occur in media with plasmalike protein concentrations, it follows that the component identified by relatively slow desorption in saline is not disqualified from a role in the reversible aggregation of red cells. However, if rouleaux-like aggregation were mediated by this component, one would expect that aggregated cells transferred to a saline medium would remain adherent. In fact, disaggregation occurs almost immediately in most cases (E. Evans, personal communication). Thus it appears unlikely that the slowly eluting component of protein adsorption can by itself account for red cell aggregation.

The remaining protein component found in cell pellets was taken to be intercellular. The mean trapped volume fractions calculated for albumin and fibrinogen are, respectively, 0.001 and 0.0002 for the given centrifugation conditions [54] and apparently imply that a greater intercellular volume is accessible to albumin. Although this agrees with the trend outlined in the trapping studies, this result is tentative, as these values are interpretations of wash data and are significantly lower than those of other workers.

The close average proximity of cells in pellets suggests that the component we have interpreted as intercellular could instead be reversibly adsorbed and available to form cross-bridges between cells. However, are there enough such molecules to explain spontaneous aggregation by this mechanism? The surface concentrations associated with intercellular fibrinogen and albumin at 1 g/L solution concentration are, respectively, 0.3 and 6.5 molecules/μm^2 (Table 9). In our opinion, such values, multiplied by the solution concentration in grams per liter at which a protein first initiates spontaneous aggregation, represent the best available estimate of the minimum surface concentration associated with aggregation. For fibrinogen this implies that aggregation would be caused by less than 1 molecule/μm^2.

Red cell aggregation in purified fibrinogen begins at a solution concentration of about 2 g/L. The measured surface affinity [27] at this concentration is about 1.5×10^{-3} erg/cm^2 (unpublished observations); red cell surface affinities in plasma are similar [38]. If the cross-bridging mechanism is assumed, the affinity per molecule at a surface concentration of 0.6 molecule/μm^2 is 360 kcal/mol. This value is extremely high when put in context with other bond energies. Physical adsorption of gases to solids is associated with free energies between about 1.5 and 9 kcal/mol, chemisorption involves about 20 kcal/mol [80], and covalent bonds are of the order of 50 kcal/mol or more. Since the calculated affinity per molecule exceeds the free energy associated with physical adsorption, it implies that surface attachment of the macromolecules involved must occur via the order of 100 binding sites/molecule. However, such multipoint adsorption is unlikely to be reversible on the time scale of cellular aggregation since all bonds would have to break at the same instant to allow release of the macromolecule upon dilution. Moreover, aggregation induced by this small number of cross-bridges is inconsistent with the observed aggregation behavior of red cells in fibrinogen. Microscopically, cell-cell adhesion is seen to form smoothly and continuously, yet the bending stiffness of the red cell membrane is too small to produce a smooth contour if adhesion were caused by only a few points of interaction [81]. It is therefore extremely difficult to rationalize our best estimates for fibrinogen adsorption with the bridging model of rouleaux formation.

Lack of spontaneous aggregation in the presence of albumin is viewed in the cross-bridging scheme as due to the greater electrostatic repulsion associated with the smaller surface-to-surface separation required by smaller cross-bridging species [12]. On the other hand, if reversibly adsorbed albumin formed cross-bridges, there would be ample binding energy available to cause adhesion. If the free energy of albumin adsorption were only 1.5 kcal/mol, the estimated minimum exchangeable surface concentration at 45 g/L albumin concentration (about 290 molecules/μm^2) would imply a surface affinity for aggregation of 3×10^{-3} erg/cm^2. This is twice the minimum affinity required to cause aggregation in the presence of fibrinogen. Hence, although it is possible that electrostatic stabilization plays a role in counteracting the effects of albumin adsorption, there would appear to be sufficient potential for cross-bridging that it is surprising that conditions have not been found where albumin induces aggregation.

It is not yet known whether the depletion hypothesis distinguishes fibrinogen from albumin with respect to the cellular aggregation produced. A quantitative theory has been reported describing depletion-dependent aggregation of particles coated with anchored polymer in polymer solutions [82]. This theory requires values for the sizes of the monomers and the number of monomers per polymer. However, no explicit expressions for the dependence of surface affinity or the range of the depletion force on polymer molecular weight are available. A comprehensive theory for proteins is even more remote, as these are structurally more complex; however, intuitively, one would expect that lower affinities and shorter interaction distances will be associated with smaller molecules. Protein adsorption plays different roles in the depletion and bridging models. In the context of the depletion mechanism, surface adsorption (without bridging) is expected to reduce surface depletion and thus counteract rather than assist aggregation. To promote aggregation, surface adsorption would ideally be zero. Bridging, on the other hand, would be promoted by the reversible (weak) binding of relatively large numbers of macromolecules. In our opinion, a definitive proof of the mechanism of reversible red cell aggregation in protein solutions is not yet available, but there seems little to support the bridging model as the predominant mechanism. Aggregation behavior at fibrinogen concentrations above 20 g/L has not been documented and may be of help in this regard. Possibly of greater interest is the potential theoretical development of both the bridging and nonadsorption hypotheses and studies on model systems. These might eventually allow the prediction of free energies of interaction from molecular characteristics.

3. Dextran Adsorption and Aggregation

The method used by Chien et al. [10] to determine apparent dextran adsorption employed a correction for intercellular trapping and a moderately high centrifugation rate, while that used by Brooks [7,9] and Brooks et al. [74] employed ultracentrifugation and no correction for trapping. At concentrations below about 100 g/L the apparent adsorption estimated by these two methods are in substantial agreement. However, at higher concentrations, the results diverge (Fig. 5). The difference could be explained by the presence of an intercellular component that was insignificant under conditions where cells are aggregated, but which contributed significantly to apparent adsorption where cells are disaggregated.

Wash analyses of dextran 36.5 and 150 apparent adsorption suggest that an intercellular component with this general behavior exists. When the intercellular component is subtracted from the total apparent adsorption reported for dextran 36.5 and 150, the apparent adsorption falls to values (open symbols, Fig. 5) generally lower than those reported by Chien et al. [10]. Whether the difference between the corrected values reported by Brooks et al. [74] and those reported by Chien et al. [10] are significant is difficult to judge. Under conditions where cells are disaggregated, the intercellular component is as much as 75% of the total apparent adsorption of dextran 150; thus the difference is likely to be imprecisely known. Similarly, Chien [12] noted that the intercellular volume fraction determined via [^{14}C]carboxydextran was less than 0.01; however, the adsorption reported is equivalent to a distribution ratio of no more than 0.015.

The isotherms reported by Chien et al. [10] have been interpreted as approaching saturation at dextran concentrations above about 100 g/L. This is suggestive of surface adsorption, which approaches a steric limit determined by the surface area of the adsorbent, the size of the adsorbed molecules, and the assumption of monolayer adsorption. From the radius of gyration determined from data reported by Granath [83], the red cell surface area, and assuming a rectangular packing arrangement, the adsorption reported by Chien et al. [10] at 200 g/L was found to be equivalent to about 30% surface coverage, independent of molecular weight. Without measurements in another suitable system or theoretical developments capable of predicting surface coverage by adsorbed dextrans, this plausible value does not prove or disprove the mechanism of dextran-associated red cell aggregation. In a similar vein, Chien [12] interpreted the lower saturating adsorption associated with neuraminidase-treated cells as being due to the aggregated state of these cells, in contrast to normal cells, which are disaggregated. He assumed that 35% of the cell surface is involved in contact between pairs of cells in rouleaux. This explanation is a starting point; however, the adsorption estimates are derived from cell pellets, where one might guess that closer to 100% of a given cell would be within cross-bridging range of an adjacent cell. Thus the saturation behavior of apparent dextran adsorption is suggestive of adsorption and cross-bridging, but does not prove the mechanism.

Wash analyses reveal that part of the reported dextran adsorption is due to a component that has been described as slowly eluting and is characterized by distribution ratios of about 1×10^{-3} (Table 10). For dextran 36.5 and 150 the distribution ratios are constant to in excess of 120 g/L. Thus on a weight basis this component of dextran adsorption is very similar to that of the slowly eluting component of protein adsorption. In addition, it has been argued on the basis of the amounts adsorbed [74] that slowly eluting dextran adsorption, like its protein analog, is not responsible for red cell aggregation. However, it may be that the interactions of these classes of macromolecules with the cell surface are different in some respects.

Slowly eluting protein can be rapidly eluted when free protein is present in solution, whereas dextran does not appear to exchange from the cell surface. Although a direct test of this has not been made, indirect evidence exists. In their experiments Rothman et al. [73] washed cells with 1% nonlabeled dextran, the same concentration as that used during adsorption. The similarity of the surface concentrations reported by these authors and by Brooks et al. [74] for the slow site after saline washing suggests that dextran does not desorb significantly by the exchange mechanism. The mechanism by which bound dextran is removed from the cell surface is not known.

The remaining component of apparent dextran adsorption has been referred to as rapidly eluting material. Because of its ready reversibility this component is viewed as the primary candidate for the role as cross-bridging agent in dextran-mediated red cell aggregation. Most of the pellet-associated radiolabel is due to this material. It is represented by distribution ratios which at concentrations below 50 g/L range from a low of 2.7×10^{-3} to a high of 15×10^{-3} (Table 10). A similar component associated with protein adsorption does not appear to exist, as the rapidly eluting counts were found to be free radioiodide. However, it is very unlikely that the dextran results could be due to free label since Chien's group determined the total amount eluted after several washes by a chemical assay and obtained results in substantial agreement with the results obtained by Brooks [9] using isotope methods. Only if the majority of the dextran in the whole suspension was hydrolyzed at the reducing end by an unrecognized reaction to produce labeled free glucose that could equilibrate with intracellular glucose, then wash out as observed, would the eluted material be identified incorrectly. The quantities involved make this possibility extremely remote.

Aggregation of red cells in dextran 150 begins at about 3 g/L and the adhesion corresponds to a surface affinity of about 2×10^{-3} erg/cm^2 (Buxbaum et al. [38]). The minimum estimate for the dextran 150 surface concentration at this bulk concentration would be about 23 molecules/μm^2 (rapidly eluted component) [74], while the maximum estimate would be 86 molecules/μm^2 [10]. Assuming the cross-bridging mechanism, the calculated affinity per mole adsorbed dextran is therefore between 12.4 and 3.3 kcal/mol. Red cell adhesion in dextran 150 becomes stronger with increasing concentration to approximately 30 g/L, reaching an affinity of 2.2×10^{-2} erg/cm^2 and then falls to zero at approximately 80 g/L [38]. The calculated affinity per adsorbed molecule is thus approximately constant up to a bulk dextran concentration of 30 g/L and decreases above this level. Presumably since adsorption is linear with concentration in this range the affinity per mole is independent of dextran concentration. The affinity per mole calculated from adhesion measurements decreases, perhaps because not all of the adsorbed molecules are contributing to adhesion. Unlike aggregation in the presence of plasma proteins, therefore, there appears to be no difficulty with the measured amounts of adsorbed dextran and the molecular affinities implied by the cross-bridging mechanism.

No conclusive evidence is available proving that cross-bridging is the mechanism of reversible red cell aggregation in dextran. However, circumstantial evidence, in particular the apparent saturation of dextran adsorption, suggests that the cross-bridging mechanism is probably the correct one in this case. Whether or not the depletion mechanism plays a significant role in erythrocyte aggregation associated with dextran depends on the extent to which adsorption counteracts the depleted concentration associated with the presence of the surface. It appears likely to us that fibrinogen and dextrans cause aggregation by different mechanisms, suggesting that these macromolecules interact differently with the surface of red cells.

ACKNOWLEDGMENTS

We thank our colleague Dr. Evan Evans for his forceful, enlightening discussion of the depletion model and for access to some unpublished results. The financial support of the Medical Research Council of Canada through award MT-5759 to D.E.B. is gratefully acknowledged.

SYMBOLS

C_c	final adsorbed amount per unit of cells
C_c'	initial adsorbed amount per unit of cells
C_p	amount per unit of cell pellet
C_s	final amount per unit of solution
C_s'	initial amount per unit of solution
C_s''	amount per unit of cell-free solution
H_e	final cell volume (or mass) fraction of cell suspension
H_i	initial cell volume (or mass) fraction of cell suspension
$\underline{H}_i$	maximum initial cell volume (or mass) fraction of cell suspension calculated assuming that P = 0
H_t	intercellular volume (or mass) fraction of a cell pellet
$\underline{H}_t$	maximum intercellular volume (or mass) fraction of a cell pellet calculated assuming that P = 0
p	volume ratio of diluent to cell water
P	final apparent adsorption distribution ratio
P'	initial apparent adsorption distribution ratio
P_j	adsorption distribution ratio of the jth species
r_{js}	fractional solution concentration of the jth species
R	parameter relating cell volume to solution tonicity
T	solution tonicity: ratio of freezing-point depressions
V_c	final cell volume (or mass)
V_c'	initial cell volume (or mass)
V_e	final cell suspension volume (or mass)
V_i	initial cell suspension volume (or mass)
V_p	cell pellet volume (or mass)
V_s	final solution volume (or mass)
V_s'	initial solution volume (or mass)
V_s''	initial cell-free solution volume (or mass)
w	erythrocyte fractional water volume

REFERENCES

1. R. Fähraeus, The suspension stability of the blood, *Physiol. Rev.* *9*:241–274 (1929).
2. K. Meyer, E. Hahnel, and R. R. Feiner, Experiments on erythrocyte sedimentation rate, *Proc. Soc. Exp. Biol. Med.* *58*:36–40 (1945).
3. G. Thorsén and H. Hint, Aggregation, sedimentation and intravascular sludging of erythrocytes, *Acta Chir. Scand. Suppl.* *154*:1–50 (1950).
4. J. Hardwicke and J. R. Squire, The basis of the erythrocyte sedimentation rate, *Clin. Sci.* *11*:333–355 (1952).
5. E. W. Merrill, G. C. Cokelet, A. Britten, and R. E. Wells, Non-Newtonian rheology of human blood: effect of fibrinogen deduced by "subtraction," *Circ. Res.* *13*:48–55 (1963).
6. S. Chien and K.-M. Jan, Ultrastructural basis of the mechanism of rouleaux formation, *Microvasc. Res.* *5*:155–166 (1973).
7. D. E. Brooks, The effect of neutral polymers on the electrokinetic potential of cells and other charged particles. IV. Electrostatic effects in dextran-mediated cellular interactions, *J. Colloid Interface Sci.* *43*:714–726 (1973).
8. G. I. Bell, M. Dembo, and P. Bongrand, Cell adhesion: competition between nonspecific repulsion and specific bonding, *Biophys. J.* *45*:1051–1064 (1984).
9. D. E. Brooks, The effect of neutral polymers on the electrokinetic potential of cells and other charged particles. III. Experimental studies on the dextran/erythrocyte system, *J. Colloid Interface Sci.* *43*:701–713 (1973).
10. S. Chien, S. Simshon, R. E. Abbott, and K.-M. Jan, Surface adsorption of dextrans on human red cell membrane, *J. Colloid Interface Sci.* *62*:461–470 (1977).
11. D. E. Brooks, Rheology interpreted through flow properties, *Microcirculation* *1*:33–52 (1976).
12. S. Chien, Aggregation of red blood cells: an electrochemical and colloid chemical problem, in *Bioelectrochemistry: Ions, Surfaces, Membranes*, Advances in Chemistry Series 188 (M. Blank, ed.), American Chemical Society, Washington, D.C., 1980, pp. 3–38.
13. B. R. Monaghan and H. L. White, Factors influencing adsorption at surfaces of red cells, *J. Phys. Chem.* *40*:1063–1070 (1936).
14. E. A. Brown, The adsorption of serum albumin by human erythrocytes, *J. Cell. Comp. Physiol.* *47*:167–176 (1956).
15. S. M. Bychkov and S. A. Kuz'mina, Role of glycosaminoglycans in red cell aggregation and adhesion, *Bull. Exp. Biol. Med. USSR* *83*:320–324 (1977).

16. J. E. Morris, Steric exclusion of cells: a mechanism of glycosaminoglycan-induced cell aggregation, *Exp. Cell. Res. 120*: 141–153 (1979).
17. D. R. Forsdyke, R. G. E. Palfree, and A. Takeda, Formation of erythrocyte rouleaux in preheated normal serum: roles of albumin polymers and lysophosphatidylcholine, *Can. J. Biochem. 60*:705–711 (1982).
18. P. A. W. Edwards, Differential cell adhesion may result from nonspecific interactions between cell surface glycoproteins, *Nature 271*:248–249 (1978).
19. S. Asakura and F. Oosawa, On interaction between two bodies immersed in a solution of macromolecules, *J. Chem Phys. 22*: 1255–1256 (1954).
20. F. Oosawa and S. Asakura, Surface tension of high-polymer solutions, *J. Chem. Phys. 22*:1255 (1954).
21. J. F. Joanny, I. Leibler, and P. G. De Gennes, Effects of polymer solutions on colloid stability, *J. Polym. Sci. Polym. Phys. Ed. 17*:1073–1084 (1979).
22 G. J. Fleer, J. H. M. H. Scheutjens, and B. Vincent, The stability of dispersions of hard spherical particles in the presence of nonadsorbing polymer, in ACS Symposium Series 240 (E. D. Goddard and B. Vincent, eds.), American Chemical Society, Washington, D.C., 1984, pp. 245–263.
23. E. Evans, D. Needham, and J. Janzen, Non-specific adhesion of phospholipid bilayer membranes in solutions of plasma proteins: measurement of the free energy potentials and theoretical concepts, in *Proteins at Interfaces: Physiochemical and Biochemical Studies*, ACS Sympsoium Series 343 (J. L. Brash and T. A. Horbett, eds.), American Chemical Society, Washington, D.C., 1987, pp. 88–102.
24. S. J. Rehfeld, D. J. Eatough, and L. D. Hansen, The interaction of albumin and concanavalin-A with normal and sickle human erythrocytes, *Biochem. Biophys. Res. Commun. 66*:586–591 (1975).
25. C. W. M. Grant, Lateral phase separations and the cell membrane, in *Membrane Fluidity in Biology*, Vol. 2 (R. C. Aloia, ed.), Academic Press, New York, 1983, pp. 131–150.
26. E. A. Evans, Minimum energy analysis of membrane deformation applied to pipet aspiration and surface adhesion of red blood cells, *Biophys. J. 30*:265–284 (1980).
27. E. Evans and K. Buxbaum, Affinity of red blood cell membrane for particle surfaces measured by the extent of encapsulation, *Biophys. J. 34*:1–12 (1981).
28. M. Bessis, Agglutination of erythrocytes, in *Living Blood Cells and Their Ultrastructure*, Springer-Verlag, New York, 1973, pp. 65–169.

29. O. Prokop and G. Uhlenbruch, *Human Blood and Serum Groups*, translated by J. L. Raven, 1969, Maclaren and Sons, London, 1969, pp. 256–265.
30. R. W. Samsel and A. S. Perelson, Kinetics of rouleaux formation. I. A mass action approach with geometric features, *Biophys. J. 37*:493–514 (1982).
31. R. W. Samsel and A. S. Perelson, Kinetics of rouleaux formation. II. Reversible reactions, *Biophys. J. 45*:805–824 (1984).
32. S. Rowlands and L. Skibo, The morphology of red-cell aggregates, *Thromb. Res. 1*:47–58 (1972).
33. J. S. K. Fung and P. B. Canham, The mode and kinetics of the human red cell doublet formation, *Biorheology 11*:241–251 (1974).
34. G. R. Cokelet, E. W. Merrill, E. R. Gilliland, H. Shin, A. Britten, and R. E. Wells, Jr., Rheology of human blood: measurements near and at zero shear rate, *Trans. Soc. Rheol. 7*: 303–317 (1963).
35. J. Goldstone, H. Schmid-Schönbein, and R. Wells, The rheology of red blood cell aggregates, *Microvasc. Res. 2*:273–276 (1970).
36. S. Chien, L. A. Sung, S. Kim, A. M. Burke, and S. Usami, Determination of aggregation force in rouleaux by fluid mechanical technique, *Microvasc. Res. 13*:327–333 (1977).
37. M. D. Hollenberg, and P. Cuatrecasas, Distinction of receptor from nonreceptor interactions in binding studies, in *The Receptors: A Comprehensive Treatise*, Vol. 1 (R. D. O'Brien, ed.), Plenum Press, New York, 1979, pp. 193–214.
38. K. Buxbaum, E. Evans, and D. E. Brooks, Quantitation of surface affinities of red blood cells in dextran solutions and plasma, *Biochemistry 21*:3235–3239 (1982).
39. B. R. Monaghan and H. L. White, Effect of proteins on electrophoretic mobility and sedimentation velocity of red cells, *J. Gen. Physiol. 19*:715–726 (1936).
40. R. D. O'Brien, Problems and approaches in non-catalytic biochemistry, in *The Receptors: A Comprehensive Treatise*, Vol. 1 (R. D. O'Brien, ed.), Plenum Press, New York, 1979, pp. 311–335.
41. M. Dalmark and J. O. Wieth, Temperature dependence of chloride, bromide, iodide, thiocyanate and salicylate transport in human red cells, *J. Physiol. 224*:583–610 (1972).
42. Z. I. Cabantchik and A. Rothstein, Membrane proteins related to anion permeability of human red blood cells. I. Localization of disulfonic stilbene binding sites in proteins involved in permeation, *J. Membr. Biol. 15*:207–226 (1974).
43. E. Ponder, *Hemolysis and Related Phenomena*, Grune & Stratton, New York, 1948, pp. 83–101.

44. E. Ponder, Compression of human red cells during centrifuging, *Nature 210*:527 (1966).
45. J. A. Sirs, The measurement of the haematocrit and flexibility of erythrocytes with a centrifuge, *Biorheology 5*:1–14 (1968).
46. J. W. Dellimore and R. G. Gosling, Compression of erythrocytes during centrifugation, *Biorheology 10*:325–328 (1973).
47. E. Evans and Y.-C. Fung, Improved measurements of the erythrocyte geometry, *Microvasc. Res. 4*:335–347 (1972).
48. H. Chaplin, Jr., and P. L. Mollinson, Correction for plasma trapping in the red cell column of the haematocrit, *Blood 7*: 1227 (1952).
49. J. M. England, D. M. Walford, and D. A. W. Waters, Re-assessment of the reliability of the haematocrit, *Br. J. Haematol. 23*:247–256 (1972).
50. C. J. Bellis and F. H. Scott, The alteration of protein distribution, in vitro, between corpuscles and plasma caused by isosomotic and hyperosmotic solutions, *J. Biol. Chem. 111*: 17–44 (1935).
51. Y. Kikuchi and T. Koyama, Red blood cell deformability and protein adsorption on the red blood cell surface, *Am. J. Physiol. 247*:H739–H747 (1984).
52. R. F. Furchgott and E. Ponder, Disk-sphere transformation in the mammalian red cells. II. The nature of the antisphering factor, *J. Exp. Biol. 17*:30–44 (1940).
53. D. L. Traber and S. N. Kolmen, Adsorption of fibrinogen onto erythrocytic surfaces, *Tex. Rep. Biol. Med. 23*:782–792 (1965).
54. J. Janzen, Fibrinogen adsorption to human erythrocytes. Ph.D. thesis, University of British Columbia, Canada, 1985.
55. D. L. Doxey, Haemotology of the ox, in *Comparative Clinical Haematology* (R. K. Archer and L. B. Jeffcott, eds.), Blackwell Scientific, London, 1977, pp. 215–269.
56. T. Peters, Jr., Serum albumin, in *The Plasma Proteins*, 2nd ed. Vol. 1 (F. W. Putnam, ed.), Academic Press, New York, 1975, pp. 133–181.
57. S. Chien, R. J. Dellenback, S. Usami, and M. I. Gregersen, Plasma trapping in hematocrit determinations: differences among animal species, *Proc. Soc. Exp. Biol. Med. 119*:1155–1158 (1965).
58. N. W. Spurling, The haematology of the dog, in *Comparative Clinical Haematology* (R. K. Archer and L. B. Jeffcott, eds.), Blackwell Scientific, London, 1977, pp. 365–440.
59. C. Cohen, J. W. Weisel, G. N. Phillips, Jr., C. V. Stauffacher, J. P. Fillers, and E. Daub, The structure of fibrinogen and fibrin. I. Electron microscopy and x-ray crystallography of fibrinogen, *Ann. N.Y. Acad. Sci. 408*:194–213 (1983).

60. R. C. Lief and J. Vinograd, The distribution of buoyant density of human erythrocytes in bovine albumin solutions, *Biochemistry 51*:520–528 (1964).
61. S. Margen and H. Tarver, The deiodination of proteins labelled with I^{131}, *N.Y. Acad. Sci. 70*:49–55 (1957).
62. R. E. Crandall, J. Janatova, and J. D. Andrade, The effects of radioiodination and fluorescent labelling on albumin, *Prep. Biochem. 11*:111–138 (1981).
63. O. N. Vazquez, K. Newerly, R. S. Yalow, and S. A. Berson, Determination of trapped plasma in the centrifuged erythrocyte volume of normal human blood with radioiodinated (I^{131}) human serum and radiosodium (Na^{24}), *J. Lab. Clin. Med. 39*:595–604 (1952).
64. M. Maizels and M. Remington, Percentage of intercellular medium in human erythrocytes centrifuged from albumin and other media, *J. Physiol. 145*:658–666 (1959).
65. J. Funder and J. O. Wieth, Trapping of sodium, potassium, sucrose, and albumin in the packed cell column of the hematocrit, *Acta Physiol. Scand. 71*:105–112 (1967).
66. M. A. Cohen Stuart, J. M. H. M. Scheutjens, and G. J. Fleer, Polydispersity effects and the interpretation of polymer adsorption isotherms, *J. Polym. Sci. Polym. Phys. Ed. 18*:559–573 (1980).
67. O. N. Vazquez, K. Newerly, R. S. Yalow, and S. A. Berson, Estimation of trapped plasma with I^{131} albumin; critique of methods, *J. Appl. Physiol. 6*:437–440 (1954).
68. N. C. Hughes-Jones and B. Gardner, The exchange of ^{131}I-labelled lipid and ^{131}I-labelled protein between red cells and serum, *Biochem. J. 83*:404–413 (1962).
69. H. E. Müller and F. Gramlich, Die normale zusammensetzung des plasmaproteinfilmes an der oberfläche menschlicher erythrocyten, *Hoppe-Seyler's Z. Physiol. Chem. 334*:103–115 (1963).
70. M. W. Rampling, The binding of fibrinogen and fibrinogen degradation products to the erythrocyte membrane and its relationship to haemorheology, *Acta Biol. Med. Ger. 40*:373–378 (1981).
71. S. Levine, M. Levine, K. A. Sharp, and D. E. Brooks, Theory of the electrokinetic behavior of human erythrocytes, *Biophys. J. 42*:127–135 (1983).
72. J. Viitala and J. Järnefelt, The red cell revisited, *Trends Biol. Sci. 118*:392–395 (1985).
73. S. Rothman, E. Adelson, A. Schwebel, and R. D. Langdell, Adsorption of carbon-14 dextran to human blood platelets and red blood cells in vitro, *Vox Sang. 2*:104–109 (1957).
74. D. E. Brooks, R. G. Grieg, and J. Janzen, Mechanisms of erythrocyte aggregation, in *Erythrocyte Mechanics and Blood*

Flow (G. R. Cokelet, H. J. Meiselman, and D. E. Brooks, eds.), Alan R. Liss, New York, 1980, pp. 119–140.

75. H. E. Müller and F. Gramlich, Uber den plasmaproteinfilm an der oberfläche menschlicher erythrocyten, *Acta Haematol. 34*: 239–249 (1965).
76. S. B. Hlady and T. J. Rink, The use of ion transporters, pH measurements and light scattering with red blood cells, in *Red Cell Membranes: A Methodological Approach*, Biological Techniques Series (J. C. Ellory and J. D. Young, eds.), Academic Press, London, 1982, pp. 335–358.
77. K.-M. Jan, Electrochemical basis of heparin-induced red blood cell aggregation, in *Bioelectrochemistry: Ions, Surfaces, Membranes*, Advances in Chemistry Series 188 (M. Blank, ed.), American Chemical Society, Washington, D.C., 1980, pp. 143–149.
78. J. J. Christensen, J. Ruckman, D. J. Eatough, and R. M. Izatt, Determination of equilibrium constants by titration calorimetry. I. Introduction to titration calorimetry, *Thermochim. Acta 3*:203–218 (1972).
79. J. L. Brash and Q. M. Samak, Dynamics of interaction between human albumin and polyethylene surface, *J. Colloid Interface Sci. 65*:495–504 (1978).
80. A. W. Adamson, *Physical Chemistry of Surfaces*, 3rd ed., Wiley, New York, 1976, pp. 548–552.
81. E. A. Evans, Detailed mechanics of membrane–membrane adhesion and separation. II. Discrete kinetically trapped molecular cross-bridges, *Biophys. J. 48*:185–192 (1985).
82. A. P. Gast and L. Leibler, Interactions of sterically stabilized particles suspended in a polymer solution, *Macromolecules 19*: 686–691 (1986).
83. K. A. Granath, Solution properties of branched dextrans, *J. Colloid Sci. 13*:308–328 (1958).

III

Corpuscular

10

Electroporation

JAMES C. WEAVER Harvard-MIT Division of Health, Sciences, and Technology, Massachusetts Institute of Technology, Cambridge, Massachusetts

I.	Introduction	253
II.	Overview of Electroporation	254
III.	Significance of Electroporation	258
IV.	Representative Experimental Conditions	260
	A. Experimental observations involving electrical measurements	263
	B. Experimental observations using cells	265
	Summary	269
	Symbols	270
	References	271

I. INTRODUCTION

Application of electrical fields, particularly strong, short-lasting electric fields, causes a variety of dramatic cell membrane effects. Some involve molecules in or on the membrane, but most involve transmembrane movement of ions and molecules. The transmembrane effects include a rapid discharge of the transmembrane potential, U, and the development of a longer-lasting, by transient, high-permeability state which allows movement of molecules of many sizes. Often, electroporation effects take place without prompt cell death, but delayed cell death may occur for some cells, while others recover completely. Although the basic mechanisms are not fully understood, there is considerable evidence that transient membrane openings or "pores" are involved. Further, although some electroporation phenomena such as reversible electrical breakdown (REB) can occur in pure, artificial bilayer membranes without membrane proteins and can be reasonably described by a transient pore theory, there is evidence that membrane proteins participate in cell membrane electroporation. The occurrence of electroporation has fundamental implications for understanding of cell membranes and is also likely to lead to widespread technical applications that extend well beyond introduction of genetic material into cells.

II. OVERVIEW OF ELECTROPORATION

Qualitatively, the sequence of events in Fig. 1 summarizes our present understanding of the major events relating to transmembrane movement of ions and molecules due to electroporation of a cell membrane [1,2]. A more detailed but still brief description of these events is given below, where for convenience we assume a spherical cell (a reasonable representation for suspension cultured mammalian cells, protoplasts, and some microorganisms). The literature references cited provide several illustrations of what is presently known about these events.

1. A "high-voltage" pulse of magnitude $\Delta\phi_{applied}$ is applied across two electrodes. Usually, a parallel planes configuration with separation L is used to contain a cell suspension, so that at low cell densities (so that nearby cells do not significantly perturb the field) each cell experiences a nearly uniform applied field. The resulting *nominal electric field* is $E_{nominal} \equiv \Delta\phi_{applied}/L$, but this is not the field actually experienced by the cells. Instead, the actual field is generally less. An electric current, I(t), flows within the aqueous electrolyte comprising the suspending medium. The corresponding current density is $j(t) = I(t)/A_{electrode}$, where $A_{electrode}$ is the macroscopic surface area of one of the two electrodes. In turn, there is an associated electric field, $E_e = j/\sigma_e$, where σ_e is the electrical conductivity of the medium. Because significant potential drops can occur at the electrode—electrolyte interface [4—6], a general condition is $E_e \le E_{nominal}$. If the electrode configuration is not plane parallel, interpretation of results will be difficult, as E_e will be nonuniform, with the consequence that different cells will experience different applied fields. E_e, the field within the electrolyte, is the relevant field, as E_e is the field actually applied to the cells.

2. The transmembrane potential, U(t), increases rapidly with time, but with U(t) having different values at different locations on the cell membrane. For a spherical cell at low values of U(t), the relationship between the applied field and the *change* in transmembrane

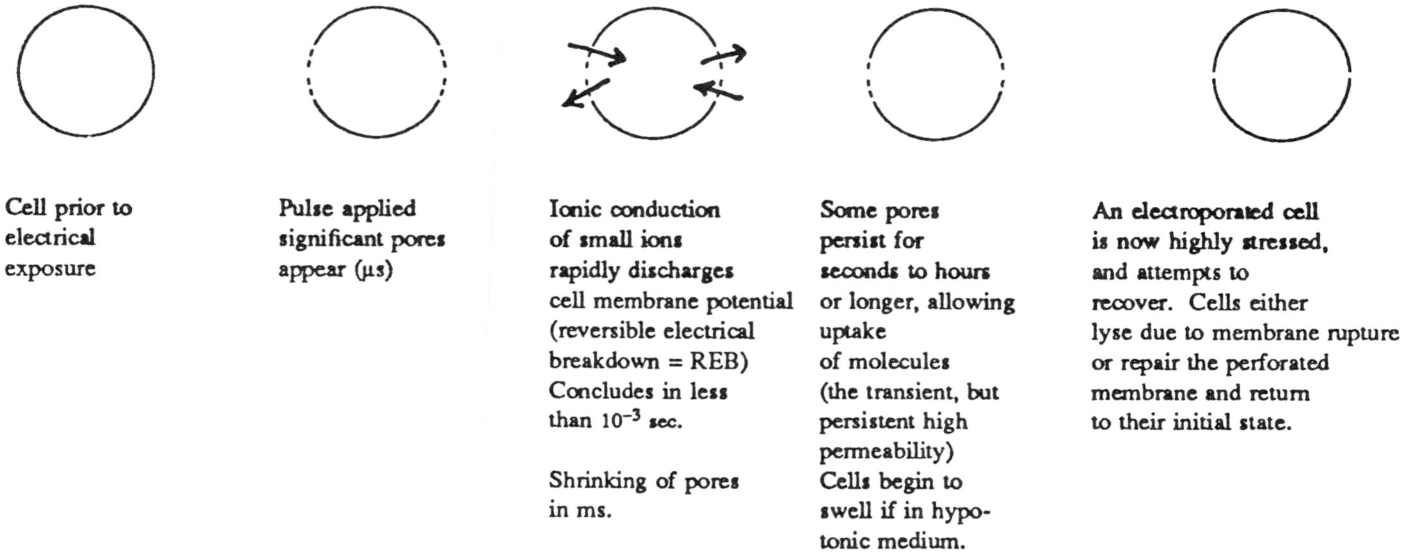

FIG. 1 Basic sequence of electroporation events involving transmembrane movement of ions and molecules. (From ref. 3.)

potential, $\Delta U(t,\theta)$ at different locations on the membrane is given by the well-known formula [7]

$$\Delta U(t,\theta) \approx 1.5E(t)_e R_{cell} \cos\theta \tag{1}$$

where R_{cell} is the cell's radius and θ is the angle between the direction of E_e and the location on the membrane. The maximum change occurs at the poles ($\cos\theta \pm 1$), so that a widely used low-field estimate is

$$\Delta U(t)_{max} \approx 1.5E(t)_e R_{cell} \tag{2}$$

If the cell resting transmembrane potential was U_0, then $U(t) \approx U_0 + \Delta U(t)$, although this approximation neglects the perturbation of the resting potential by the changes in permeability caused by electroporation. Dramatic electroporation effects are believed to occur if $\Delta U(t)_{max}$ reaches values of about 500 mV to 1500 mV for short pulses. Electroporation at smaller elevated values of U(t) which occur for longer times has not yet received much attention.

3. Pores or some type of membrane openings rapidly develop in the sense that large numbers and/or large sizes of transmembrane channels appear if $500\ \text{mV} < U(t) < 1500\ \text{mV}$. Such pores are generally capable of passing both ions and molecules, but with specificity and size cutoff that is not yet well established. Macromolecules such as proteins and DNA, or small particles such as latex beads [1] or viruses [8], begin to cross the membrane at this stage. While U(t) is elevated, charged molecules may move across the membrane by virtue of their electrical mobility and the strong electric fields within the aqueous pathways provided by pores. Molecular transport by hindered diffusion may also occur.

4. Reversible electrical breakdown (REB) occurs; that is, there is a very rapid decrease in membrane resistance, R(t), during which the membrane capacitance, C, hardly changes [9], such that the membrane is rapidly discharged with a RC time constant controlled by the changing R(t). This discharge is usually completed rapidly, within 10^{-3} s and sometimes within 10^{-6} to 10^{-5} s.

5. The temperature T_e of the medium rises at a tremendous rate (e.g., 10^5 degs s^{-1}), but the net temperature change is often negligible (e.g., 1°C) because of the short duration of the applied electric field. Clearly, the occurrence of significant temperature rises can greatly complicate the interpretation of results, as many proteins denature at temperatures in excess of about 45°C, and a heat shock response may also be induced. The use of a low initial temperature (e.g., slightly above 0°C) is an obvious way to accommodate modest ohmic heating. The use of special, low-electrical-conductivity media is also possible, but the associated cell stress (see below) due to

extracellular–intracellular molecular exchange during and after electroporation must also be considered.

6. A transient high-permeability state can exist (Fig. 2), which persists for a much longer time, often seconds to minutes to hours, depending on cell type and on temperature. This high permeability appears to allow transport of all types of ions and molecules, ranging from the ubiquitous small ions to macromolecules such as proteins and nucleic acids. The maximum size of transported entities is not yet known, nor is the time range at which high permeability for the larger molecules vanishes.

7. An electroporated cell becomes highly stressed because of a loss of intracellular compounds and/or an influx of extracellular compounds [5,10,11]. As a result of such chemical imbalances, a variety of cellular functions can become temporarily or permanently disrupted. Electroporation-induced cell stress is poorly understood at present.

8. Recovery of the membrane occurs following cessation of the electric pulse and after elevated transmembrane potentials have returned to normal or low values. Pores gradually shrink and disappear so that the cell membrane excludes progressively smaller

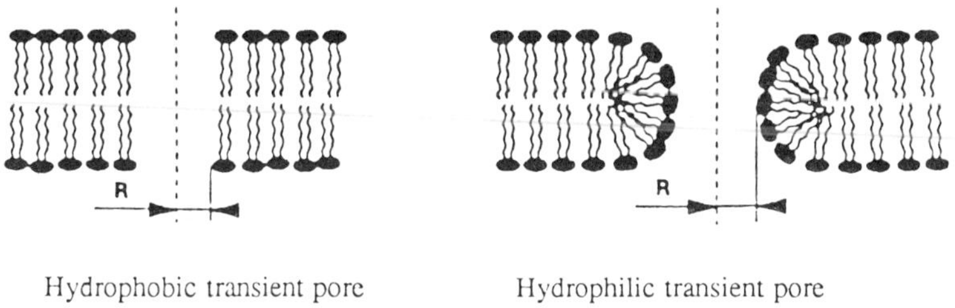

FIG. 2 Hypothetical structures of transient pores in the bilayer portion of a cell membrane. As suggested by Chizmadzhev and coworkers [22,23], the hydrophobic pores can be hypothesized to be of any size, with radii from zero upward, whereas in contrast, the hydrophilic pores are believed to have a minimum size that is related to the phospholipid head group size and the need to involve several head groups to obtain a pore. Other possible transient structures for hydrophilic pores involve a block complex [24], or in the case of cell membranes may involve membrane macromolecules.

molecules and ions at smaller and smaller rates. Persistent pores, even very small pores that can pass small ions, may result in long-term cell stress and cell lysis [5,10,11]. The kinetics of membrane recovery or "pore resealing" is poorly understood at present.

9. Cell recovery and survival can occur in a significant percentage of electroporated cells, while the others die, apparently because of osmotic lysis, unrelieved taxing of metabolic systems by the membrane shunting of persistent small pores, or other uncorrected imbalances [10,12–14]. Surviving cells retain the electroporatively introduced molecules if the molecules do not permeate the recovered membrane and are not degraded intracellularly [15–17]. These events are inferred from some of the reported electroporation studies, which are few in number compared to the general cell membrane literature. Perhaps significantly, many of the electroporation studies have been carried out on erythrocytes or their ghosts, or on artificial planar bilayer membrans [16,18–20], whilc the application of electroporation to gene transfer (which typically determines only the biological endpoint of transformation, and therefore reveals relatively little about early electroporation events), has utilized a much wider range of cell and protoplast types [1,2,21].

III. SIGNIFICANCE OF ELECTROPORATION

New electrical phenomena with application to biology and medicine are discovered only occasionally, but once understood and methodically applied, they can have major impacts. Naturally occurring phenomena of established significance include neural action potentials, electrocardio potentials, myoelectric action potentials, and other active cellular potentials. Important electrical phenomena caused by applied electrical stimuli include electrophoresis, microelectrodes, electrochemical sensors (e.g., pH, O_2), and membrane potential fluorescent dyes. A tremendous effort has gone into the investigation and application of these electrical phenomena, so that it is worthwhile to ask briefly why externally stimulated electrical phenomena are important. Each phenomenon has its individual attributes, but generally electric currents and potentials have the feature of being relatively easy to apply, manipulate, measure, and control. In short, in comparison to chemical phenomena, electrical phenomena are readily caused and measured.

The external fields that cause dramatic electroporation effects can generate an electric field in excess of the nominal dielectric strength of the cell membrane. However, that actual energy available is zeU, which is seldom more than 1 eV, and therefore incapable of causing a classic avalanche breakdown [3,25,26]. Further, the membrane of a cell in a pathological state may be much more susceptible, with the consequence that even a normal physiological transmembrane potential may result in electroporation. Relatively little investigation of the involvement of electroporation of pathologic cell membranes in disease states of an organism has been carried out (Tsong, priv. com.).

At present, electroporation is poorly understood in detail, but appears to occur universally in cell and artificial bilayer membranes, and therefore can probably be used almost universally to temporarily elevate cell membrane permeability. For this reason, electroporation is likely to have a significant impact on:

1. *Understanding of basic cell membrane behavior*, as the very existance of electroporation phenomena poses challenges to understanding of fundamental cell and artificial bilayer membranes. Knowledge of the properties of membranes which allow electroporation to occur, and to vary, will be important.
2. *Introduction and expression of foreign genes in cells*, as electroporation is used increasingly to insert genes in cells difficult to transform by other means [1,2,21], and these early applications have demonstrated several advantages that may make electroporation suitable as a general method, including use in gene therapy.
3. *Possible historical role for gene transfer early in evolution*, as primitive cells in close proximity may have experienced electroporation because of nearby lightning strikes [1]. This is a tantalizing possibility, and it would be of interest to estimate the frequency of occurrence of such transfers, to determine whether this could have been significant.
4. *Introduction of other important macromolecules into cells*, such as enzymes [15], other proteins [27,28], fluorescence-labeled probes [5,28–30], antibodies [31], and subcellular particles is clearly important [1,8]. The ability to carry out intracellular assays, interventions, and manipulations by adding molecules to cells has the potential to be broadly important in biological research and in applications.
5. *Tissue electroporation*, as this has been implicated as a damage mechanism in muscle injury associated with electrical trauma [32–34], but it is not clear what the relative roles of prompt and delayed membrane rupture may be [26]. A simplified approach to tissue electroporation considers sheets of cells connected by tight junctions. Pores develop because of transtissue electric fields, which in turn allows transport of ions and molecules. A number of early studies observed a reversible electrical breakdown in frog skin, an almost monolayer tissue model system [35,36]. Tissue electroporation allows large numbers of cells to be studied simultaneously, with all cells exposed to

similar transcellular potentials. This allows average electrical behavior to be examined and allows estimates of the frequency of occurrence of some possible types of damage [37].

6. *Electrical control of cellular function* may be possible by several general mechanisms: (1) preexisting channels [38], (2) electroconformation of membrane macromolecules [39,40], (3) surface redistribution of membrane molecules [41], and (4) pores that are created spontaneously by the combined influence of thermal fluctuations and increased U(t) [42]. The latter is electroporation. Although the emphasis to date has been on strong field electroporation, the possibility of causing electroporation by prolonged application of moderate to weak fields [43] deserves greater emphasis.
7. *Electroinsertion of macromolecules into cell membranes* has recently been demonstrated [44], under suboptimal conditions for transporting molecules across membranes. A tentative explanation is that smaller hydrophobic pores are involved [22,23], which have been hypothesized to be precursors of the hydrophilic pores, which presumably allow ion and molecular transport. This explanation is consistent with an absence of cell damage, because if predominantly hydrophobic pores are involved, negligible molecular transport across the membrane is expected.

Overall, electroporation appears to be short-term reversible, often nonlethal to cells, and a general way of transiently manipulating cell membrane structure by conveniently applied electrical fields, so that the long-term potential for applications is enormous.

IV. REPRESENTATIVE EXPERIMENTAL CONDITIONS

Typical conditions are described for the usual strong dc-field, short-duration cases, although RF (radio-frequency) fields may also be important [43,45]. Cells to be electroporated are usually prepared in suspension, allowing better control of applied field conditions, but cells can also be present on insulating solid surfaces such as plastic or glass [46]. Usually, cells are exposed briefly to a large applied electric field of magnitude and direction $\vec{E}_e$. The applied field is thus a pulsed electric field (PEF), generally caused by applying a large potential difference ("high voltage"), $\Delta U_{applied}$, across two metal electrodes. Usually it is desired to expose all of a suspension of cells to the same field, so that some approximation to parallel plane electrodes is used. For example, flat stainless steel or platinum electrodes separated by a distance $L \approx 0.1$ to 1 cm can be used, so that at first glance the applied field magnitude is $E_{nominal} = \Delta U_{applied}/L$. However, the field within the aqueous electrolyte of the suspending medium E_e is generally smaller, due to significant potential drops at the two electrode–electrolyte interfaces. For example, in the case of stainless steel electrodes, the electrolyte phosphate buffered saline (PBS; pH 7) and the geometry of a small electroporation chamber, approximately 40% of $\Delta U_{applied}$, can appear across the electrode–electrolyte interface region [6], so that in this case cells are exposed to 60% of $E_{nominal}$. Although often neglected, it can often be important to determine the actual applied field by additional measurements: for example, if electroporation conditions are to be duplicated or controlled, and the geometry, electrode material, or suspending medium composition is to be significantly altered.

As an example, consider the square-wave pulse. The total current, I, passing through the electroporation chamber can be measured by placing a small (e.g., 20 Ω) resistor in series with the chamber and using an oscilloscope to record the voltage pulse across this resistor. Then, by using the chamber geometry and the medium conductivity, $\sigma_e = 1/\rho_e$, where ρ_e is the medium resistivity, the actual field E_e is calculated. Specifically, for a parallel-plane chamber with electrodes of area $A_{electrode}$ and separation L, the chamber resistance is $R_{chamber} = \rho_e L/A_{electrode}$. Thus the potential drop ΔU_e across the electrolyte is $\Delta U_e = IR_{chamber}$, and the associated field is $E_e = \Delta U_e/L$, that is,

$$E_e = \frac{I}{\sigma_e A_{electrode}} \tag{3}$$

which is independent of L and is therefore obtained by measuring I, $A_{electrode}$, and σ_e, and then computing E_e.

The field E_e causes an ohmic current of density J_e to flow, which can usually be regarded as instantaneously following E_e because the charge relaxation time [47], $\tau_Q = K_e \varepsilon_o/\sigma_e$, is very short. For example, in physiologic media τ_Q is of order 10^{-9} s, which is small even compared to the shortest pulses used to cause electroporation. Also, it is generally acceptable to use low field conductivity values, as E_e tends to be in the range 1 to 10 kV/cm (10^5 to 10^6 V/m), whereas the (first) Wien effect, in which the conductivity σ_e increases in high fields, occurs at fields of about 100 kV/cm (10^7 V/m) [48]. Typical media have electrical conductivity close to that of physiological saline, that is, $\sigma_e \approx 1.43 \times 10^{-2}\ \Omega^{-1}\ cm^{-1} = 1.43$ S [49], so that the corresponding resistivity is $\rho_e = 1/\sigma_e = 7 \times 10^{-1}\ \Omega$ m. For applied field magnitudes of $E_e \approx 1$ to 10 kV cm^{-1} (100 to 1000 kV m^{-1}) the resulting electrical power density $\rho_e J^2 = \sigma_e E_e^2$ is very large, about 1×10^{10} W $m^{-3} = 1 \times 10^4$ W cm^{-3} for 1 kV cm^{-1}, and a factor of 100 larger for 10 kV cm^{-1}. This average power dissipation occurs in the extracellular medium through the entire cell suspension, with significant heat removal possible only at the surfaces of the suspension. Slightly larger power densities occur near cells, as the consequence of cells perturbing the otherwise uniform field. As a result there is a rapid average temperature rise rate

$$\overline{\frac{dT_e}{dt}} = \frac{\sigma_e E_e^2}{\rho_w c_w} \tag{4}$$

where for aqueous media $\rho_w \approx 10^3$ kg m^{-3} is the mass density, and $c_w \approx 4.18 \times 10^3$ J kg °C is the specific heat. For 10 kV cm^{-1} the temperature rises at the tremendous rate of about 2×10^5 °C s^{-1}, but is 100-fold less at 1 kV cm. As an example, for a 10-kV/cm, 50-μs "square-wave" pulse in which the applied field rises essentially instantly from zero to magnitude E_e, remains for 50 μs and drops instantly again to zero, the resulting temperature rise, ΔT_e, is about 10°C, but is only 0.1°C for E_e = 1 kV cm^{-1}. Many proteins denature upon prolonged exposure to $T_e > 45$°C, which is also an approximate threshold for hyperthermic killing of mammalian cells. Depending on the initial cell suspension temperature and the type of cell used, the higher value of 10°C may or may not be damaging, and therefore describes approximately a thermal limit for the applied field pulse.

A similar limit exists for any shape electrical pulse. Two types of pulses are generally used, the square-wave pulse described above, and an exponentially decaying pulse for which the applied potential has the behavior

$$\Delta U(t)_{applied} = \Delta U_{applied,0} e^{-t/\tau_{RC}} \tag{5}$$

where the time constant, τ_{RC}, is a constant for any one pulse, but can often be varied prior to initiating the pulse. Usually, C can be varied by adjustment of the pulse generator, but R is the total electrical resistance driven by the generator, and therefore determined by the resistance of the electroporation chamber, which in turn is dependent on the medium and cell suspension. The average cell suspension heating associated with an exponential pulse is

$$\overline{\Delta T_e} = \int_0^{\infty} \frac{\sigma_e E_e^2(t)}{\rho_w c_w} \, dt \tag{6}$$

Provided that the local potential drop at the metal electrode/suspending medium interface is either negligible (often not true) or is a constant fraction of the total applied potential, $\Delta U(t)_{applied}$, the decaying field magnitude, $E_e(t)$, is proportional to $\Delta U(t)_{applied}$. Assuming for simplicity that $E_e(t) \approx \Delta U(t)_{applied}/L$, we find that

$$\overline{\Delta T_e} \approx \frac{\sigma_e E_0^2 \tau_{RC}}{2\rho_w c_w} \tag{7}$$

where the factor $\tau_{RC}/2$ replaces the pulse width, Δt, of the square wave. As an example, mammalian cell electroporation used for DNA introduction typically involves maximum applied fields of $E_0 \approx 1$ kV cm^{-1} or less, with maximum time constants of about $\tau_{RC} \approx 10$ ms = 10^{-2} s, so that a typical maximal temperature rise is about 0.1°C. Thus, although long-lasting high-intensity fields could cause a significant temperature rise, usually the combination of initial temperature and small $\overline{\Delta T_e}$ results in insignificant heating, and for this reason electroporation phenomena are generally referred to as nonthermal. Clearly, protocols that involve long pulses at high field strengths should be explicitly examined to determine whether a significant $\overline{\Delta T_e}$ occurs.

A. Experimental Observations Involving Electrical Measurements

Based on experiments with artificial planar bilayers [50–54], with isolated nodes of Ranvier [55], with electrode-impaled algal cells [52], the squid axon [56], and with isolated frog skin [35–37] it is believed that the cell membrane responds to the marked increase in U by rapidly decreasing its electrical resistance, often resulting in a very rapid discharge of the membrane. This inability of the membrane to support a large value of U has generally been referred to as *reversible electrical breakdown*, although the term *dielectric breakdown* has also been used. However, actual dielectric breakdown is impossible, as conventional dielectric breakdown involves the creation of ionized molecules which then conduct, often by a cascade ionization and amplification process. For this reason "dielectric breakdown" is misleading terminology. There is already a well-established usage of "dielectric breakdown" which relates to ion-pair avalanche breakdown phenomena (e.g., gas discharge breakdown at high field strengths). This is a very different phenomenon which involves significantly larger energy [i.e., about 10 V (1.6×10^{-18} J) per monovalent ion] [25]. In contrast, the about 500- to 1500-mV transmembrane potential associated with REB corresponds to about 0.5 to 1.5 eV per monovalent ion, about an order of magnitude less energy, and this is generally insufficient for ion-pair formation. Rather than "breakdown," the idea of structural membrane "rearrangement" [57,58] is probably a more useful concept.

Formation, growth, shrinkage, and destruction of transient pores all form a general membrane rearrangement possibility. Theoretical models based on transient aqueous pores provide an explanation of rupture and REB [22,23,59–62], so that appeal to "dielectric breakdown" appears unnecessary.

One hypothesis is that during electroporation (1) many significant (in size and number) aqueous openings or pores are either created and then rapidly expand because of a large ΔU, or (2) many very small pores are always present (e.g., as the result of thermal fluctuations) and these preexisting pores expand rapidly because of the electrical pressure associated with a large ΔU. In either case, the result of a large ΔU value is the rapid appearance of many large pores, which provide a pathway for the movement of large numbers of small ions, a discharge current that is responsible for REB. The appearance of ionically conducting large pores renders the membrane incapable of supporting the large ΔU. Or put another way, the appearance of many large pores results in a sudden change in membrane resistance from high to low, and the membrane then promptly discharges with a "time constant" governed by the membrane resistance R and membrane capacitance C [51,61,63]. Based on careful experiments with artificial planar bilayer membranes, it was found that C changes negligibly [64], and therefore that essentially all of the dramatic reversible electrical breakdown behavior is due to a tremendous, rapid decrease in membrane electrical resistance R.

Biological membranes are structurally much more complex than artificial bilayer membranes, and for this reason electroporation mechanisms in cell membranes are probably more complex. Electroporation occurring in the lipid domain will probably closely resemble that in lipid bilayers and in lipid vesicles [20,65], although it is not yet resolved whether the prompt rupture observed in planar bilayer membranes can occur in the closed membranes of vesicles and cells [24]. However, in a cell membrane electric activation of channels and pumps may also lead to electroporation, as the local membrane field, $E_m \equiv U/d$, exceeds the normal value associated with the resting potential [39,66,67]. It is highly suggestive that some experiments have shown that 30% of the membrane conductance due to electroporation in human erythrocytes was blocked by oubain, a potent inhibitor of [Na,K] = ATPase [43,65,66]. Additional studies have shown that this enzyme can be activated by an applied field, although if the field strength exceeds normal physiological values, the enzyme may be denatured. Such electroactivation of membrane channels and pumps may involve denaturation, which in turn is involved in the formation of semipermanent pores [68–70]. These investigations have led to the suggestion that *electroconformational change* of membrane proteins and lipids plays a role in the pore initiation process, and the thermal effect arising from the subsequent intense transmembrane current is responsible for pore expansion. This hypothetical process is represented by

$$P \longleftrightarrow P^* \longrightarrow [P]$$

where P is a membrane protein under normal physiologic conditions, P* is an electrically altered conformation of the protein which aids pore formation, and [P] is the protein in a denatured state due to microscopically local temperature rises due to electrical conduction through the pore. In this scheme, the first step is an electric field–induced reversible channel opening, and the second step is a thermally induced irreversible denaturation of the proteins. Thus the first step is a physiological event, while the second is artificial, caused by the applied field.

B. Experimental Observations Using Cells

To our knowledge, the first report of an observation relating to electroporation appeared in 1958 and consisted of finding reversible electrical breakdown in the node of Ranvier [55]. Other early reports of a breakdown leading to very low electrical membrane resistance used systems such as *Nitells* [71] and frog skin [35,36]. Later it was reported that a cell suspension experienced cell killing upon application of short but intense electric fields, but without significant temperature rise [12]. This nonthermal cell death is now believed to be due to electroporation, most likely the case in which a very large field magnitude and/or long pulse duration results in the occurrence of persistent pores that lead to cell stress and eventual cell death.

The first report of a direct nonelectrical consequence of electroporation described molecular release across vesicular membranes [18]. Subsequent analysis of these experiments supports the interpretation that transmembrane potential changes of order 1 V had occurred [72,73]. Other experiments with red blood cells (RBCs) in a modified Coulter counter subsequently showed that the cells experienced a dramatic decrease in transcellular resistance as the cells passed through the strong electric field of the apparatus, and this was interpreted as being due to a significant change in cell membrane resistance, and therefore as evidence that ubiquitous small ions such as Na^+ and Cl^- then readily cross the RBC membrane [19]. Subsequent studies with RBCs using a different experimental configuration showed that macromolecules such as albumin (about 70 kDa) and the enzyme urease (about 400 kDa) were taken up and retained by RBC ghosts [15,27].

The use of strong, short-lasting electric fields to form RBC ghosts, the biological vesicles that result from emptied RBCs, was also reported in several experiments [13–15,27,74]. An explicit interpretation based on pore formation, as opposed to dielectric breakdown, was then advanced [10]. In this study the extent of hemolysis was compared to the electrical heating of the medium and also to the field strength, with the conclusion that E_e caused large changes in U(t), which then resulted in lysis. The possibility that bulk Joule heating had caused a large $\overline{\Delta T_e}$ was ruled out. This study also showed that a pulse of several kV/cm applied to an isotonic RBC suspension resulted in a dramatic increase in permeability to Na^+ and K^+ ions while remaining impermeable to sucrose and hemoglobin. Surprusingly, however, when essentially the same experiment was carried out with a different composition isotonic medium (mixture of isotonic saline and isotonic sucrose), the RBCs became permeable to sucrose. Except for extreme (very large or very long duration) pulses, the RBC membranes remained impermeable to hemoglobin and enzymes. In contrast to the then often-reported dielectric breakdown [15,27,75–79], these permeability differences led these investigators to propose that pores of limited size had been created [10]. The implausibility of true dielectric breakdown, on simple energetic grounds, was emphasized earlier in our discussion. Further, as was first qualitatively conjectured [59], it has now been shown quantitatively that a theory based on transient aqueous pores can account for the very rapid electrical discharge termed *reversible electrical breakdown*, and that this is a relatively low energy event [61–63,80,81].

Additional support for the creation of some types of aqueous pores was provided in this and related studies [10,16,82]. The studies demonstrated that the PEF-induced hemolysis was the result of cell swelling and subsequent cell lysis due to overswelling. An osmotic imbalance and corresponding loss of the normal transmembrane potential was invoked. It was also inferred that some pore sizes changed as the field strength, pulse duration, and ionic strength of the suspending medium were varied. Further, if the osmolality of the medium was balanced to that of the intracellular (cytoplasmic) fluid by adding molecules larger than the pore size to the extracellular medium, cell swelling was prevented and cell membrane recovery subsequently occurred. For example, it was possible to obtain electroporated RBCs that had taken up small carbohydrates of various sizes (e.g., radioactive sucrose) but without loss of hemoglobin [82]. The biologically stringent test of introducing these loaded RBCs into mice demonstrated that membrane and cell recovery was essentially complete, as a significant circulating half-life in a laboratory animal model was found. This success led to the suggestion that electric field pulses could be used to introduce genetic material into living cells, not ghosts [79], and this accomplishment was subsequently reported [17,83]. Success with other cells quickly followed [1,2,21,84]. Presently, there is an almost explosive growth in the use of electroporation to introduce genetic material into cells.

The inferred average kinetics of pore formation and subsequent membrane recovery have been partially investigated by monitoring the pulsed-field-induced membrane permeability [85] and comparing it to the decrease in permeability of Rb^+ as a function of pulse duration [86]. In human RBCs, pore initiation occurred for pulse durations in the microsecond range, and pore expansion occurred in the 50-μs range. During the membrane recovery phase, pores shrink significantly within 10 ms, but complete recovery to Rb^+ transport requires 20 min to several hours at 37°C [16]. In other experiments, the kinetics of the alteration of U(t) in sea urchin eggs was measured using time-resolved fluorescence imaging microscopy with a membrane potential dye and was found to be in the 10-μs range [87]. Earlier the rise time of the membrane potential in human RBCs was estimated to be 0.1 to 0.5 μs [85].

In 1982 it was demonstrated that cells can take up, incorporate, and express DNA following electroporation [17,83]. More difficult-to-transform cells were subsequently also transformed using electroporation [84], and this has since been followed by a rapid increase in the use of electroporation for gene transfer, particularly for cells that cannot readily be transfected by conventional methods [1,2,21]. Such gene transfer experiments are usually scored by counting transformed cells, and because transformation requires DNA uptake, intracellular transport, into the nucleus, incorporation, and expression, the number (percentage) of cells experiencing electroporation could not actually be determined. Presently, it appears that electroporation offers an essentially universal method for the introduction of genetic material. However, because the separate events of uptake, intracellular transport, and so on, are not understood, it is not clear which phenomena govern the overall efficiency. If electroporation proves to be a universal cell membrane phenomenon, mammalian cells, and microorganisms with walls that permit passage of DNA, should be capable of transformation by electroporation. Additional experiments with RBCs suggested that cells may acquire zero or one significant and persistent pore, and that "flip-flop" (the transmembrane location of certain membrane molecules) rates increase following electroporation [88]. This experimental observation, in combination with the basic assumed pore structure in quantitative electroporation theories [22,23,59,60,62,80], is consistent with the early idea that the structure of transient aqueous pores suggests that they should be preferred sites of flip-flop [89,90].

The electroporative transport of relatively uncharged molecules across membranes is in contrast to ion transport, which is acutely sensitive to the Born energy [91–93]. This is particularly important to understand for macromolecules. Several experiments have demonstrated that electroporation caused uptake of fluorescent molecules [5,11,28–30,46,94]. One investigation studied individual RBC ghosts, using fluorescent or fluorescent-labeled molecules to determine the uptake or release of different size and charge fluorescent molecules. For example, such experiments showed that lucifer yellow (457 D; charge ≈ -2e) and larger molecules, such as FITC-dextran (broad range of sizes; nominally 10,000 to 156,000 D; relatively uncharged), myoglobin (about 17,000 D; negatively charged), and FITC-labeled bovine serum albumin (BSA) (about 60,000 D; negatively charged) were released by electroporation following preloading by osmotic swelling [28]. It has also been shown that electroporative molecular transport preferentially occurs at the cell's poles [95]. Similar experiments have been used to estimate uptake with other types of cells [11,46]. Use of fluorescent probes is particularly attractive because transport of molecules of different sizes, shapes, and charges can be studied.

A closely related and complementary approach also uses fluorescent molecules, but instead of monitoring the fluorescence of one or a small number of cells as a function of time, "snapshots" of the population distribution of cell fluorescence are obtained at different times by using flow cytometry. Such experiments reveal the variation of macromolecule uptake and of membrane recovery within a cell population, and can employ statistically significant nubmers (e.g., 10,000) of individual cell measurements [5,6,30,94,96]. For example, flow cytometry revealed that for a large fraction of RBCs (20%), the intact yeast *Saccharomyces cerevisiae* (35%) and the intact yeast *S. pombe* (75%) take up significant amounts of FITC-dextran (70 kD) following a single 50-μs square-wave pulse. For the yeasts, membrane recovery 25 min after the electroporation pulse was tested by exclusion of PI, with cell viability tested by plating [5]. The numbers of molecules taken up by cells in the participating subpopulations can be large. For example, intact *S. cerevisiae* took up on the order of 10^5 FITC-dextran molecules (70 kD) when suspended in 500 μM FITC-dextran, or took up 10^8 propidium iodide molecules (660 D) when suspended in an 80 μM solution [30]. Overall, such electroporation phenomena population distribution experiments show that although large percentages of cells can participate, not all cells do, and several significant subpopulations are often found.

Based on both electrical and molecular transport studies, it is clear that the cell membrane becomes highly conductive, and also tremendously increases its permeability to water-soluble molecules, whether or not electrically charged (e.g., glucose, sucrose, dextrans, albumin, enzymes, antibodies, and nucleic acids). Although apparently universal, the electrical behavior accompanying electroporation is short lived. If U(t) causes reversible electrical breakdown, the discharge of membrane is completed within a few microseconds, and if rupture occurs, the discharge time appears to be significantly longer, but still is typically completed within 10^{-4} to 10^{-6} s. Thereafter, the membrane conductance may remain fairly high, often for seconds or more [97].

In contrast to the very rapid characteristic times for membrane discharge, the transient high-permeability state can have a long lifetime. Presently it appears that if $P(t)_{d,m}$ denotes the diffusive permeability as a function to time, there may be both fast and slow components. That is, during times when U(t) is large, $P(t)_{d,m}$ may be extremely large, and this may be the time when most of the macromolecular transport occurs, probably due to electrical drift. Nevertheless, following a rapid decrease in $P(t)_{d,m}$, there also appears to be residual high permeability that persists for seconds to minutes to hours. Further, the residual high permeability appears to have a lifetime that is significantly temperature dependent. For example, pulsed RBCs remain highly permeable to K^+ and Na^+ for about 30 min at 37°C, but for days at 0°C [10,16]. Similar ion permeability behavior has been observed by others [98]. Overall, the existence of a transient high-permeability state with complex time behavior appears to be a major, but still poorly characterized, feature of electroporation. This and other key attributes of molecular transport due to electroporation remain to be elucidated.

SUMMARY

Electroporation-related phenomena appear to be universal, and to involve transient pores that occur when the bilayer portion experiences large transmembrane potentials. Both ions and molecules can cross the membrane, and some types of membrane-adsorbed molecules can be electro-inserted into the cell membrane. Prompt processes (those that occur while the potential is still elevated) and delayed processes (those that occur well after the potential has decayed) are both significant. The fate of cells experiencing electroporation is complex and can include the absence of an effect, reversible cell stress, and irreversible cell stress leading to cell death. In the case of cell damage, it is also likely that both relatively prompt and delayed cell membrane rupture can occur. The complex interactions between pores, the elevated transmembrane potential, membrane macromolecules, cellular structures, and the molecules and ions that enter pores remain to be understood, but appear to be fundamental to cell membranes.

ACKNOWLEDGMENTS

This work supported partially by ONR Contract N00014-87-K-0479. Discussions with many individuals, particularly T. Y. Tsong, are gratefully acknowledged.

SYMBOLS

$A_{electrode}$	macroscopic area of an electrode
C	electrical capacitance of the membrane
d	thickness of the lipid portion of the membrane
D_p	diffusion constant governing pore radius
E_e	actual electric field in the electrolyte (medium) which cells experience
$E_{nominal}$	nominal electric field (neglects interfacial effects at electrodes)
G(t)	conductance of the membrane as a function of time
h	thickness (total) of the membrane, including both lipid and head group regions
J_e	average current density in the electrolyte (medium)
K_M	dielectric constant of the lipid portion of the membrane
K_W	dielectric constant of bulk water
L	separation of the two electrodes in a parallel plane geometry
r	pore radius r_W; effective radius of a water molecule
T	temperature (absolute), kelvin
U_C	critical transmembrane potential associated with membrane rupture
U(t)	transmembrane potential as a function of time, t
U_0	value of transmembrane potential at the end of a square pulse
$U_{0,C}$	maximum transmembrane potential at the end of a square pulse
$\Delta E(r,U_p)$	pore free energy as a function of radius and local transmembrane potential
Δt	pulse width applied by external potential (pulse) generator
$\Delta U(t)_i$	pulse (potential) generator output as a function of time
$\Delta\phi$	potential difference (voltage) applied across electrodes
η_i	electrical mobility of the ith ionic species
σ_e	electrical conductivity of bulk electrolyte
σ_p	electrical conductivity of electrolyte within a pore

REFERENCES

1. U. Zimmermann, Electrical breakdown, electropermeabilization and electrofusion, *Rev. Physiol. Biochem. Pharmacol. 105*:175–256 (1986).
2. E. Neumann, A. Sowers, and C. Jordan (eds.), *Electroporation and Electrofusion in Cell Biology*, Plenum Press, New York, 1989.
3. J. C. Weaver, Electroporation: a new phenomenon to consider in medical technology, in *Emerging Electromagnetic Medicine* (M. E. O'Connor, R. H. C. Bentall, and J. C. Monahan, eds.), Springer-Verlag, New York, 1990, pp. 81–102.
4. L. A. Geddes, *Electrodes and the Measurement of Bioelectric Events*, Wiley, New York, 1972.
5. J. C. Weaver, G. I. Harrison, J. G. Bliss, J. R. Mourant, and K. T. Powell, Electroporation: high frequency of occurrence of the transient high permeability state in red blood cells and intact yeast, *FEBS Lett. 229*:30–34 (1988).
6. J. G. Bliss, G. I. Harrison, J. R. Mourant, K. T. Powell, and J. C. Weaver, Electroporation: the distribution of macromolecular uptake and shape changes in red blood cells following a single 50-μsecond square wave pulse, *Bioelectrochem. Bioenerg. 19*:57–71 (1988).
7. K. R. Foster and H. P. Schwann, Dielectric properties of tissues, in *CRC Handbook of Biological Effects of Electromagnetic Fields* (C. Polk and E. Postow, eds.), CRC Press, Boca Raton, Fla., 1986, pp. 27–96.
8. J. W. Watts, J. M. King, and N. J. Stacey, Inoculation of protoplasts with viruses by electroporation, *Virology 157*:40–46 (1987).
9. O. Alvarez and R. Latorre, Voltage-dependent capacitance in lipid bilayers made from monolayers, *Biophys. J. 21*;1–17 (1978).
10. K. Kinosita, Jr. and T. Y. Tsong, Hemolysis of human erythrocytes by a transient electric field, *Proc. Natl. Acad. Sci. USA 74*:1923–1927 (1977).
11. M. R. Michel, M. Elgizoli, H. Koblet, and Ch. Kempf, Diffusion loading conditions determine recovery of protein synthesis in electroporated P3X63 Ag8 cells, *Experientia, 44*:199–203 (1988).
12. A. J. H. Sale and W. A. Hamilton, Effects of high electric fields on microorganisms. III. Lysis of erythrocytes and protoplasts, *Biochim. Biophys. Acta 163*:37–43 (1968).
13. T. Y. Tsong and E. Kingsley, Hemolysis of human erythrocytes induced by a rapid temperature jump, *J. Biol. Chem. 250*:786–789 (1975).

14. U. Zimmermann, G. Pilwat, Chr. Holzapfel, and K. Rosenheck, Electrical heymolysis of human and bovine red blood cells, *J. Membr. Biol. 30*:135–152 (1976).
15. U. Zimmermann, F. Riemann, and G. Pilwat, Enzyme loading of electrically homogeneous human red blood cell ghosts prepared by dielectric breakdown, *Biochim. Biophys. Acta. 436*:460–474 (1976).
16. K. Kinosita Jr., and T. Y. Tsong, Formation and resealing of pores of controlled sizes in human erythrocyte membrane, *Nature 268*:438–441 (1977).
17. E. Neumann, M. Schaefer-Ridder, Y. Wang, and P. H. Hofschneider, Gene transfer into mouse lyoma cells by electroporation in high electric fields, *EMBO J. 1*:841–845 (1982).
18. E. Neumann and K. Rosenheck, Permeability changes induced by electric impulses in vesicular membranes, *J. Membr. Biol. 10*:279–290 (1972).
19. U. Zimmermann, J. Schultz, and G. Pilwat, Transcellular ion flow in *Escherichia coli* B and electrical sizing of bacterias, *Biophys. J. 13*:1005–1013 (1973).
20. E. M. El-Mashak and T. Y. Tsong, Ion selectivity of temperature-induced and electric-field induced pores in dipalmitoylphosphatidylcholine vesicles, *Biochemistry 24*:2884–2888 (1985).
21. H. Potter, Electroporation in biology: methods, applications, and instrumentation, *Anal. Biochem. 174*:361–373 (1988).
22. I. G. Abidor, V. B. Arakelyan, L. V. Chernomordik, Yu. A. Chizmadzhev, V. F. Pastushenko, and M. R. Tarasevich, Electric breakdown of bilayer membranes. I. The main experimental facts and their qualitative discussion, *Bioelectrochem. Bioenergy. 6*:37–52 (1979).
23. L. V. Chernomordik and Y. A. Chizmadzhev, Electrical breakdown of lipid bilayer membranes: phenomenology and mechanisms, in *Electroporation and Electrofusion in Cell Biology* (E. Neumann, A. E. Sowers, and C. A. Jordan, eds.), Plenum Press, New York, 1989, pp. 83–95.
24. I. P. Sugar and E. Neumann, Stochastic model for electric field-induced membrane pores: electroporation, *Biophys. Chem. 19*:211–225 (1984).
25. E. U. Condon and H. Odishaw, *Handbook of Physics*, McGraw-Hill, New York, 1958.
26. J. C. Weaver, Cell membrane rupture by strong electric fields: prompt and delayed processes, in *Electrical Trauma: Pathophysiology and Clinical Management* (R. C. Lee, J. F. Burke, and E. G. Cravalho, eds.), Cambridge University Press, Cambridge, 1991 (submitted).

27. U. Zimmermann, G. Pilwat, and F. Riemann, Preparation of erythrocyte ghosts by dielectric breakdown of the cell membrane, *Biochim. Biophys. Acta 375*:209–219 (1975).
28. A. E. Sowers and M. R. Lieber, Electropore diameters, lifetimes, numbers, and locations in individual erythrocyte ghosts, *FEBS Lett. 205*:179–184 (1986).
29. W. Mehrle, U. Zimmermann, and R. Hampp, Evidence for asymmetrical uptake of fluorescent dyes through electro-permeabilized membranes of *Avena* meosphyll protoplasts, *FEBS Lett. 185*:89–94 (1985).
30. D. C. Bartoletti, G. I. Harrison, and J. C. Weaver, The number of molecules taken up by electroporated cells: quantitative determination, *FEBS Lett. 256*:4–10 (1989).
31. I. Uno, K. Fukami, H. Kato, T. Takenwa, and T. Ishikawa, Essential role for phosphatidylinositol 4,5-biphosphate in yeast cell proliferation, *Nature 333*:188–190 (1988).
32. R. C. Lee and M. S. Kolodney, Electrical injury mechanisms: electrical breakdown of cell membranes, *Plast. Reconstr. Surg. 80*:672–679 (1987).
33. R. C. Lee, D. C. Gaylor, D. Bhatt, and D. A. Israel, Role of cell membrane rupture in the pathogenesis of electrical trauma, *J. Surg. Res. 47*:709–719 (1988).
34. D. L. Bhatt, D. C. Gaylor, and R. C. Lee, *Rhabdomyolysis Due to Pulsed Electric Fields* (submitted).
35. H. M. Fishman and R. I. Macey, The N-shaped current-potential characteristic in frog skin. I. Time development during step voltage clamp, *Biophys. J. 9*:127–139 (1969).
36. H. M. Fishman and R. I. Macey, The N-shaped current-potential characteristic in frog skin. II. Kinetic behavior during ramp voltage clamp, *Biophys. J. 9*:140–150 (1969).
37. K. T. Powell, A. W. Morgenthaler, and J. C. Weaver, Tissue electroporation: observation of reversible electrical breakdown in viable frog skin, *Biophys. J. 56*:1163–1171 (1989).
38. W. S. Agnew, T. Claudia, and F. J. Sigworth (eds.), Molecular Biology of Ionic Channels, Current Topics in Membranes and Transport, 33, Adacemic Press, New York, 1989.
39. E. H. Serpersu and T. Y. Tsong, Activation of electrogenic Rb^+ transport of (NaK)-ATPase by an electric field, *J. Biol. Chem. 259*:7155–7162 (1984).
40. R. D. Astumian, P. B. Chock, T. Y. Tsong, and H. V. Westerhoff, Effects of oscillations and energy-driven fluctuations on the dynamics of enzyme catalysis and free-energy transduction, *Phys. Rev. A 39*:6416–6435 (1989).
41. R. B. Gennis, *Biomembranes: Molecular Structure and Function*, Springer-Verlag, New York, 1989.

42. J. C. Weaver and A. Barnett, *A Theoretical Model of Electroporation Mechanism: Membrane Electrical Behavior and Molecular Transport* in Handbook of Electroporation and Electrofusion, Academic Press (D. C. Chang, B. M. Chassy, J. A. Saunders, and A. E. Sowers, eds.) (submitted).
43. J. Teissie and T. Y. Tsong, Voltage modulation of Na^+/K^+ transport in human erythrocytes, *J. Physiol. Paris. 77*:1043–1053 (1981).
44. Y. Mouneimne, P.-F. Tosi, Y. Gazitt, and C. Nicolau, Electroinsertion of xeno-glycophorin into the red blood cell membrane, *Biochem. Biophys. Res. Commun. 159*:39–40 (1989).
45. D. C. Chang, Cell poration and cell fusion using an oscillating electric field, *Biophys. J. 56*:641–652 (1989).
46. H. Liang, W. J. Purucker, D. A. Stenger, R. T. Kubiniec, and S. W. Huie, Uptake of fluorescence-labeled dextrans by 10T 1/2 fibroblasts following permeation by rectangular and exponential-decay electric field pulses, *BioTechniques 6*:550–558 (1988).
47. J. O'M. Bockris and A. K. N. Reddy, *Modern Electrochemistry*, Plenum Press, New York, 1970.
48. W. J. Moore, *Physical Chemistry*, 3rd ed., Prentice-Hall, Englewood Cliffs, N.J., 1962.
49. A. P. Mazzoleni, B. F. Sisken, and R. L. Kahler, Conductivity values of tissue culture medium from 20°C to 40°C, *Bioelectromagnetics 7*:95–99 (1986).
50. H. T. Tien, *Bilayer Lipid Membranes (BLM): Theory and Practice*, Marcel Dekker, New York, 1974.
51. R. Benz, F. Beckers, and U. Zimmermann, Reversible electrical breakdown of lipid bilayer membranes: a charge-pulse relaxation study, *J. Membr. Biol. 48*:181–204 (1979).
52. R. Benz and U. Zimmermann, Relaxation studies on cell membranes and lipid bilayers in the high electric field range, *Bioelectrochem. Bioenerg.* 7:723–739 (1980).
53. R. Benz and U. Zimmermann, Pulse-length dependence of the electrical breakdown in lipid bilayer membranes, *Biochim. Biophys. Acta 597*:637–642 (1980).
54. R. Benz and U. Zimmermann, The resealing process of lipid bilayers after reversible electrical breakdown, *Biochim. Biophys. Acta 640*:169–178 (1981).
55. R. Stämpfli, Reversible electrical breakdown of the excitable membrane of a Ranvier node, *An. Acad. Bras. Cienc. 30*:57–63 (1958).
56. R. Benz and F. Conti, Reversible electrical breakdown of squid giant axon membrane, *Biochim. Biophys. Acta 645*:115–123 (1981).
57. E. Neumann, The electroporation hysteresis, *Ferroelectrics 86*:325–333 (1988).

58. E. Neumann, in *Electroporation and Electrofusion in Cell Biology* (E. Neumann, A. Sowers, and C. Jordan, eds.), Plenum Press, New York, 1989, pp. 61–82.
59. J. C. Weaver and R. A. Mintzer, Decreased bilayer stability due to transmembrane potentials, *Phys. Lett. 86A*:57–59 (1981).
60. I. P. Sugar, The effects of external fields on the structure of lipid bilayers, *J. Physiol. Paris 77*:1035–1042 (1981).
61. K. T. Powell, E. G. Derrick, and J. C. Weaver, A quantitative theory of reversible electrical breakdown, *Bioelectrochem. Bioelectroenerg. 15*:243–255 (1986).
62. A. S. Barnett and J. C. Weaver, Electroporation: a unified, quantiative theory of reversible electrical breakdown and rupture, *Bioelectrochem. Bioenerg.* (1991) (in press).
63. J. C. Weaver and K. T. Powell, Theory of electroporation, in *Electroporation and Electrofusion in Cell Biology* (E. Neumann, A. Sowers, and C. Jordan, eds.), Plenum Press, New York, 1989, pp. 111–126.
64. L. V. Chernomordik, S. I. Sukharev, I. G. Abidor, and Yu. A. Chizmadzhev, The study of the BLM reversible electrical breakdown mechanism in the presence of $UO_2{}^{2+}$, *Bioelectrochem. Bioenerg. 9*:149–155 (1982).
65. J. Teissie and T. Y. Tsong, Electric field induced transient pores in phospholipid bilayer vesicles, *Biochemistry 20*:1548–1554 (1981).
66. J. Teissie and T. Y. Tsong, Evidence of voltage-induced channel opening in Na/K ATPase of human erythrocyte membrane, *J. Membrane Biol. 55*:133–140 (1980).
67. E. H. Serpcrsu and T. Y. Tsong, Stimulation of Rb^+ pumping activity of (Na,K)-ATPase, *J. Membr. Biol. 74*:191–201 (1983).
68. T. Y. Tsong and R. D. Astumian, Absorption and conversion of electric field energy by membrane bound ATPases, *Bioelectrochem. Bioelectroenerg. 15*:457–476 (1986).
69. T. Y. Tsong and R. D. Astumian, Electroconformational coupling and membrane protein function, *Prog. Biophys. Mol. Biol. 50*:1–45 (1987).
70. R. D. Astumian, P. B. Chock, F. Chauvin, and T. Y. Tsong, Electroconformational coupling and the effects of static and dynamic electric fields on membrane transport, in *Electromagnetic Fields and Biomembranes* (M. Markov and M. Blank, eds.), Plenum Press, New York, 1988, pp. 59–77.
71. H. G. L. Coster, A quantitative analysis of the voltage–current relationships of fixed charge membranes and the associated property of "punch-through," *Biophys. J. 5*:669–686 (1965).

72. E. Neumann and K. Rosenheck, Potential difference across vesicular membranes, *J. Membr. Biol. 14*:194–196 (1973).
73. P. Lindner, E. Neumann, and K. Rosenheck, Kinetics of permeability changes induced by electric impulses in chromaffin granules, *J. Membr. Biol. 32*:231–254 (1977).
74. T. Y. Tsong, T. T. Tsong, E. Kingsley, and R. Siliciano, Releaxation phenomena in human erythrocyte suspension, *Biophys. J. 16*:1091–1104 (1976).
75. U. Zimmermann, G. Pilwat, and F. Riemann, Dielectric breakdown of cell membranes, *Biophys. J. 14*:881–899 (1974).
76. F. Riemann, U. Zimmermann, and G. Pilwat, Release and uptake of haemoglobin and ions in red blood cells induced by dielectric breakdown, *Biochim. Biophys. Acta 394*:449–462 (1975).
77. G. Pilwat, U. Zimmermann, and F. Riemann, Dielectric breakdown measurements of human and bovine erythrocyte membranes using benzyl alcohol as a probe molecule, *Biochim. Biophys. Acta 406*:424–432 (1975).
78. H. G. L. Coster and U. Zimmermann, Dielectric breakdown in the membranes of *Valonia utricularis*: the role of energy dissipation, *Biochim. Biophys. Acta. 382*:410–418 (1975).
79. D. Auer, G. Brandner, and W. Bodemer, Dielectric breakdown of the red blood cell membrane and uptake of SV 40 DNA and mammalian cell RNA, *Naturwissenschaften 63*:391 (1976).
80. K. T. Powell and J. C. Weaver, Transient aqueous pores in bilayer membranes: a statistical theory, *Bioelectrochem. Bioelectroenerg. 15*:211–227 (1986).
81. J. C. Weaver, R. A. Mintzer, H. Ling, and S. R. Sloan, Conduction onset criteria for transient aqueous pores and reversible electrical breakdown in bilayer membranes, *Bioelectrochem. Bioelectroenerg. 15*:229–241 (1986).
82. K. Kinosita, Jr. and T. Y. Tsong, Survival of sucrose-loaded erythrocytes in circulation, *Nature 272*:258–260 (1978).
83. T.-K. Wong and E. Neumann, Electric field mediated gene transfer, *Biochem. Biophys. Res. Comm. 107*:584–587 (1982).
84. H. Potter, L. Weir, and P. Leder, Enhancer-dependent expression of human K immuinoglobulin genes introduced into mouse pre-B lymphocytes by electroporation, *Proc. Natl. Acad. Sci. USA 81*:7161–7165 (1984).
85. K. Kinosita, Jr. and T. Y. Tsong, Voltage induced conductance in human erythrocyte membranes, *Biochim. Biophys. Acta 554*:479–497 (1979).
86. E. H. Serpersu, K. Kinosita, and T. Y. Tsong, Reversible and irreversible modification of erythrocyte membrane permeability by electric field, *Biochim. Biophys. Acta 812*:779–795 (1985).

87. K. Kinosita, Jr., I. Ashikawa, N. Saita, H. Yoshimura, H. Itoh, H. Nagayama, and A. Ikegami, Electroporation of cell membrane visualized under a pulsed-laser fluorescence microscope, *Biophys. J. 53*:1015–1019 (1988).
88. V. Dressler, K. Schwister, C. W. M. Haest, and B. Deuticke, Dielectric breakdown of the erythrocyte membrane enhances transbilayer mobility of phospholipids, *Biochim. Biophys. Acta 732*:304–307 (1983).
89. C. Taupin, M. Dvolaitzky, and C. Sauterey, Osmotic pressure induced pores in phospholipid vesicles, *Biochemistry 14*:4771–4775 (1975).
90. A. G. Petrov, S. A. Seleznev, and A. Derzhanski, Principles and methods of liquid crystal physics applied to the structure and functions of biological membranes, *Acta Physiol. Pol. A55*: 385–405 (1979).
91. V. A. Parsegian, Energy of an ion crossing a low dielectric membrane: solutions to four relevant electrostatic problems, *Nature 221*:844–846 (1969).
92. P. C. Jordan, Effect of pore structure on energy barriers and applied voltage profiles. I. Symmetrical channels, *Biophys. J. 45*:1091–1100 (1984).
93. V. F. Pastushenko and Yu. A. Chizmadzhev, Stabilization of conducting pores in BLM by electric current, *Gen. Physiol. Biophys. 1*:43–52 (1982).
94. J. C. Weaver, J. G. Bliss, G. I. Harrison, J. R. Mourant, and K. T. Powell, Electroporation in individual cells: measurements using light scattering and fluorescence by flow cytometry, in *Proc. 9th IEEE/Engineering in Medicine and Biology Society*, 1987, pp. 708–709.
95. A. E. Sowers, Fusion events and nonfusion contents mixing events induced in erythrocyte ghosts by an electrical pulse, *Biophys. J. 54*:619–626 (1988).
96. M. J. O'Hare, M. G. Ormerod, P. R. Imrie, J. H. Peacock, and W. Asche, Electropermeabilization and electrosensitivity of different types of mammalian cells, in *Electroporation and Electrofusion in Cell Biology* (E. Neumann, A. Sowers, and C. Jordan, eds.), Plenum Press, New York, 1989, pp. 319–330.
97. L. V. Chernomordik, S. I. Sukharev, S. V. Popov, V. F. Pastushenko, A. V. Sokirko, I. G. Abidor, and Y. A. Chizmadzhev, The electrical breakdown of cell and lipid membranes: the similarity of phenomenologies, *Biochim. Biophys. Acta 902*: 360–373 (1987).
98. K. Schwister and B. Deuticke, Formation and properties of aqueous leaks induced in human erythrocytes by electrical breakdown, *Biochim. Biophys. Acta 816*:332–348 (1985).

11

Dynamic Mechanism for Cytoskeleton Structures

E. DEL GIUDICE Istituto Nazionale Fisica Nucleare (INFN), Milano, Italy

S. DOGLIA and M. MILANI Dipartimento di Fisica, Università degli Studi di Milano, Milano, Italy

G. VITIELLO Dipartimento di Fisica, dell'Università di Salerno, Salerno, Italy

I.	Introduction	279
II.	Physical Model	280
III.	Process of Filamentation	281
IV.	Conclusions and Outlook	284
	References	284

I. INTRODUCTION

In recent years cytoskeleton has been widely investigated and many important features have been elucidated. It is a complex system that is able to perform in a flexible way [1]. Theoretical understanding of cytoskeleton structure and functions is still an open challenge. In this chapter we suggest that standard quantum field theory could give helpful insights into the problem. The present analysis aims to fit the main qualitative features of the system and to discover its own inherent limitations. Actually, this problem is a part of a more general one in physics: *the description of the growth of an ordered state* out of a nonordered one. This transformation is a nonequilibrium transition in a thermodynamic sense. Actually, the thermodynamics of nonequilibrium states was proposed some decades ago as a useful tool for understanding living systems [2], but no dynamic description has supplemented this scheme. Recently, a theoretical approach [3] has been proposed that could fit the concent of thermodynamic nonequilibrium in the framework of quantum field theory (QFT).

II. PHYSICAL MODEL

We assume that the living system is the endpoint of an evolution whose starting point is a set of many electric dipoles (biomolecules) open to environmental influence. Usually, an organized extended object is described through a macroscopic variable, called an *order parameter*. Examples of order parameters are the density of magnetization in ferromagnets and the orientation angle in nematic crystals. An organized set of electric dipoles admits the vector P (density of electric polarization) as an order parameter. We have discussed elsewhere [4,5] the physical evolution of a set of electric dipoles open to environmental influence. Let us briefly recall the main points:

1. Electric dipoles, interacting through short-range forces (e.g., hydrogen bonds, van der Waals forces), rotationally invariant, get phase-correlated if their evolution occurs in the presence of a nonvanishing electric polarization P_0. The statement above is a particular case of the Goldstone—Nambu theorem [6]. In that case a coherent polarization mode emerges that forces the component dipoles to oscillate coherently [7].
2. If an electromagnetic field (either externally supplied or produced by the nearby molecules) is applied on a set of correlated dipoles, three possibilities can occur:
   a. The field is stronger than the correlation. The field is then able to disregard the correlation and propagate as in a noncorrelated medium (usual Maxwellian propagation).
   b. The field is weaker than the correlation. Then the field cannot enter the region where the correlation holds.
   c. The field is as strong as the correlation. Then the field is prevented from propagating apart from a number of singular pathways, where the correlation is broken.
3. Case 2c is possible only if the rotational invariance of the short-range dynamics of the electric dipoles is completely broken (see ref. 5). That means that a left-handed rotation must be physically distinguishable from a right-handed one. As a consequence, the electromagnetic field, propagating along a singular pathway, acquires a preferred helicity.

The propagation of the electromagnetic field according to cases 2c and 3 is not governed by the Maxwell equations but by

$$(\Box + M^2)A_\mu = j_\mu \tag{1}$$

where $\square$ is the usual differential d'Alembert operator

$$\square = \frac{1}{c^2}\frac{\partial^2}{\partial t^2} - \frac{\partial^2}{\partial x^2} - \frac{\partial^2}{\partial y^2} - \frac{\partial^2}{\partial z^2}$$

A_μ is the potential four-vector, j_μ is the source distribution, and M is a parameter (which is zero in the Maxwell case) connected to the radius r of the singular filament where the field is trapped:

$$M = \frac{\hbar}{c}\frac{1}{r} \tag{2}$$

The parameter M is zero in the Maxwell case. This parameter is dimensionally a mass; an observer completely immersed within the filament would actually measure a nonzero mass of the photon. Real observers are external to the filament and the physical meaning of the mass term is completely exhausted by Eq. (2), that is, the electromagnetic field is confined inside a filament. In ref. 5, we have calculated, in the simplified case of a completely aqueous medium, r = ~150 Å and consequently M = 13.6 eV. The theory described here is analogous to the Anderson theory of conventional superconductors [8], where the correlation holds among the Cooper electron pairs. The main difference is that conventional superconductors are systems at equilibrium whereas here we are considering systems out of equilibrium.

III. PROCESS OF FILAMENTATION

We discuss now how the trapping of the electromagnetic field into a filament induces a rearrangement in the molecule concentrations in the cell medium. On the side boundary of each filament there is a strong field gradient. It can be shown that the dominant force acting on the molecules (the "gradient force") is [9]

$$F = \text{const}\cdot\text{grad } E^2\left[\Sigma_k \frac{\omega_k^2 - \omega^2}{(\omega_k^2 - \omega^2) + \Gamma_k^2}\right] \tag{3}$$

where ω is the electromagnetic field frequency and ω_k is the frequency of the kth molecular mode whose damping factor is Γ_k. According to Eq. (3), molecules with different vibrational frequencies are acted on differently and that allows "recognition" of different molecules by the field. A molecular pattern recognition can then be construed as a frequency pattern. In particular, if there is an index k such that $\omega \sim \omega_k$, a very strong force develops (positive or negative according to the sign of $\omega - \omega_k$); if the sign is correct, the side boundary of the filament gets coated with molecules, which oscillate with a frequency ω_k. These molecules wind up around the filament according to the helicity chosen by the electromagnetic field as a consequence of the original symmetry breaking. Notice that very small changes of the frequency can transform an attractive force into a repulsive one. Modifications in the sources of the field (e.g., the charged subgroups in DNA) could then produce, through the subsequent frequency modification, breakdown of the filament structure. Notice that sudden filament disappearances, termed *microtubule catastrophes* in ref. 10, could be traced back to this instability against small frequency fluctuations.

Let us allow a final consideration of filament structure. We have said above that the electromagnetic field gets propagated along the filament by breaking all the correlations there. Actually, the cell's main component, water, has a very high density since many transient correlations ($\tau \sim 10^{-11}$ s) are present. Those correlations produce a zero average on significant time intervals, so that the medium appears uncorrelated with an incoming electromagnetic field. Correlations then produce a sharp decrease in the average distance between molecules (i.e., an increase of density). Since the process of correlation is depressed by Eq. (1), we expect a decrease in water density within the filament. This is a way to test the predictability of the present approach.

The force given by Eq. (3) must be supplemented by the so-called *radiation pressure force*, whose density (force per unit volume) is

$$\tilde{F}_v = \frac{1}{c}\,\frac{\partial \tilde{P}}{\partial t} \times \tilde{H} \tag{4}$$

This force has two components: the component orthogonal to the field propagation that sums up to a small correction to the gradient force of Eq. (3); the longitudinal component pushes particles along the filament, giving rise to a squeeze among the molecules assembled on the side boundary and to the transport of small molecules inside. The combined effect of Eqs. (3) and (4) would then be the attraction of selected molecules on the boundary, their orientation through both the field direction and the water density gradient across the side boundary, and finally, the binding induced by the squeezing produced by the radiation pressure. In this way we get a sort of "dynamic" polymerization [11] where energy is provided by the field.

A definite prediction that follows from the foregoing approach is the "treadmilling" of the polymer chain [12]. Monomers, which are squeezed along the filament, would be released at the end of the correlation range, while new monomers would fill the gap produced at the chain beginning with the forward motion induced by the radiation pressure. Moreover, dynamics, rather than a chemical connection among monomers, is able to avoid stable attachments of the filaments to organelles and membranes, letting them slide on the surfaces. Relative rotation of the cytoskeleton with respect to the membrane, under externally applied magnetic fields, is a test of the prediction.

It has been discussed in refs. 5 and 13 that Eq. (1) implies a kind of superconductive property of the medium. The microscopic carriers of this peculiar superconductivity, which is quite different from the standard superconductivity of metals at low temperature, are not yet recognized; the discussion in ref. 13 suggests that the electric charge of the carrier would be twice the electron charge. This point requires further research. In any case, a consequence that can be derived at this stage is that ions could travel along the *superconductive pathways*, which are the filaments, without energy losses.

Let us make a last point. Suppose that heat is released at some point of the chain (e.g., the energy output of some chemical reaction). Usually, heat propagates in a diffusive way, à la Fourier, since in a noncorrelated medium energy is transferred at the microscopic level through the scattering of accelerated particles; as a consequence, the temperature of the region traversed increases. In our case, however, a different method of energy transfer is available. Molecules are correlated so that the oscillation of one component induces the oscillation of all the other components of the string. Wave propagation, à la d'Alembert, replaces Fourier propagation. Heat propagation in wave form has been observed in all the correlated media, such as superfluid helium and extremely pure crystals; sometimes this heat wave is called *second sound* [14]. Phenomenologically, this collective oscillation of dipoles appears as an electric polarization wave which travels in the cylindrical mantle encircling the filament and produces ponderomotive forces similar to those discussed above. The frequency of this heat wave has been shown to depend on the strength and the range of correlation. Since these parameters can be assumed to be time dependent, a time-dependent frequency of the heat wave results and due to Eq. (3), a time-dependent pattern of forces emerges.

This last feature is interesting since the energy output of a chemical reaction occurring at a definite time may trigger, without any losses, a sequence of more chemical reactions governed by the time-dependent frequency of the associated heat wave propagating along the filament. The concept of cytoskeleton as a privileged place for biochemical reactions is then vindicated. The feature above is the possible source of a biochemical feedback scheme.

IV. CONCLUSIONS AND OUTLOOK

In this chapter we have sketched a scheme, where, by using the conceptual framework of modern quantum field theory, a filamentous structure emerges from the collective dynamics of microscopic components. In the living cell this structure could be the cytoskeleton. The derivation of the filamentation process from basic principles of physics is an important result. A more quantitative approach to the fundamental interaction between matter and electromagnetic field in the cell and to the cytoskeleton formation would require a detailed knowledge of the excitation frequencies propagating in the cell and of the proper frequencies of the relevant biomolecules. As a matter of fact, the amount of available information is at present very poor. Application of this model to phenomena such as magnetotaxis could produce helpful insights.

REFERENCES

1. I. I. Wolosewick and K. B. Porter, *J. Cell Biol. 82*:114 (1979); J. S. Clegg, *Collect. Phenomena 3*:289 (1981); J. S. Clegg, in *Coherent Excitations in Biological Systems* (H. Fröhlich and F. Kremer, eds.), Springer-Verlag, Berlin, 1983, p. 189.
2. P. Glansdorff and I. Prigogine, *Thermodynamic Theory of Structure, Stability and Fluctuations*, Wiley, New York, 1971; I. Prigogine and G. Nicolis, *Self-Organization in Nonequilibrium Systems: From Dissipative Structures to Order Through Fluctuations*, Wiley, New York, 1977.
3. H. Matsumoto, M. Tachiki, and H. Umezawa, *Thermofield Dynamics and Condensed States*, North-Holland, Amsterdam, 1982.
4. E. Del Giudice, S. Doglia, M. Milani, and G. Vitiello, *Nuclear Physics B 251*(FS 13):375 (1985).
5. E. Del Giudice, S. Doglia, M. Milani, and G. Vitiello, *Nuclear Physics B 275*(FS 17):185 (1986).
6. C. Itzykson and J. B. Zuber, *Quantum Field Theory*, McGraw-Hill, New York, 1980.
7. H. Fröhlich, *Adv. Electron. Electron Phys. 53*:85 (1980).
8. P. W. Anderson, *Phys. Rev. 110*:827 (1958); *Phys. Rev. 130*:439 (1963).
9. G. A. Askar'yan, *Sov. Phys. Usp. 16*:680 (1974).

10. J. R. McIntosh, *Nature 312*:196 (1984); T. Mitchison and M. Kirschner, *Nature 312*:232, 237 (1984).
11. E. Del Giudice, S. Doglia, and M. Milani, *Phys. Lett. 90A*:104 (1982); in *The Application of Laser Light Scattering to the Study of Biological Motion* (J. C. Earnshaw and M. W. Steer, eds.), Plenum Press, New York, 1982, p. 493.
12. R. L. Margolis and L. Wilson, *Nature 298*:705 (1981).
13. E. Del Giudice, S. Doglia, M. Milani, and G. Vitiello, in *Coherence in Biology and Response to External Stimuli* (H. Fröhlich, ed.), Springer-Verlag, Berlin, 1988.
14. J. Wilks, *Properties of Solid and Liquid Helium*, Oxford University Press, Oxford, 1967.

IV

Interactions

12

Surface Activity of Proteins

SHLOMO MAGDASSI and NISSIM GARTI Casali Institute of Applied Chemistry, The Hebrew University of Jerusalem, Jerusalem, Israel

I.	Introduction	289
II.	Adsorption at Fluid–Fluid Interfaces: Kinetics and Mechanism	290
III.	Factors Affecting Surface Activity	292
IV.	Adsorption at Solid–Liquid Interfaces	294
V.	Proteins at Interfaces of Biological Interest	295
	A. Enzymes at interfaces	295
	B. Cell attachment onto surfaces	296
	C. Emulsification	297
	Summary	297
	References	297

I. INTRODUCTION

Proteins are macromolecules that are composed of a mixture of polar and nonpolar side chains and may have a net electrical charge. Therefore, these molecules tend to adsorb strongly to various interfaces, such as liquid–liquid and solid–liquid, hence decrease the surface energy of the system. This reduction in surface energy (i.e., the surface activity) is utilized in many natural processes as well as in many technological applications, mainly in the pharmaceutical and food industries (emulsion and foam stabilizers). However, sophisticated applications take advantage of both the surface activity of the proteins and the specific biological activity, especially in biotechnology and medical applications. For example, immobilization of enzymes and antibodies can be achieved by adsorption onto microcarriers [1]. The adsorption of protein molecules onto surfaces may alter the nature of the surface and could lead to adhesion of blood platelets, followed by thrombus formation [2], or enhance the adhesion of mammalian cells in some biotechnological applications [3].

The adsorption of proteins differs from the adsorption of monomeric surfactants, due mainly to the attachment of one protein molecule to the surface through several segments and to the possible alteration in protein structure. These differences could result in significant changes in the biological activity of the adsorbed protein.

A large number of publications dealing with the adsorption of proteins to various surfaces is available, describing both the mechanism of adsorption and the structure of the molecules adsorbed [4–7]. In this chapter we describe the adsorption process of proteins at fluid–fluid and solid–liquid interfaces and the parameters that affect the adsorption. This part will be followed by examples of systems in which protein adsorption plays an important role.

II. ADSORPTION AT FLUID–FLUID INTERFACES: KINETICS AND MECHANISM

The adsorption process can be described by three consecutive or concurrent processes: (1) the diffusion of the protein molecule from the bulk to the interface and attachment at the interface; (2) penetration of new molecules, when an energy barrier exists; and (3) molecular rearrangements of the adsorbed molecules. The adsorption process could be followed by measurements of the amount of adsorbed protein by various techniques [5] and is reflected in a reduction of the interfacial pressure [8]. It is assumed that any process responsible for the time dependence of surface tension should involve an increase in the number of adsorbed segments per unit area with time [9,10].

At the first stage of the adsorption process, it is assumed that there are no energy barriers to the adsorption. Therefore, in the absence of convection and desorption, the surface concentration Γ at time t is given according to Ward and Tordai [11] by the following equation, which is based only on a diffusion-controlled process:

$$\Gamma = 2C\left(\frac{Dt}{\pi}\right)^{1/2} \tag{1}$$

where C is the concentration of the protein in the bulk and D is the diffusion coefficient.

Since a reduction in interfacial tension is related to the number of active groups of protein segments that are adsorbed, the number of adsorbed molecules per unit area Γ will be replaced instead by the number of adsorbed active groups [8]. At low surface pressure, Eq. 1 can be combined with the simple two-dimensional equation of state, $\pi = \Gamma RT$, and we obtain the correlation between surface pressure and time [12], while a plot of the surface pressure against $t^{1/2}$ will be linear, as was shown for various proteins [8]. It should be emphasized that in many cases the diffusion coefficient calculated by the correlations above is higher than the value obtained by ultracentrifuge technique, probably due to convection [13].

After the initial adsorption of molecules at the interface, an energy barrier to adsorption will exist and the penetration rate of the protein molecules (or segments) will be rate limiting. It could be assumed that for a molecule to adsorb, it must compress molecules already adsorbed against an interfacial pressure π to create an area of interface ΔA. If this step is sufficiently rapid, the amount of work needed for penetration is $\pi\Delta A$ [5], and the rate of adsorption, if the adsorption is irreversible, is given by

$$\frac{d\Gamma}{dt} = k_a c_0 e^{-\pi \Delta A/KT} \tag{2}$$

where k_a is the rate constant for adsorption, π the surface pressure, C_0 the bulk concentration, K is Boltzman's constant, and T is the absolute temperature. Plotting $\log(d\Gamma/dt)$ against π has been found to give straight lines for a number of proteins [14], from which ΔA could be calculated. It was found that the ΔA values are much smaller than those expected from the dimensions of the proteins and they all fall between narrow limits of 100 to 200 Å^2. For example, ΔA for bovine serum albumin (BSA) was found to be 135 Å^2, while calculations based on its dimensions in solution yield the smallest molecular cross-sectional area of about 1100 Å^2. π—A curves also lead to molecular cross-sectional area, which is much higher (i.e., 10,000 Å^2) [5].

Similar observations were obtained for lysozyme and ovalbumin, from measurements of surface tensions and using a correlation between the rate of change in surface pressure and the surface pressure [8,42] as presented in

$$\ln \frac{d\pi}{dt} = \ln B - \frac{\pi \Delta A}{KT} \tag{3}$$

where π is the surface pressure and B is a constant that depends on protein concentration in the bulk and on a rate constant of the process.

Surface viscosity data of protein monolayers also indicate that the flow unit is much smaller than the molecular size and is in the range of 6 to 8 amino acid residues [15]. Since the surface pressure usually continues to increase even after the surface concentration has reached a constant value, it appears that only conformational changes take place at the last stage of adsorption.

The conclusions that can be drawn so far are: (1) only a small portion of the protein molecule needs to enter the interface, in order for the adsorption to proceed spontaneously [5], and (2) the spreading and conformational changes in the protein molecule occurs by movement of only small segments of the molecule, several amino acid residues each.

The adsorption process could lead to a drastic unfolding of the protein at the interface, a phenomenon that is termed *surface denaturation.* Under certain conditions the protein present at the interface may coagulate, and again drastic changes in protein properties are observed. An important implication of surface denaturation and coagulation is that the proteins may lose their specific biological activity, an important factor in enzyme immobilization or drug targeting.

III. FACTORS AFFECTING SURFACE ACTIVITY

In view of the adsorption mechanism above, it is possible to point out the factors that can affect the surface activity of the protein and which lead to the significant differences between various proteins. Generally, the surface activity is derived from the amino acid sequence of each protein and from the possible attachement of various segments to the interface. Therefore, the main parameters that result in variance in surface activity, as reviewed by Horbert and Brash [6], are:

1. *Molecule size.* It is assumed that larger molecules may have more contact points, thus increasing the total adsorption energy (which leads to irreversible adsorption). However, it appears that this parameter is less significant than the protein structure; for example, hemoglobin is far more surface active than fibrinogen, although it has much lower molecular weight (65,000 and 330,000, respectively).
2. *Amphipathicity and hydrophobicity.* It is expected that increased hydrophobicity of the protein would lead to increased surface activity, as indeed was observed: A native ovalbumin

was denatured for various periods of time, hence yielding proteins with various degrees of hydrophobicities [16]. It was concluded that increasing the hydrophobicity leads to greater reduction in interfacial tension and better emulsifying properties. The significance of protein hydrophobicity can be demonstrated by the surface tension reduction obtained by ovalbumin that was covalently attached to several hydrophobic chains, compared to that of the native ovalbumin (Fig. 1). It is clear that the surface tension is reduced faster and to lower values by the modified ovalbumin. It was also found that modified ovalbumins are much better emulsifiers than the native molecule, even upon attachment of six C_8 chains [17].

3. *Flexibility.* It is very likely that proteins which can unfold to a greater degree, or that unfold more rapidly, would be more surface active due to increased adsorbed sites at the interface, and due to entropic gain caused by the unfolding of the molecule. Thus the flexible β-casein molecules are more surface active than the globular BSA and lysozyme molecules, which contain many disulfide bonds [18].
4. *Charge.* It is frequently observed that the highest adsorption and surface pressure are obtained near the isoelectric point of

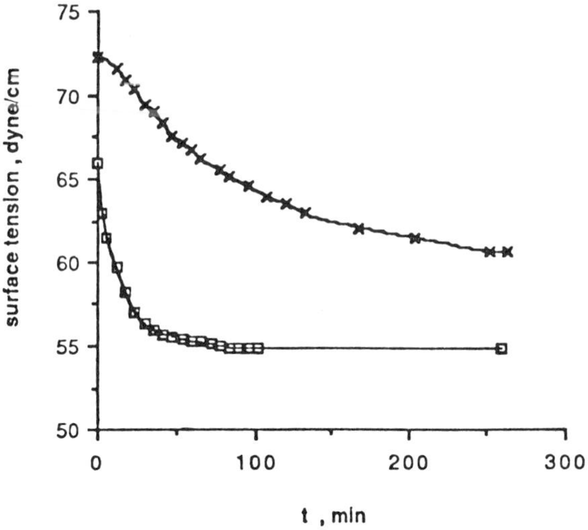

FIG. 1 Effect of attachment of caproyl groups to ovalbumin (pH = 10, 0.01 mg/ml). X, native ovalbumin; □, modified ovalbumin.

the protein. This observation results from decreased electrical barrier to adsorption, decreased repulsion among adsorbed molecules, and minimal solubility of the proteins in the bulk phase.

IV. ADSORPTION AT SOLID–LIQUID INTERFACES

A large number of applications in medicine, biology, and biotechnology are based on the adsorption of proteins onto solid surfaces, a subject that has been investigated extensively [7,19–22]. The colloid chemical approach has been discussed in an excellent review by Norde and Lyklema, in which the protein molecules as a whole were considered [28]. In this section the general features of the adsorption isotherms and the factors that affect the adsorption are described briefly in order to evaluate the role of the proteins at various interfaces of interest.

The surface activity at the solid–liquid interface is usually described by isotherms showing the adsorbed amount as a function of protein concentration in the bulk, at equilibrium. For most proteins at low bulk concentration (<1 mg/ml), a plateau in the adsorption isotherm is observed. The adsorbed amounts in the plateau region are usually in the range of several mg/m^2. This coverage is close to the expected amounts in a close-packed monolayer of molecules in the native configuration. However, in some cases, as for human plasma albumin adsorbed on polystyrene latex, it was observed that the concentration at the surface, at saturation, is considerably lower than the expected value. This was attributed to the spreading and unfolding of the molecules at the interface [26,27], in a similar way to that described for fluid–fluid interfaces. It also appears that the protein structure has a dominant role in the adsorption process compared to the substrate; for nonrigid molecules, the maximal adsorption is obtained near the isoelectric point (IEP) irrespective of the charge and nature of the substrate [28,29]. It was suggested that below and above the IEP the molecules may expand at the surface more readily than at the IEP, thus having a larger area per molecule [7]. This finding is of great importance in immobilization techniques: Adsorption that would be performed at the IEP would favor the preservation of the native conformation, hence could retain the biological activity. For rigid molecules such as RNase, unfolding is more difficult, and therefore the adsorption is dominated mainly by the electrostatic interactions between the surface and the protein molecules [23].

Thermodynamic analysis of the adsorption reveals several parameters that contribute to the adsorption energy: hydrophobic dehydration, ion incorporation, electrostatic interactions, and structural changes in the protein molecule. It was concluded that in most cases the adsorption is entropically driven and that the adsorption is usually irreversible. The irreversibility results from the high adsorption energy due to the large number of adsorbing segments and from the unlikely event of simultaneous desorption of all the segments. This issue has some implications on protein chromatography and purification, since desorption can be achieved only by changing the solution conditions or by using displacers.

V. PROTEINS AT INTERFACES OF BIOLOGICAL INTEREST

The physicochemical phenomena described in previous sections (i.e., adsorption kinetics, desorption, conformational changes, surface denaturation, and coagulation) may occur in both natural and technological systems. In this section we demonstrate the relevance of the colloid chemical viewpoint of the surface activity of proteins, to several selected systems.

A. Enzymes at Interfaces

The adsorption of enzymes at interfaces is common in two cases: (1) enzymatic reactions that occur at the interface between two immiscible liquid phases, as in the hydrolysis of fats by lipase [30–32], and (2) immobilization of enzymes, by adsorption onto microcarriers, in order to achieve a heterogeneous catalytic reactor [1, 30,33]. It appears that the main problem related to these applications is the conformational changes of the enzymes, due to unfolding at the interface. As shown by Sandwick et al. [34], the enzymatic activity of all the enzymes studied was lost upon adsorption to hydrophobic polystyrene latex (alkaline phosphatase, β-galactosidase, lysozyme, horseradish peroxidase, and catalase).

Studies performed with enzymes adsorbed on porous glass concluded that inactivation of the enzymes varied from one enzyme to another and did not follow a common pattern [24,35]. The activities of the free plasmin, trypsin, and phosphatase were greater than those in the adsorbed state, while the opposite was observed for thrombin. However, the activities of all adsorbed enzymes deceased with increasing adsorption time, a result that could be explained by the possible unfolding of the enzymes at the interface as described in previous section. It should be emphasized that there are also reports on immobilized enzymes which retained their activity [1,30,33], but the conditions in which the adsorption was performed were not reported.

An interesting concept was suggested by Torchilin et al. [36]: formation of a more surface-active enzyme by covalent attachment of hydrophobic groups to the protein molecule. The surface-active enzyme, chymotrypsin, was successfully introduced into a phospholipid membrane and retained its activity. The increase in surface activity of the modified protein could also be used to increase its adsorption onto various carriers, as shown for modified ovalbumins adsorbed onto hydrophobic silica particles.

The appealing feature of adsorption immobilization is its simplicity. Therefore, this technique is widely used in industry, by adsorption onto ion exchangers such as DEAE-cellulose [1]. However, the conditions that are essential to retain enzyme activity are not well established. From previous sections it could be speculated that the best result could be achieved by performing the adsorption at conditions that favor the rapid formation of a close-packed monolayer (e.g., high bulk concentration and pH equal to the isoelectric point of the protein).

B. Cell Attachment onto Surfaces

Cell attachment onto surfaces has both positive and negative aspects, and is usually initiated by adsorption of protein to the surface. The best known examples, adhesion of blood cells onto biological and synthetic surfaces and adsorption onto dental surfaces, are discussed elsewhere in this volume. Generally, the initial step of adhesion involves the adsorption of a protein layer at the surface, which in turn enhances the adsorption of various cells. The interaction could be either specific, in which a recognition mechanism is involved, or nonspecific, in which various proteins could lead to similar adhesion.

Cell attachment is of great importance in biotechnology, in work based on tissue cultures. Unlike bacterial cultures, which can grow in suspensions, most mammalian cells, such as fibroblasts, require attachment to a surface in order to undergo mitosis [25]. The attachment is achieved upon adsorption of plasma or serum proteins onto the solid surface (glass or polystyrene). Thereafter, the cells are attached to the protein monolayer instead of the clean surface itself.

Another biotechnological application based on protein adsorption is cell separation. The concept was introduced in 1969 [38] and is based on adsorption of a protein molecule, BSA, on a carrier, which was followed by biospecific adsorption of cells that formed antibodies to BSA.

A unique approach to cell growth medium is based on attachment of cells at the oil–water interface. By this method, cells are attached to fluorocarbon or silicon oil droplets, hence allowing new manipulations of the cell cultures [25,39]. It was suggested that the attachment of the cells was performed via adsorption of BSA at the oil–water interface, but no adsorption data are given.

C. Emulsification

The role of proteins in formation and stabilization of emulsions was investigated extensively [40,41]. The most common naturally occurring emulsion is milk, which is composed of fat globules surrounded by casein molecules and other components, such as cholesterol. As stated earlier, casein is a surface-active molecule, which can easily unfold at the interface, thus reducing significantly the interfacial tension and may form a stable film that is irreversibly adsorbed. These features are of great importance in formation and stabilization of emulsions: The protein molecule should diffuse rapidly to the interface, reduce the interfacial tension, and lead to formation of steric and electrical barriers against coalescence. Due to their surface activity, the proteins are widely used in commercial food emulsions, such as dairy products, mayonnaise, and so on, and their role in the various emulsions is similar to that of raw milk, namely to form a membranelike structure, which surrounds the droplets. It should be noted that the transport of lipids and fats in the blood circulation is performed by "pseudoemulsions," the lipoproteins. The lipoproteins are composed of a core of triglycerides and esterified cholesterol, and an outer shell composed of cholesterol and several apoproteins. The protein content may vary and therefore the lipoproteins are divided into classes based on their density (HDL, LDL, VLDL, VHDL).

SUMMARY

The proteins play an important role in many other biological systems, such as incorporation in cell membranes, lung surfactants, drug targeting, and immunodiagnostics. It is expected that better understanding of the amphipathic nature and hence the surface activity of proteins would contribute to more sophisticated applications in which the biological activity is combined with the surface activity of the proteins.

REFERENCES

1. A. M. Klibanov, *Science 219*:722 (1983).
2. R. E. Baier and R. C. Dutton, *J. Biomed. Mater. Res. 3*:191 (1969).

3. I. Giaver and E. Ward, *Proc. Natl. Acad. Sci. USA 75*:1366 (1978).
4. R. E. Baier, ed., *Applied Chemistry of Protein Interfaces*, Advances in Chemistry Series 145, American Chemical Society, Washington, D.C., 1973.
5. F. Macritchie, *Adv. Protein Chem. 32*:283 (1978).
6. J. L. Brash and T. A. Horbett, eds., *Proteins at Interfaces*, ACS Symposium Series 343, American Chemical Society, Washington, D.C., 1987.
7. J. Lyklema, *Colloids Surf. 10*:33 (1984).
8. E. Tornberg, *J. Colloid Interface Sci. 64*:391 (1978).
9. J. M. C. Lankveld and J. Lyklema, *J. Colloid Interface Sci. 41*:454 (1972).
10. H. L. Frish and R. Simha, *J. Chem. Phys. 24*:652 (1956).
11. A. F. Ward and L. Tordai, *Recueil, 71*:572 (1952).
12. J. T. Davies, *Biochem. Biophys. Acta 11*:165 (1953).
13. D. E. Graham and M. C. Phillips, *J. Colloid Interface Sci. 70*:403 (1979).
14. F. MacRitchie and F. Alexander, *J. Colloid Sci. 18*:458 (1963).
15. F. MacRitchie, *J. Macromol. Sci. Chem. A4*:1169 (1970).
16. A. Kato and S. Na Kai, *Biochim. Biophys. Acta 624*:13 (1980).
17. S. Magdassi and A. Stawski, *J. DIspersion Sci. Technol. 10*: 213 (1989).
18. D. E. Graham and M. C. Phillips, *J. Colloid Interface Sci. 70*: 427 (1979).
19. J. L. Brash and D. J. Lyman, in *The Chemistry of Biosurfaces* (M. L. Hair, ed.), Marcel Dekker, New York, 1971, p. 177.
20. W. Norde, *Adv. Colloid Interface Sci. 25*:267 (1986).
21. B. Ivarsson and I. Lundstrom, *CRC Crit. Rev. Biocompat. 2*:1 (1986).
22. J. D. Andrade, ed., *Surface and Interfacial Aspects of Biomedical Polymers*, Vol. 2, Plenum Press, New York, 1985, p. 2.
23. W. Norde and J. Lyklema, in *Proteins at Interfaces*, ACS Symposium Series (J. L. Brash and T. A. Horbert, eds.), American Chemical Society, Washington, D.C., 1987, p. 36.
24. T. Mizutani, in *Proteins at Interfaces*, ACS Symposium Series (J. L. Brash and T. A. Horbert, eds.), American Chemical Society, Washington, D.C., 1987, p. 63.
25. I. Giaver and R. Kees, in *Proteins at Interfaces*, ACS Symposium Series (J. L. Brash and T. A. Horbert, eds.), American Chemical Society, Washington, D.C., 1987, p. 583.
26. W. Norde and J. Lyklema, *J. Colloid Interface Sci. 66*:266 (1978).
27. W. Norde and J. Lyklema, *J. Colloid Interface Sci. 71*:350 (1979).

28. B. Matuuszewska, W. Norde, and J. Lyklema, *J. Colloid Interface Sci. 84*:403 (1981).
29. P. G. Koutsoukos, W. Norde, and J. Lyklema, *J. Colloid Interface Sci. 95*:385 (1983).
30. C. Brady, L. Metcalf, D. Slaboszewski, and D. Frank, *J. Am. Oil Chem. Soc. 65*:917 (1988).
31. Y. Gorgouri, G. Pieroni, C. Riviere, L. Sarda, and R. Verger, *Biochemistry 25*:1733–1738 (1986).
32. K. Oyama, S. Nishimura, Y. Nonaka, K. Kihara, and T. Hashimoto, *J. Org. Chem. 46*:5241 (1981).
33. B. Huge-Jensen, D. R. Galluzzo, and R. G. Jensen, *J. Am. Oil Chem. Soc. 65*:905 (1988).
34. R. K. Sandwick and K. J. Schray, *J. Colloid Interface Sci. 115*:130 (1987).
35. S. Magdassi, M. Kleinberg, and S. Braun, unpublished results.
36. V. P. Torchilin, V. G. Omelyaneko, A. L. Klibanov, A. I. Mikhailov, V. I. Goldanski, and V. N. Smirnov, *Biochim. Biophys. Acta 602*:511 (1980).
37. S. Magdassi, D. Leibler, and S. Braun, *Langmuir 6*:376 (1990).
38. H. Wigzell and B. Anderson, *J. Exp. Med. 129*:23 (1969).
39. M. D. Rosenberg, in *Cellular Control Mechanisms and Cancer* (P. Emmelot and O. Muhlbock, eds.), Elsevier, Amsterdam, 1964, p. 146.
40. P. J. Halling, *Crit. Rev. Food Sci. Nutr. 15*:155 (1981).
41. M. C. Phillips, *Food Technol. 35*:50 (1981).
42. F. Macritchie and A. E. Alexander, *J. Colloid Sci. 18*:458 (1963).

13

Supercomputer Simulations of the Interaction of Biomolecules

ENRICO CLEMENTI, STEVEN CHIN, GIORGINA CORONGIU, JAMES GIVEN*, GEORGE C. LIE†, MICHELE MIGLIORE‡, and PIERO PROCACCI§ IBM Corporation, Kingston, New York

I.	Global Simulations and Productivity	301
II.	Example of Global Simulation	310
III.	Water as a Solvent: Simple Ions, DNA, Proteins, and Gramicidin as Solute	316
IV.	Conclusions	322
	References	329

I. GLOBAL SIMULATIONS AND PRODUCTIVITY

A basic goal in science is to achieve universal understanding, thereby providing quantitative models for any event, in specific as well as for all events in general. Engineering attempts to put this understanding to work in practical applications and realizations of ever-increasing complexity and boldness. These goals and attempts have been with humankind since the beginning of our social evolution and

---

Current affiliations:
*Department of Chemistry, State University of New York, Stony Brook, New York.
†Dow Corning Corporation, Midland, Michigan.
‡Institute for Interdisciplinary Applications of Physics, National Research Council, Palermo, Italy.
§Dipartimento di Chimica, Università di Firenze, Firenze, Italy.

today seem to be more and more at hand. A turning point on the learning curve is the ability to perform global simulations, which, as we shall show, can be accomplished only if supercomputers are extensively available.

In the last 40 years we have witenssed the birth and evolution of computers, minicomputers, personal computers, and supercomputers. However, all the available indications point to the expectation that we are only at the beginning of a new socioeconomic era, where computing machines and signal processing machines play a determinant role. Ages are often classified in terms of some characteristic of its production tools—thus the stone, bronze, and iron ages. As we know, "tools" enhance the dimension and power of human beings; in the past the physical dimension was enhanced, but today it is the intellectual and mental dimensions. We are clearly in the computer age.

There are many ways to define a supercomputer: for example, in terms of speed, cost, I/O, and so on. However, as a tool, we prefer to define it as the necessary tool to perform global simulations. Let us now once more define what we mean by "global simulations"; beforehand, however, let us note that obviously, computers with one to a few gigaflops of peak performance can be used to carry out computations requiring megaflop (or even lower) performance and a price—performance analysis might actually suggest a move in exactly this direction. Yet it would be a pity to use such beautiful machines for such low-complexity tasks! Indeed, supercomputers should be used essentially for super computations and for the exploration of new avenues. Achieving the ability to simulate complex laboratory experiments realistically is one such goal.

We have been advocating for some time the *global simulation* approach to computer simulations in science and engineering [1]. The main idea is borrowed from history: from the events leading England to the first industrial revolution when assembly lines were introduced, thus enormously increasing productivity per work unit. In an assembly line one starts with simple raw material and, step by step without interruption, transforms the raw material into more and more finished and complex products. In the global simulation model we start with the simplest assumption: that molecules are built up of point-charge nuclei and electrons. With quantum mechanics we can then assemble simple molecules or larger ones; generally, this is done in a timeless and temperatureless environment. With statistical mechanics we can consider many molecules at a given pressure and temperature, pass from an enthalpic description to a free-energy description, and consider trajectories involving time. Finally, for even larger systems we can use fluid dynamics, where viscosity, transport coefficient, convection, turbulence, and so on, can be analyzed. Of course, all this is well known at the theoretical level, but the novel aspect we have been stressing is how to move from one mechanic to the next in such a way that the entire process is self-consistent for computer simulations at the operational level. Another novel proposal is to use molecules, Newton equations, and boundary conditions to solve problems previously considered at the fluid dynamical level only. Briefly stated, in the global simulation model we decompose a given problem into n subproblems, each corresponding to a submodel I, ..., i, ..., n. The operational rules are that the input needed in the submodel i is fully obtained as the output from submodel i-1 and that the input for submodel 1 must be very simple in a computational sense. Thus the global simulation approach is an assembly line designed to increase our productivity in generating information.

Let us now consider in some detail the relationships between the global simulation modeling and productivity; as it is known, the latter correlates both to goods and information. The main evolution of productivity in human history is sketched in Fig. 1. The productivity of information trails the main phases of the productivity of goods; it is therefore not surprising that even though production assembly lines revolutionized industrial output since the early nineteenth century, the global simulation method is only now being introduced. We recall that two conditions (market demand and a tool—the automated loom) were essential for initiating a revolution

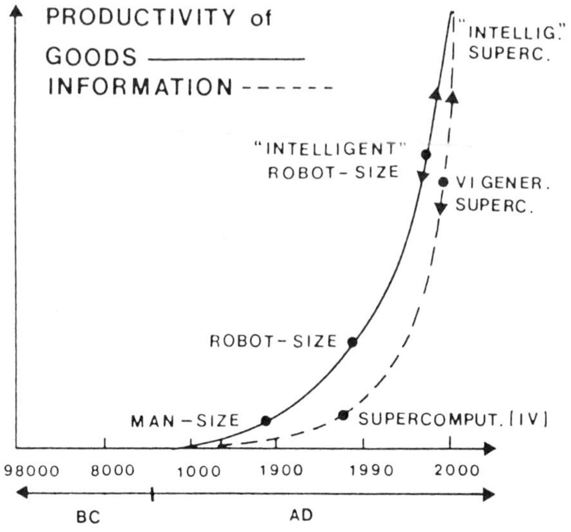

FIG. 1. Some aspects in the evolution productivity. Landmark dates for productivity of goods and information are shown on the time axis.

in the textile industry. Equivalently, the need and demand for interdisciplinarity of information, essential to avoid expensive errors in our highly technical and thus risk-prone society, had to wait for the appropriate tool—the supercomputer—before we could set up a production assembly line for information, namely the global simulation approach.

Advances in material sciences, brain research, networking, signal processing, cybernetics, and artificial intelligence are sufficiently reliable pointers for confidently predicting intelligent robots for productivity of goods, and intelligent supercomputers for productivity of information (see Fig. 1). An intelligent supercomputer, in this context, is an engine of one to a few gigaflops sustained performance, thus with much higher peak performance (up to about 100 gigaflops) and with the attributes to hear, talk, and—within the limit of expert systems—understand an extremely rudimental language (e.g., Fortran), and to see, display, and understand a few two-dimensional patterns such as digits and logical-mathematical symbols. We recall that pattern and voice recognition and artificial intelligence are today capable of performing these tasks and could notably increase the user friendliness of computer systems and thus enhance the user's productivity. The raw speed we have forecasted for the intelligent supercomputer of the next decade is conservative. Indeed, an lCAP/3090 with 30 nodes (five clusters of IBM-3090/600) would exhibit a peak performance of over 3 gigaflops; the same architecture and software with a cycle time of 2 ns would yield a performance increase of about a factor of 10 for a total of about 30 gigaflops. In the 10 years between now and the end of this century we can expect three new generations with a speed up to a factor of 3 each time; thus one would expect a performance not far from a teraflop. The forecast of an intelligent supercomputer within this century is most reasonable.

The tool to perform global simulations must be a supercomputer with super scalar, vector, and parallel characteristics. Let us recall that when we operate on numbers we can select between either the scalar (S), vector (V), or parallel (P) mode of operation. Let us consider S, V, and P like vectors of a three-dimensional space. Any computation is then associated to a trajectory within the SVP space, as shown in Fig. 2.

For complex algorithms, the corresponding trajectories will often occur within a preferential volume in the SVP space, thus defining an envelope, characteristic of the given algorithm. The ideal computer for an algorithm is the one that has high performance for the algorithm envelope in the SVP space. For global simulations the ideal computer must have superperformance along each of the three dimensions (i.e., S, V, P) it being most unlikely that any simpler choice will suffice. Alternatively stated, the envelope for global

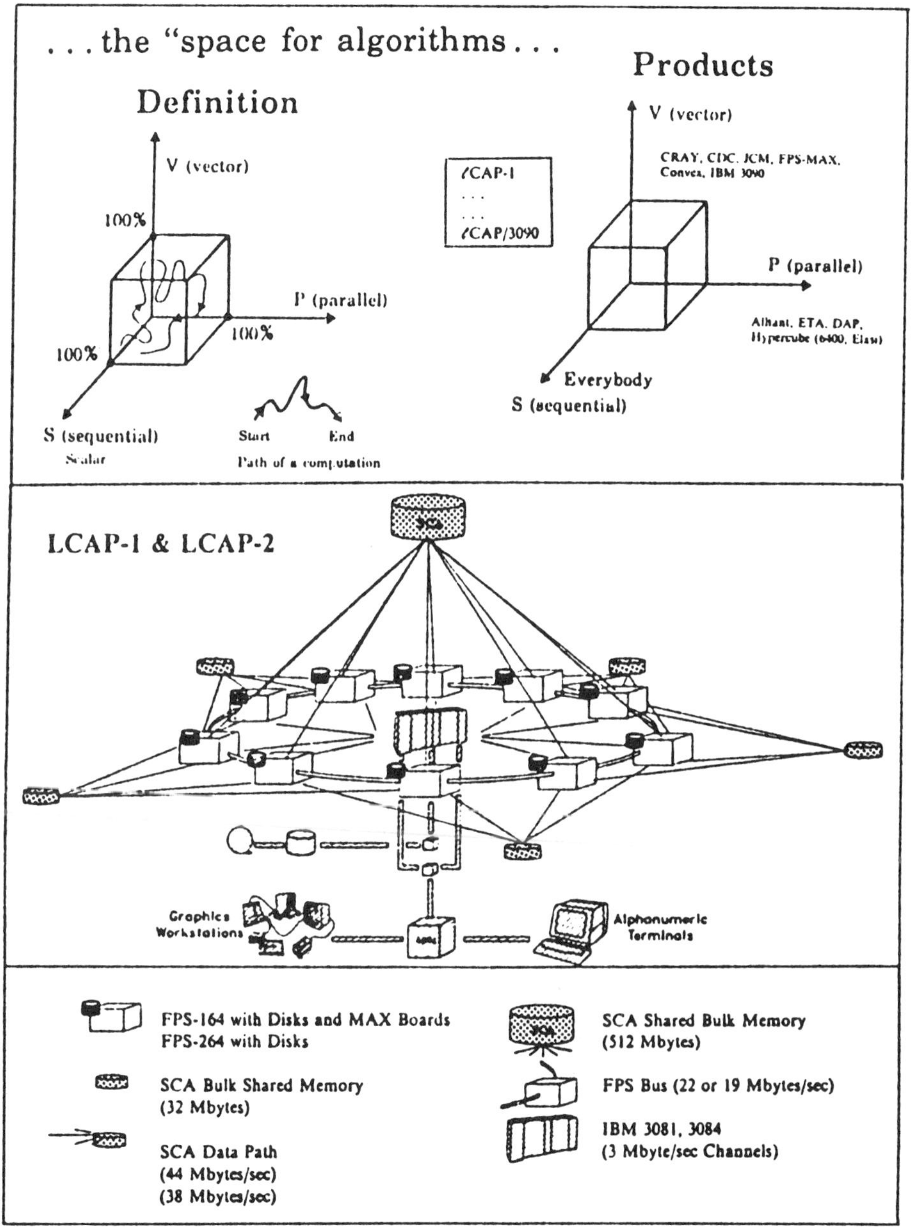

FIG. 2 LCAP-1 and LCAP-2 supercomputers (see the text) and the SVP space.

simulations extends over most of SVP space. In the early 1980s when we extended our simulations to molecular dynamics and fluid dynamics, there was no such supercomputer; the need therefore existed for us to build one. Our first attempt resulted in the so-called loosely coupled array of processors number 1 (LCAP-1, for short), consisting of ten FPS-164 array processors, each with local disk packs and eight megabytes of random memory; two vector accelerators (Max-Boards) brought the FPS-164 peak performance from 11 megaflops to 55 megaflops. These 10 nodes were individually connected to an IBM mainframe via standard IBM channels. In addition, five 32-megabyte memories connected (with a transmission rate of 44 megabytes/s) four successive nodes, forming a double-ring connectivity; further, a large memory of 512 megabytes connected each of the 10 nodes. Our computer had about 550 megaflops peak performance. A second system, LCAP-2 was constructed with FPS-264s replacing the FPS-164s; this second system had about 380 megaflops peak performance. The two supercomputers were connected for experimentation in cluster-to-cluster parallel computing and the resulting system, with about a 1-gigaflop performance, was called lCAP-3 (see Fig. 3).

The power of LCAP-3 can be exemplified by recalling the colossal ab initio SCF computation for gramicidin and a K^+ ion which required only about 11 h of elapsed time. We note that (1) this system has 276 atoms with 1010 electrons, (2) we used 2736 primitive Gaussian functions, (3) the SCF threshold was set at 10^{-5} atomic units, and (4) only integrals smaller than 10^{-5} atomic unit were neglected. All other integrals, a total of 433,646,989, were computed in double precision and stored on disk packs (working in parallel). For additional details on the LCAP system and performance relative to other commercial supercomputers, we refer to reader to refs. 2, 3, and 4.

Presently, we are assembling one more supercomputer: lCAP-3090. In this system each cluster is an IBM-3090 system that can have up to six nodes (IBM-3090/600); we are considering up to five clusters. In Fig. 3 we have shown an lCAP-3090 built up by IBM-3090/400 clusters. The advantages of LCAP-3090 over LCAP-3 are (1) one single vendor for all hardware and software; (2) very high hardware and system software reliability; (3) portability of application programs relative to LCAP-1 and LCAP-2; (4) peak performance up to 3 gigaflops; (5) growth opportunity starting from a single node; and (6) vector hardware, system software, and mathematical library availability.

In Fig. 4 we report on the efficiency in parallelization for several (12) application areas. The optimum line (straight line) states that if we use N nodes, we expect at best an N-fold speedup relative to N = 1. Of course, one would expect some degradation due to overhead at the system software, hardware, and application program levels. Still, for the many applications reported, the ideal

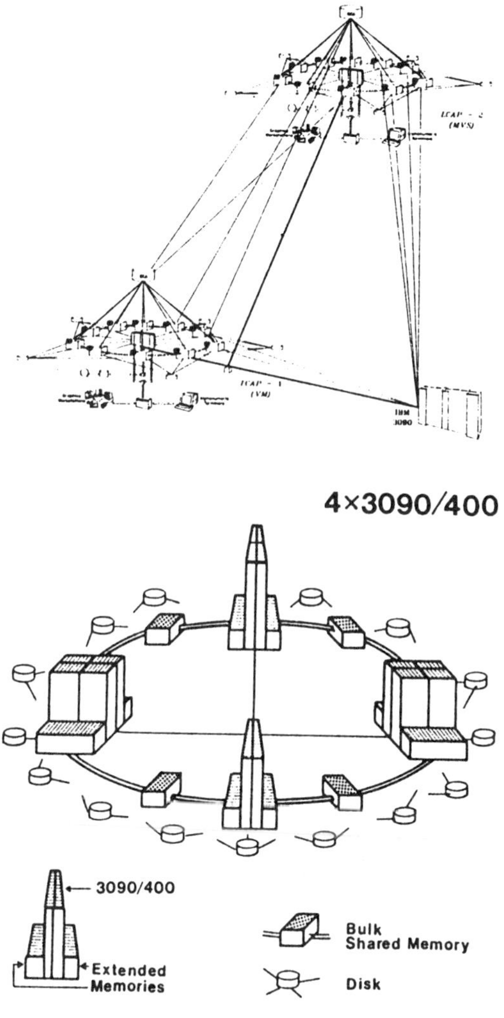

FIG. 3 LCAP-3 (a) and LCAP-3090 (b): the 10 nodes in each one of the two clusters of LCAP-3 are explicitly drawn; for LCAP-3090 the nodes of each cluster are compacted into one box.

goal is reached to a very considerable extent. We expect that the LCAP-3090 will reduce the overhead even more, thus yielding even better performance. For more details on the 12 examples reported in Fig. 4, see refs. 5 to 9. Notice that most of computational chemistry and very important areas in physics and engineering

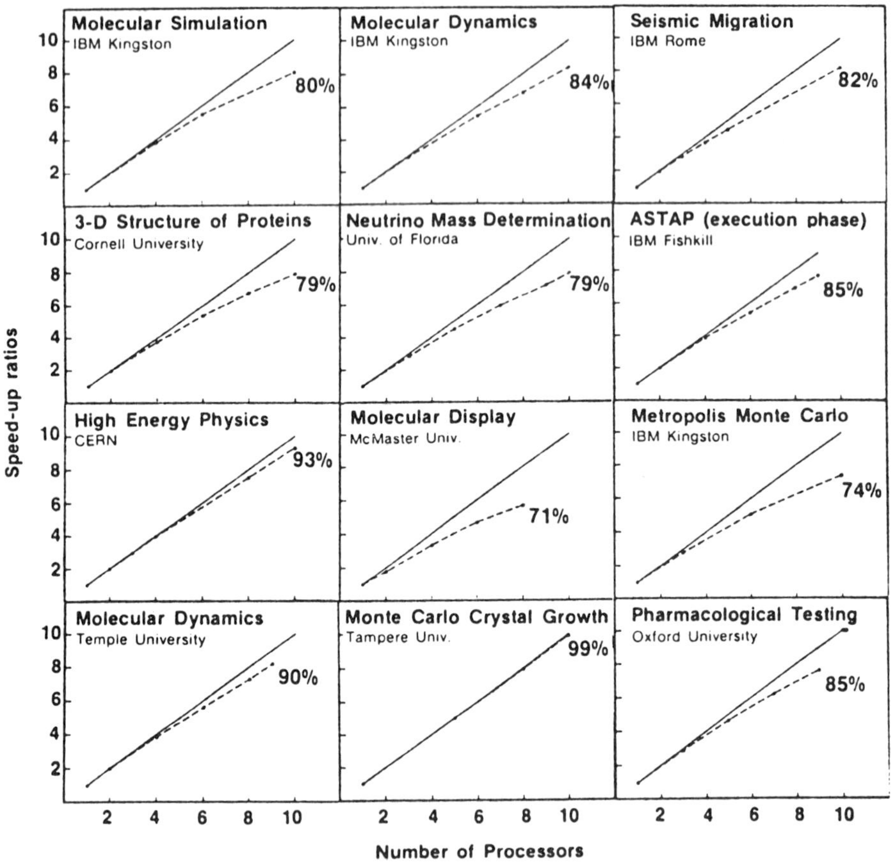

FIG. 4 Percent parallelization in simulations for science and engineering.

are essentially represented in the examples of Fig. 4. The 12 examples reported in Fig. 4 have some characteristics in common. To start with, the codes were originally written for sequential computers and the parallelization was obtained using our first-generation precompilers (1983 to 1985 vintage). The reported speedup [namely, the ratio T_n/T_1, where T_1 is the elapsed time for execution on 1 node or array processor (AP), and T_n is the elapsed time for the same job executed in parallel on n nodes] is obtained considering the full job, not only kernels; thus input—output timing is included. Alternatively stated, the examples correspond to computations executed to elicit interest in a scientist and/or an engineer rather than in a computer-scientist or a computer-hardware architect.

The 12 examples above did require a little code modification for the translation from sequential to parallel and no algorithm modification. Indeed, the examples above can be characterized as coarse-grain parallelism and correspond to what we refer to as *natural parallelism*. Often the structure of a coarse-grain parallel code is composed by a set of nested DO loops; the parallelization is done in the outermost loop and the innermost loop is often vectorizable; in this case we have optimal computational conditions, that is, vectorization and parallelization. Quantum and classical many-body dynamics and also quantum or classical stochastic problems are most often naturally parallelizable. Other natural parallel codes are those that simulate computations easily partitionable into subtasks. An interesting example of the latter is graphics, where clearly the rotation and hidden-line computing of a complex image can just as well be obtained by operating with many parcels of the image all at the same time.

Natural parallelism is often found in complex solutions since it tailors well the scientific method. Indeed, in the case of a complex problem we must often decompose complexity by solving first for only some of the variables (the most relevant ones) and then coupling the partial solutions obtained either with perturbation or relaxation techniques or appropriate many-body expansion. However, codes and algorithms are not always naturally parallel! In such cases it is basic that each node communicates efficiently with other (either a few or many) nodes. This can be only poorly accomplished with today's slow channels, but rather efficiently with buses and bulk shared memories. For this reason we have devoted much attention in the 1985–1987 period to these communcation devices, which move our lCAP toward a tCAP (i.e., from "loose" to "tight" coupling).

The FPS-BUS software is based on a message-passing scheme; the user has to specify from which and to which processor data are sent and received. If the same data have to be sent from one processor to all the others, the broadcast mode can be used. The facilities provided by the precompiler for using the bulk memories offer a choice of two modes: the shared and the message-passing mode. In the shared mode the data to be shared are stored in the memory at locations known to all the processors, and it is passed between processors by specifying a write of data from one processor and by a read of data by another one. In the message-passing mode, messages in the form of contiguous strips of data are passed between processors via a network of paths transparent to the user, who specifies in his code the send and receive transactions along with the targeted processors.

Performance measuremens have been done using either the FPS-164 or the FPS-264 as processors, exploiting the three communications schemes described above, and choosing different patterns of data transmission. We refer the reader elsewhere (refs. 4 and 10) for details of performance; here we report on an interesting example of the use of these communication devices for a Metropolis Monte Carlo study of water using two- and three-body potentials for a box containing 512 particles [10]. The net conclusion of this study is summarized in Fig. 5; loose parallelism is insufficient for this case and, all in all, the bulk shared memory is preferable to the BUS. We stress, however, that these timing examples should not be extrapolated to machines and software much different than the one used. Indeed, for example, the latency time for transmission of data depends very much on the specific system software and hardware that one has [11].

II. EXAMPLE OF GLOBAL SIMULATION

Let us consider, as an example, the simulation of some chemical in the liquid phase and, to be more specific, let us consider water, a rather complicated but most interesting liquid. To start with (the input for the first submodel), all one needs to know is that one molecule of water has the stoichiometric composition of two hydrogen atoms and one oxygen atom, which, to fit our submodel, we

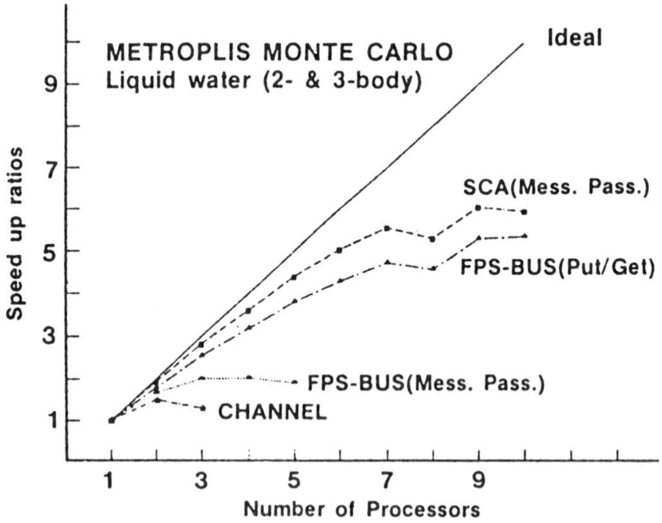

FIG. 5 Parallelization with either channels of FPS-BUS or SCA bulk shared memories. (From ref. 10.)

translate into the notion that our molecule is made up of 10 electrons and has two one-charge-point-like nuclei plus one nucleus with eight positive charges. Standard quantum chemistry yields a stable molecule; within the Hartree–Fock approximately [12] the simulated water molecule has an HOH angle slightly too wide relative to experimental data [13]. The correct (within experimental accuracy) value is easily obtained by correlating the electrons' motions [14]. Thus we now know the geometry and the energetics of the water molecule from simulations; it is also not difficult to obtain the vibrational motions by mapping the intraatomic potential surface in the vicinity of the equilibrium configuration [14].

Let us move from one molecule of water to two, three, ..., several [15], and finally many. Computations of the interaction between two molecules even at the near to Hartree–Fock limit [16] yield the well-known hydrogen-bond dimer (see Fig. 6). This picture

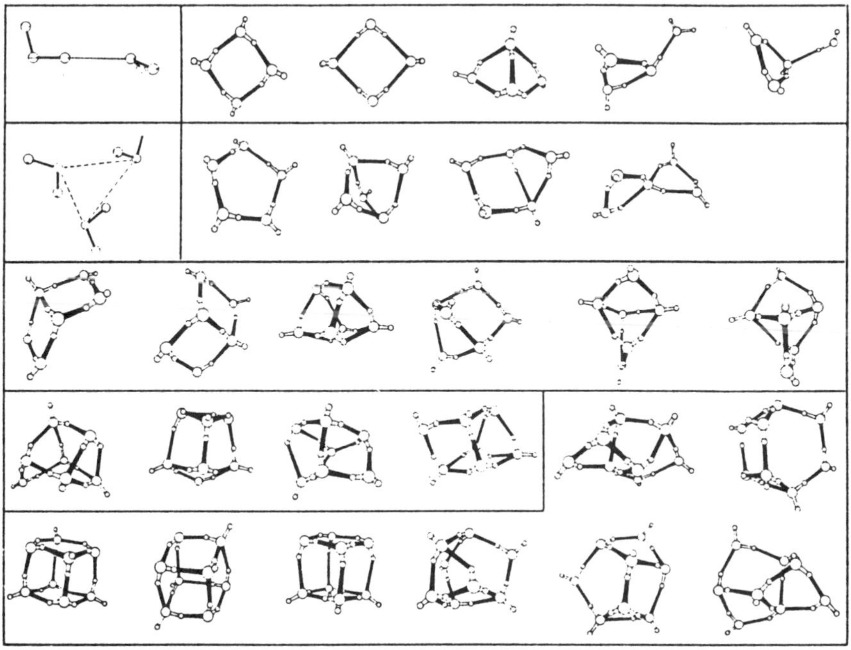

FIG. 6 Clusters of water molecules $(H_2O)_n$. For $N \geq 4$ the most stable configuration is given at the left, the less stable at the right. For details, see ref. 25.

can be improved by using correlated wave functions and energies [17] (e.g., configuration interaction approach). The computed energy points, sampling the potential energy surface, can be fitted to some analytical expression, within the guidelines of a simple physical model. In this way we obtained the two-rigid-waters intermolecular potential (two-body), known as MCY [17], or, by relaxing the water rigidity and allowing vibrational flexibility, the MYCL two-flexible body potential [18]. In the same way one could proceed and consider trimers, tetramers, and so on. A study of small water clusters [15] reminds us that even for a few water molecules, the cluster can have many stable geometries, all very close in energy: namely, that we must face a difficult minimization problem. It should be noted that the cluster study also provides us with very interesting geometrical arrangements of hydrogen-bonded structures, leading to clathrate-type complexes. Obviously, one should not build up the cluster size indefinitely, not only because it would become computationally impractical and eventually impossible, but mainly because it would be the logically incorrect approach to a problem whose answers are to be found in an understanding of statistical fluctuations.

It is time to move to the second submodel—equilibrium statistical mechanics—limiting ourselves to two main techniques: metropolis Monte Carlo [19] and molecular dynamics [20]. In the former, N molecules are placed in a volume V at a given temperature T and many (millions) geometrical configurations are randomly generated, their energy computed with the interaction potentials obtained from quantum mechancs. Properties are statistically averaged in the Boltzmann sense. Periodic boundary conditions are imposed on the N particles. In molecular dynamics again we place N molecules in a given volume V and let them interact according to the quantum-mechanically derived potential. For each particle the Newton's equations of motion are solved; we compute molecular trajectories (i.e., each molecule will modify its translational and rotational motion reacting to the field exerted by the N - 1 surrounding molecules). Only motions corresponding to a very small temperature range (ideally, one temperature value) are considered. The motions are discretized in time steps generally of 0.5×10^{-15} s for rigid molecules and 0.125×10^{-15} s for vibrating molecules.

Figure 7 reports some of the Monte Carlo results [18,21–23]. At the top of the figure we compare the computed and experimental scattering intensity for liquid water at room temperature using either x-rays (left) or neutron beams (right), comparing simulations at the two-body level with those that include three- and four-body corrections. As we can see, the latter brings about a marginal improvement in an already good match. However, for enthalpy the many-body corrections are very important. Thus static and

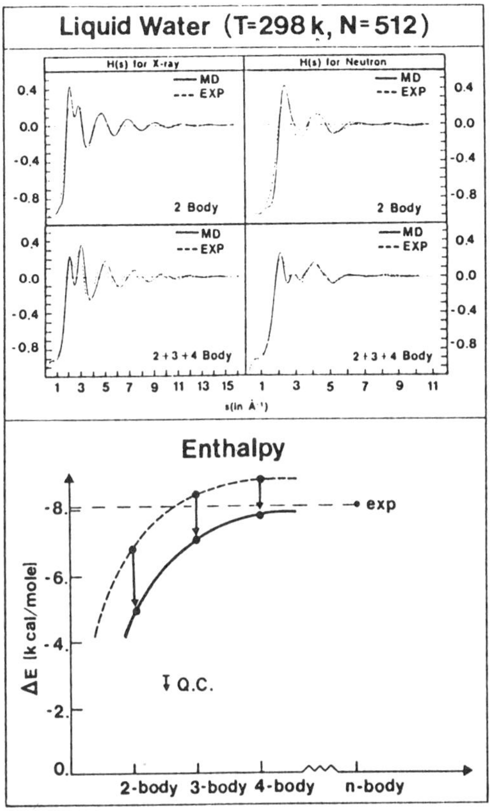

FIG. 7 Top: x-ray and neutron-beam spectra of liquid water at room temperature either from experiment or from simulations with or without three- and four-body corrections. Bottom: Monte Carlo enthalpy of liquid water from two-, three- and four-body potentials.

structural properties are expected to be well described at the two-body approximation (and even better agreement can be obtained using well-calibrated effective potentials), but this might not be the case for dynamical properties. The latter types of properties can be subdivided into two groups: those relating the motions (translations and rotations) of one water molecule within its solvation shell [24] and those describing the behavior of many water molecules [25] within a box of size L and with wavenumber $K = 2\pi/L$.

From Fig. 8 we learn that rotational motions [angular correlation functions (ACF)] are nearly five times faster than translational motions [velocity correlation functions (VCF)] for rigid molecules [24]. These motions and the corresponding orientational correlation functions are at the origin of the infrared and Raman spectra for liquid water. The computed spectra are reported at the bottom of Fig. 8; note that the intermolecular vibrational infrared (IR) spectra are incorrectly shifted to low values (bottom left insert), while the intramolecular motions (bottom right insert) are computed in remarkable agreement with experiment [18]. For the intramolecular motions we have also reported equivalent spectra for the gas phase. The spectral shifts, comparing gas to the liquid, are very accurately reproduced in our simulation. In Fig. 9 we report the dynamical structure factor, a collective property, computed from molecular dynamics and from hydrodynamics (i.e., where the value of K approaches zero). Note that the shape of the structure factor for $K = 0.319$ is not too dissimilar from that predicted from hydrodynamics [25].

The simulated results presented in Figs. 7, 8, and 9 show that within the spirit of global simulation we can indeed progress from nuclei and electrons, to a few water molecules, to the hydrodynamic limit. Let us now progress into the area hitherto reserved exclusively for fluid dynamics (i.e., for solutions of Navier–Stokes equations). We ask the question: Can we trust the Navier–Stokes equations for very small systems a few hundred angstroms in size and for events having a lifetime in the pico- or nanoseond range? Alternatively stated, the question concerns the description of systems generally larger than those analyzed by molecular dynamics but much smaller relative to fluid dynamics. We answered these basic questions by performing molecular dynamic computations with molecules or atoms interacting via intermolecular potentials, but in an open (nonequilibrium) system and with those boundary conditions that are typical in fluid dynamics. For example, in Fig. 10 we consider a flow past a plate [26] (upper insert) or past a cylinder [27–29] (bottom insert). Our simulations were performed on about 10^5 particles for events in the picosecond to nanosecond range over a sale of a few hundred angstroms. We have reported that such systems (amazingly well and very accurately) scale from results and laboratory experiments of systems at the cm^2 size and with a lifetime of seconds! For example, the Reynolds number at which we observe wakes past the cylinder is about 25 ± 10 (i.e., essentially equal to 35), the Reynolds number for which laboratory data observed the same behavior [27].

Equally gratifying are the results of Fig. 11, where we report on Benard-type flow obtained by considering 200,000 argon atoms [30]. This rather astonishing accomplishment, even if preliminary since we have not obtained true Benard's patterns, can be

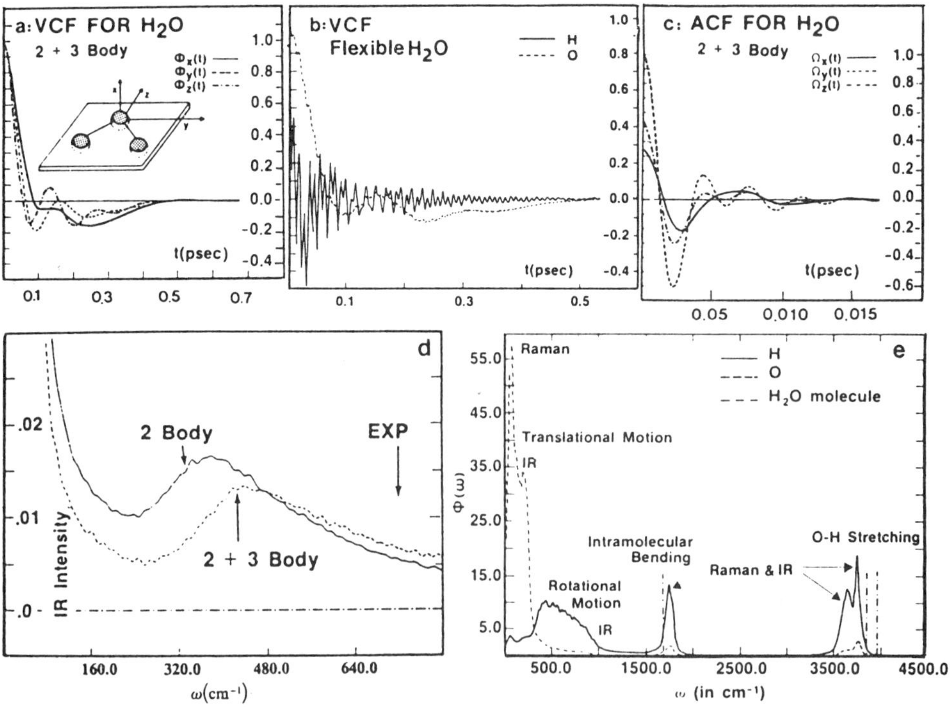

FIG. 8 (a–c): angular, ACF, and translational, VCF, velocity correlation functions for liquid water at room temperature from molecular dynamics simulations with three-body corrections and with flexible water (middle). (d) and (e): computed IR and Raman spectra.

appreciated by comparing our simulation with Benard patterns observed in laboratories at a trillionth larger scale. We returned to the problem focusing on the vertical boundaries since we noticed that the particles, when reflected, entered the fluid with excess energy, namely in a destructive way, and thus prevented the formation of convective flows. In Fig. 12 we report such flows, which exhibit the correct sinusoidal (rather than linear) temperature profile. For this and other details, see ref. 31.

In closing we note that the flow and the pipes in our simulations are about 1500 Å large and are thus of definitive biochemical

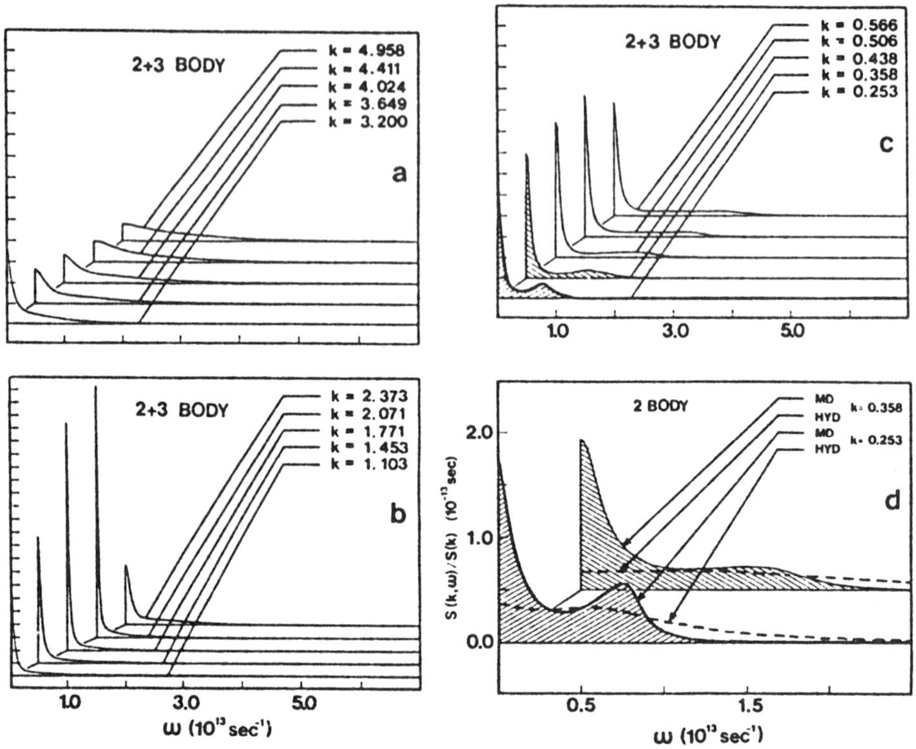

FIG. 9 Dynamical structure factor from molecular dynamic simulations (a–c) and from hydrodynamics (d).

interest. These results are exciting, but in the context of this chapter, the most relevant aspect is that indeed our global method is verified, starting from systems with a few nuclei and electrons up to vastly larger systems described previously only with fluid dynamics.

III. WATER AS A SOLVENT: SIMPLE IONS, DNA, PROTEINS, AND GRAMICIDIN AS SOLUTE

The study on a liquid described above is cast into the form of an example of the global simulation approach. Our interest in water,

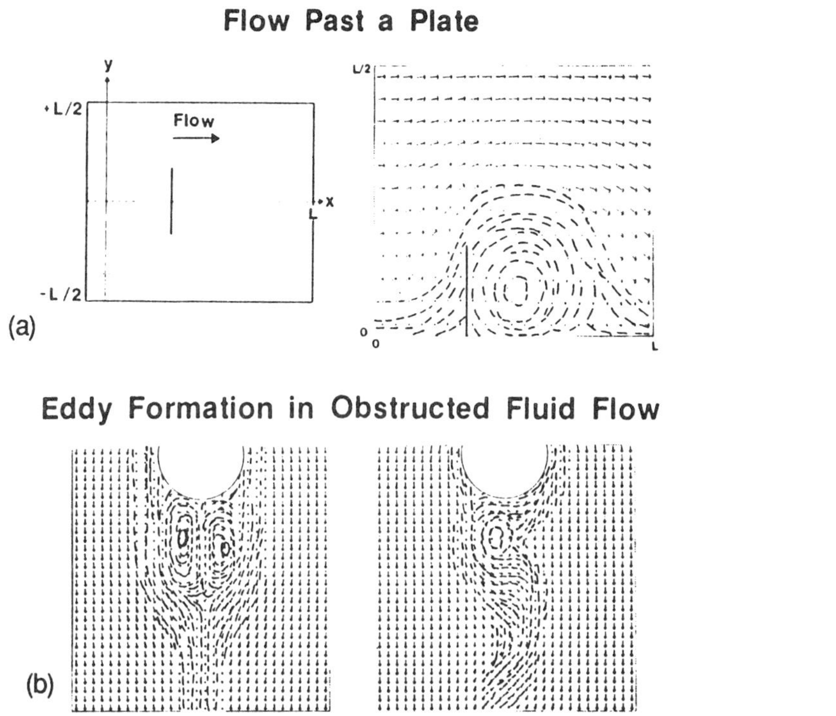

FIG. 10 (a) Schematic of a flow past a plate (left) and visualization of velocity field for steady state. Arrow lengths are scaled as the square root of the ratio of individual cell velocity to maximum cell velocity. (b) Velocity field for flow past a cylinder.

however, relates to our work in biophysics and biochemistry: as is well known, water is the most abundant solvent on earth and *the* medium to carry life. Water, as a solvent, is a network of hydrogen-bonded molecules in which rearrangement cost is low in terms of internal energy but not in terms of entropy variations. For a long time, evaluation of free energies was a difficult task; recently, this has been changing because of the use of a perturbational technique with a compling parameter [32] aimed at determining free-energy differences between solutes. In our laboratory [33] we have used these methods to evaluate the hydration free energies of monovalent ions such as Li^+, Na^+, K^+, F^-, and Cl^-. As shown in Table 1,

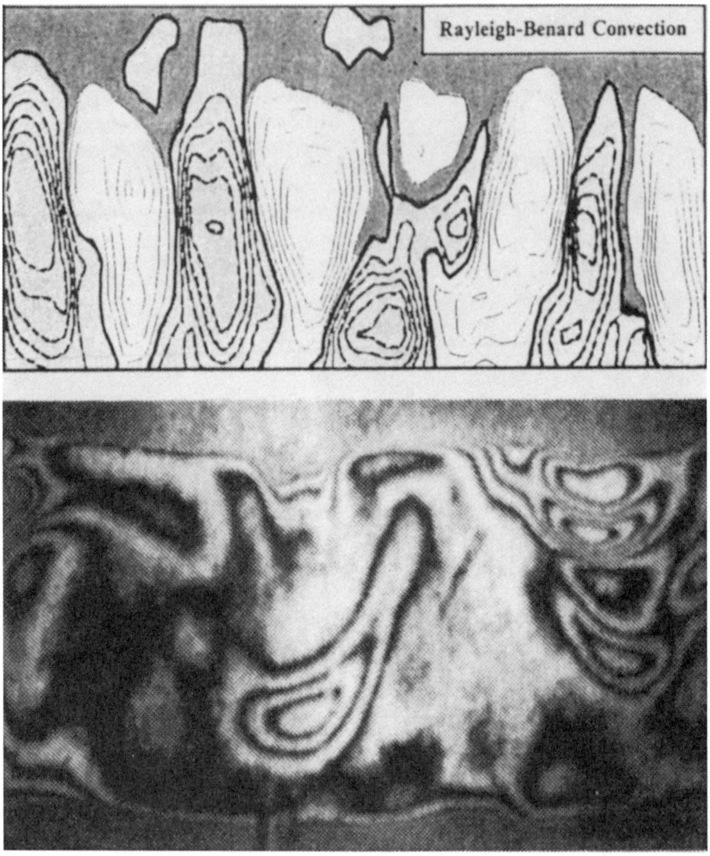

FIG. 11 Top: simulated Benard flow obtained from 200,000 neon atoms. Bottom: laboratory Benard-flow taken from a film with the kind permission of P. Bergé and M. Dubois. C.E.N. Saclay, Gif-sur-Yvette (France).

the computed energies are in relatively good agreement with experimental values, and the relative trends of positive versus negative ions correlate excellently [34]. The values reported in Table 1 have been obtained using two-body potentials. Introduction of three-body corrections brings the simulated free energy even closer to the experimental values. A simulation performed for Li^+ yielded a value of 148 ± 8 kcal/mol.

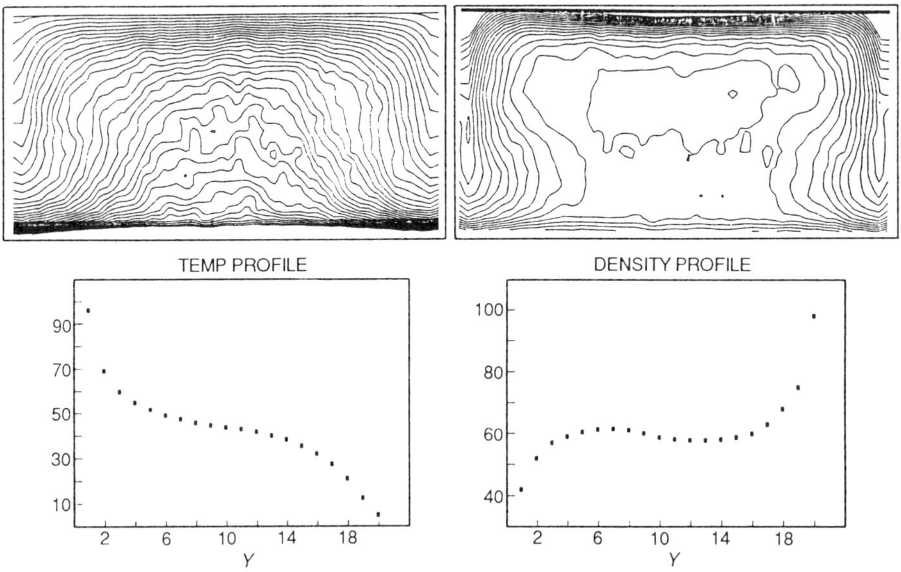

FIG. 12 Steady-state Rayleigh—Benard convective flow from J. Given and E. Clementi.

In Fig. 13 we condense aspects of our simulations of water solvating biomolecules. Specific interactions of water molecules and local hydrogen-bond rearrangements at the active site of enzymes are a concomitant aspect not discussed here (for papaine, see ref. 35; for carbonic anhydrase, see ref. 36; and for lysozyme, see refs. 37 and 38). Figure 13(a—c) shows the water network in the B-DNA grooves as first reported by our Monte Carlo simulations [39—41], later confirmed by Dickersen et al. [42], and more recently reconfirmed by our molecular dynamics study, where the interest was on the time dependency of the water motions [43,44]. Indeed, our simulation of the dipole correlation functions as computed is in good agreement with relaxation experiments by Lindsay et al. [45]. Sketches (b) and (c) report on our computed positions for the counterions in B- and Z-DNA, respectively, at a concentration of one counterion or phosphate group. We have stressed [40, 41] that a study of DNA solvation without explicit account of counterions (and vice versa) is meaningless; in addition, we should note that the exact location of the counterions around DNA is dependent on the counterions concentration (or excess).

TABLE 1 Free Energy (kcal/mol) of Hydration from Monte Carlo and Laboratory Data

Ion	MC results (ref. 33)	Experimental results	
		Ref. 34a	Ref. 34b
Li^+	140.9 ± 0.06	118.1	123.5
Na^+	109.6 ± 1.8	90.6	98.6
K^+	76.7 ± 2.7	73.1	80.8
F^-	99.8 ± 2.1	110.7	103.8
Cl^-	56.7 ± 0.6	81.5	75.8

In Fig. 13(d—f) we report a view of myoglobin, a cross section of a myoglobin unit cell with those water molecules identified in x-ray diffractions, and the much larger population of water molecules located by preliminary Monte Carlo simulations. Despite the enormous amount of experimental data available for this protein, its solvent structure is not yet well resolved: about 165 water molecules per protein have been identified from x-ray diffractions [46], but there is evidence that many more molecules are present. The aim of our study, initiated with Professor Parak, is (1) the determination of the positions of an additional 250 to 300 water moleucles presently not accounted for, (2) the determination of the number and structure of water molecules hydrating each protein residue, and (3) to find if there is any correlation between solvent mobility (obtained from our simulations) and protein dynamics (obtained from experiments). In this regard we note Frauenfelder—Parak's proposal that the measured x-ray root means displacements (rms) of the protein atoms arise from several kinds of motions, such as lattice vibrations, brownian diffusion, and fluctuations between conformational substates. At room temperature, the hydration water mediates, through its diffusive motion, the transition between conformational substates. We can attempt a correlation between water hydration and protein motion in the following way: Protein residues with experimentally high rms (and therefore with high contribution from conformational fluctuations) should be highly hydrated; protein residues with experimental low rms should either not be exposed to water or should be poorly hydrated or hydrated by highly structured water (as in the case of charged residues).

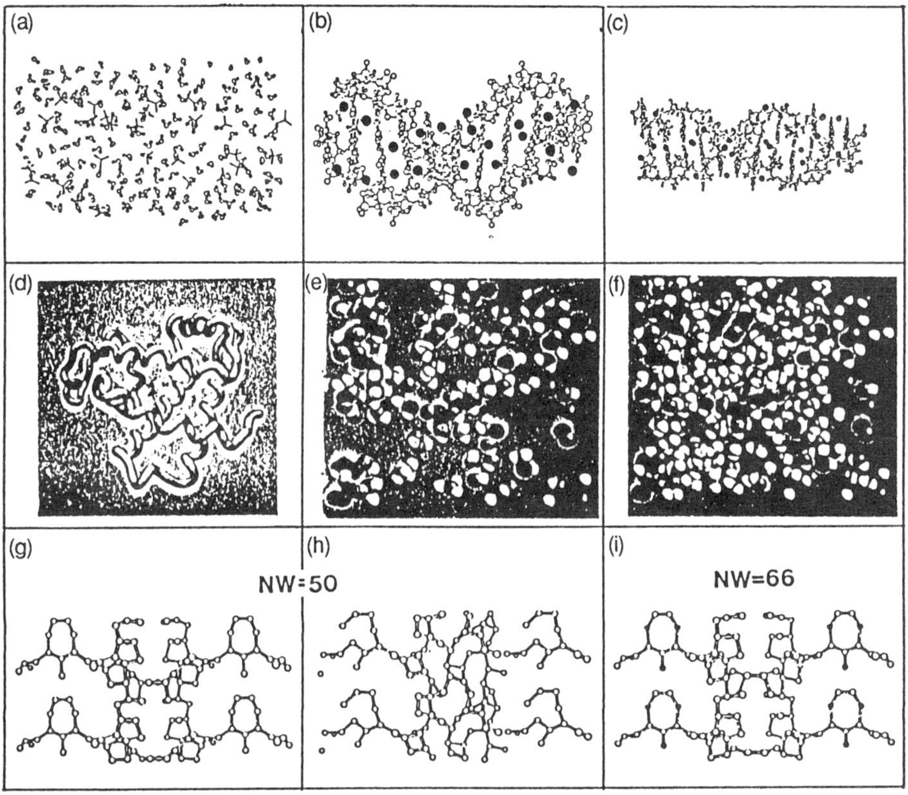

FIG. 13 (a—c) Water network in B-DNA grooves, K^+-ion positions in B-DNA and in Z-DNA: (d—f) ribbon model of x-ray crystal structure for myoglobin, and view of myoglobin crystallization molecules from x-ray diffraction and from Monte Carlo simulations; (g—i) oxygen-atom water network, from x-ray diffraction data, assuming 50 water molecules (simulated) and 66 water molecules (simulated).

A related simulation is the determination of the number and positions of the water molecules hydrating a crystal of proflavine deoxydinucleotide phosphate complex [47,48]. The oxygen-atom diffraction pattern (Fig. 13g), obtained experimentally [49] agrees well if we simulate, with Monte Carlo techniques, 61 or 66 water molecules per unit cell (Fig. 13g), but not 50 ± 2 water molecules (Fig. 13h), as assumed in the laboratory work [49].

Finally, in Fig. 14 we consider the transport of K^+ ions through a transmembrane channel, gramicidin A, using molecular dynamics. In Fig. 14(a) we report on the K^+ gramicidin interaction potential obtained by *ab initio* computations of K^+ with the entire gramicidin channel. With our lCAP-3 computer, each interaction energy evaluation requires about 4 h of computer time (however, the initial one requires 11 h). We recall that the system contains 256 atoms and 1050 electrons and is thus the largest system studied, up to now, by *ab initio* quantum-mechanical methods [50,51]. Figure 14(b) reports on the organization of water molecules outside the channel and inside when an ion is in the middle of the channel [52,53]. In Fig. 14(c) we present one frame from the computer-animated video [54]. It is very interesting to recall that upon application of a voltage differential across the channel a K^+ ion does not go through but is trapped at the mouth end unless it is sandwiched in between two water molecules. The latter undergo a continuous series of vibrations or even rotations, changing the electrostatic interaction with K^+ and thus pushing and pulling it through the channel.

IV. CONCLUSIONS

We have not mentioned the very obvious point that lCAP needs specialized system software. The master-slave and fork-joint concepts are very familiar and not new, but still much skill is required to provide for an operational system. This aspect of our department's research is discussed in a recent review [11]. The global simulation methodology is expected to become the standard approach to complex engineering problems, especially to those in which materials are subjected to stress, rapid temperature differentials, and unusual environments.

Because they are related to lCAP and to our work, two areas are worth mentioning: biophysics (with biochemistry and bioengineering) and modern organic syntheses of material for future computers. In Fig. 15 we report a matrix taken from a rather old work [55], where we summarize the pair potentials we have been and are working on. The assembly of these pair potentials is, in itself, a supercomputer task. The use of the pair potential will profoundly affect either specific problems (such as understanding liquid water) or entire fields (electrochemistry, for example), and again this will require supercomputers, possibly of the lCAP type. Presently, we have completed the pair potentials indicated in the first row of the matrix and for the first and last elements of the second row. Concerning the third row, it is our pleasure to report that the amino

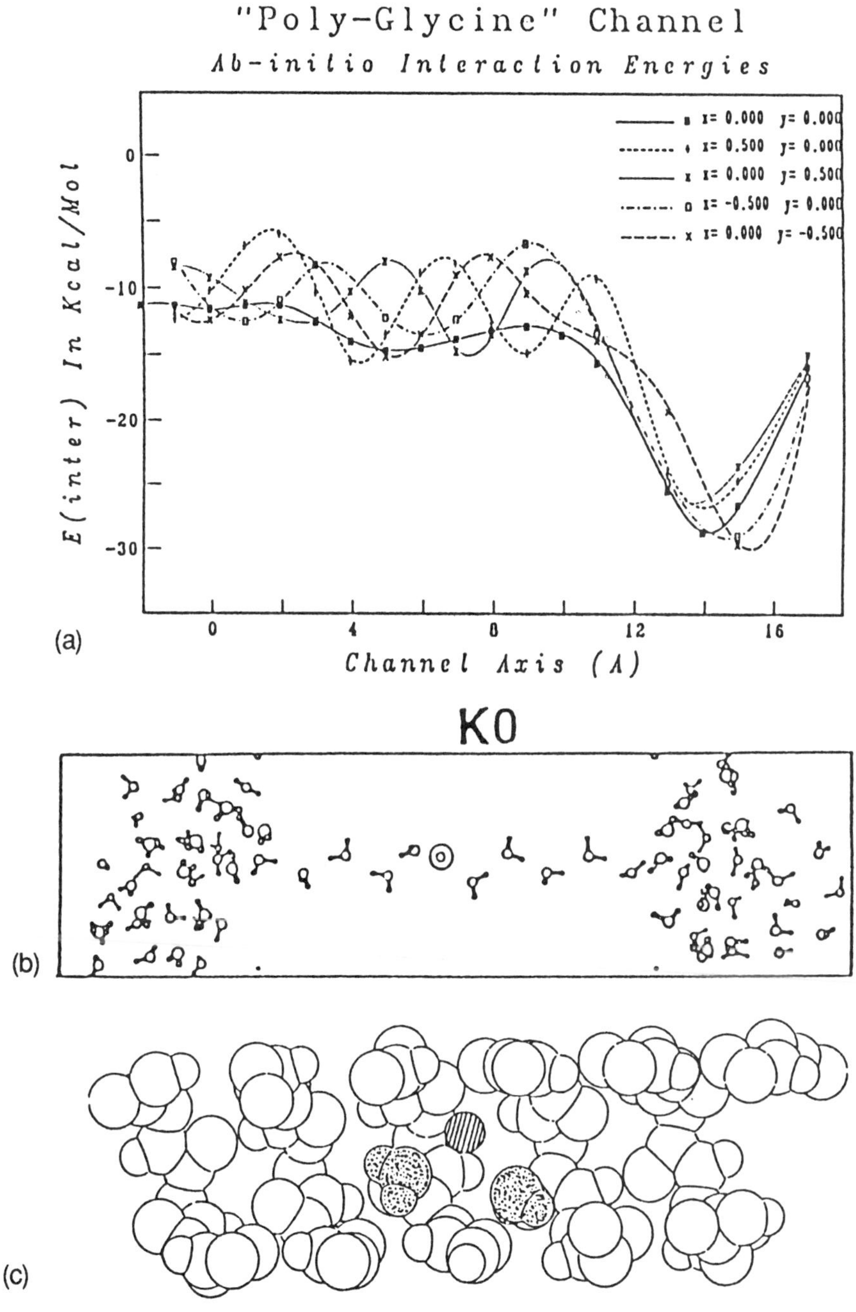

FIG. 14 (a) *Ab initio* interaction energies for K^+ with the entire gramidicin channel; (b) Monte Carlo water molecules outside the channel and around K^+ inside the channel; (c) two water molecules sandwiching K^+ inside the channel from molecular dynamics simulations. (Frame from computer-animated video, ref. 54.)

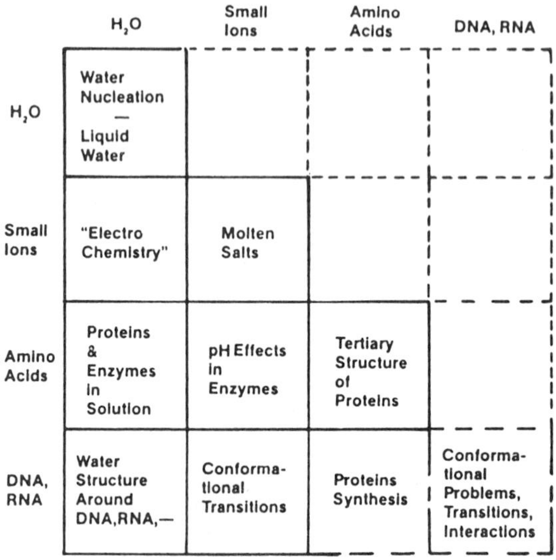

FIG. 15 Matrix of pair-potentials, and problems that can now be solved.

acid—amino acid potentials have very recently been obtained, and about 3000 interaction energies have been computed for the interaction of amino acids—DNA. The amino acid—amino acid potential required about 5000 *ab initio* computations; the amino acids—DNA will require about two to three times that number. The amino acid—amino acid pair potential library is complemented by the equivalent library for amino acid—water [56—58].

As known, today the potentials needed in Monte Carlo and in molecular dynamics to simulate protein motions and active site reactivity and to refine or even determine the protein structure are fitted from experimental data, particularly x-ray data. The number of parameters needed are about three per pair potential. Even if most atoms can be grouped under a given type of pair potential (what we call "class"; see refs. 56—58), we have to determine hundreds of parameters. The problem is that there are not sufficient numbers of mutually consistent experiments to obtain a reliable fit. Even for the water—water interactions, despite the hundreds of experiments, ambituities have plagued the literature with quantitatively unreliable pair potentials. This problem only gets more serious if we consider the pair potentials for the interaction of proteins with DNA.

Let us discuss, as an example, attempts to obtain protein structure from computer simulations. For the past several years we have been working in collaboration with Professor Scheraga's group to predict the three-dimensional structures of proteins and, in particular, the structure of interferon using the technique of direct energy minimization. Human leukocyte interferons (IFN-α) are antiviral proteins of 155 or 156 amino acid residues, which are synthesized and secreted by leukocytes in response to several types of viral or bacterial infection [60]. At least 15 different IFN-αs have been identified and sequenced [61,62]. Attempts have been made to determine the regular backbone structure with the general agreement that the protein is rich in α-helix with a small amount of β-sheet [63].

In our work on interferon we have concentrated on building the residues toward the C-terminus end of the protein. A previous publication [64] had reported the building of three polypeptide fragments of 27, 9, and 8 residues in length. We have continued this work, eventually minimizing a 98-residue fragment of the interferon, encompassing residues numbered from 52 to 149. This 98-residue fragment was built by combining two fragments, one of 47 residues and a second of 51 residues. Both fragments exhibited a good deal of α-helix content, the N-terminus fragment being somewhat of a straight α-helix, and the C-terminus fragment containing in part a β-sheet surrounded by part of a globular packing followed by an α-helix.

A ribbon model of the lowest-energy minimum for the 98-residue fragment is given in Fig. 16. The labels in this figure correspond to relative residue numbers beginning from the N-terminus residue. The picture clearly depicts the presence of local secondary structures such as α-helix and β-sheet. The figure also clearly shows the fact that the two fragments have not folded, but instead have simply been joined together. Upon closer interpretation of the structure, it also becomes obvious that certain hydrophilic residues, such as aspartic acid, glutamic acid, and threonine, labeled in Fig. 16 as residue numbers 43, 45, 55–57, and 62–74. are buried in the inside of the protein, away from the surface. This is contrary to chemical and physical expectations of protein structure, obviously pointing out limitations in the build-up procedure for predicting structures of globular proteins.

Limitations can be categorized into three areas. In the first category is the effect of the solvent. In this study we have chosen to include the solvent effect based on a shell model by Gibson and Scheraga [65a] that has been modified slightly for our work. However, the model suffers from the limitation that few

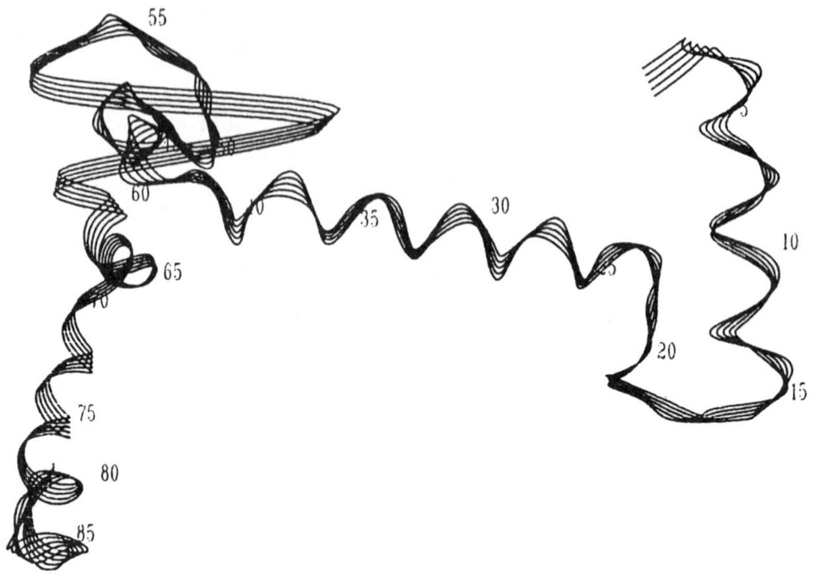

FIG. 16 Ribbon representation for the computed lowest-energy conformation of the 98-residue fragment of interferon. The labels are not actual residue numbers, but are relative to the N-terminus end of the peptide fragment.

reliable experimental data exist on prototype molecules to parametrize the model properly. Therefore, the Gibson and Scheraga caveat that the present version of the algorithm must be regarded with caution [65b], is also obvious from our results. The second area in which the above procedure fails is in the poor representation of charged amino acids, which were assumed essentially as neutral. We know that the charged amino acids play an important role in determining the interactions with the solvent as well as contributing to the overall structure of the molecule. The third area is the neglect of long-range interactions. As mentioned previously, the buildup procedure assumes that the three-dimensional structure of a protein is dominated by short-range interactions. From our results this statement appears to be true if one is interested in secondary structures. However, for predictions of full three-dimensional structures, the effect of long-range interactions should be important. Future efforts to alleviate these shortcomings are discussed below. Here, however, we stress that this type of experimentation is truly important. Notice that simulations on small polypeptides would probably tell us little about the latter two types of problems.

Molecular dynamics (MD) might provide an alternative method for studying the problem of determining protein structure. MD techniques can be used to simulate protein structure in two ways: first, to study detailed properties of particular conformations, and second, to study conformational changes in proteins. Once the secondary structures are determined, either by energy minimization or MD, it is necessary to find the protein topology: namely, trying to arrange different local structures together in the most favorable three-dimensional structure. Our strategy [66] is to implement an expert system to provide reasonable folding pathways to be analyzed by numerical techniques. The idea is to take advantage of the huge amount of information that is available in the protein data banks and to design a program capable of using statistical methods to select the best local secondary structures to be combined into supersecondary structures. These supersecondary structures should be good starting points for investigating final folding of the protein by additional simulation, using either energy minimization or molecular dynamics.

It seems clear from the above that the problem of simulating the three-dimensional structure of protein is still open, but perhaps we can see some light at the end of the tunnel. First, one will have to clean up the methodological problems mentioned above and then address the problem of the reliability of present-day empirical potentials [67–72] and, of course, improve on the *ab initio* one. Notice that for water–water the *ab initio* potentials are at a rather accurate level; for water DNA we are now recomputing the previous potentials [73] at the Hartree–Fock level, and even for the amino acid–amino acid potentials [59] we have included the superposition correction for the small basis set. We refer the reader interested in this problem to a recent paper by Scheraga [74]. Clearly, the many-dimensional space to be searched to find the minima corresponding to the stable tertiary structure can be substantially reduced if we impose experimental constraint as done, for example, in K. Wuthrich's group. The importance of this simplification—as a learning step—is stressed.

In Fig. 17 we plot and extrapolate the expected miniaturization of electronic components (given in a number of atoms, 10^N, rather than in units of volume) and the molecular complexity one will be able to deal with quantum- and statistical-mechanical simulations. Each curve is drawn three times; since we wish to give lower and upper error bounds on our extrapolations. Notice that the lower bound for molecular simulations (the year 2000) is nearly met today;

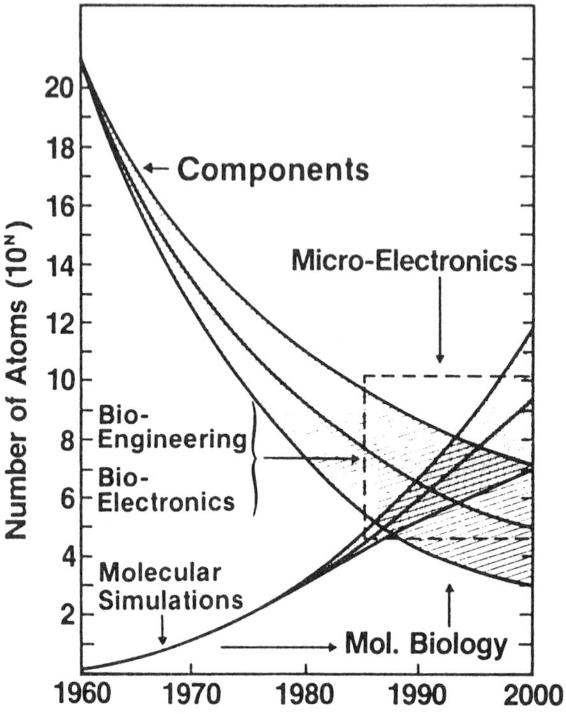

FIG. 17 Expected miniaturization of electronic components, and its impact on the computer industry.

for example, in an unpublished study at our laboratory, we are expanding from about 200,000 atoms to 200,000 × m atoms (with 15 < m < 25) for the Benard problem. Therefore, Fig. 17 is somewhat conservative, possibly because it is by now 6 to 7 years old. The molecular simulations we are particularly interested in are for electronic devices at the molecular level. We must recall and stress that modern organic chemistry deals increasingly with the problem of synthesizing molecular structures for chemicals with very specific functionality, such as for transport, conduction of electrons or protons, optical effects, pharmacological effects, and so on. If we draw a simple and classical electric circuit with resistances, impedances, condensers, and wires, we can ask the question: Which chemical system will represent the resistance, or the condensors, or the wire? In time, today's obvious but naive answers (e.g., saturated hydrocarbons, dimers of planar molecules with π electron mobility and charge transfer, and unsaturated olephins with proper terminals) will become more and more detailed, concrete, and subject to theoretical and laboratory verification. Slowly complicated circuits will have optimal organic molecule analogs [75]. The syntheses of these exact chemicals will probably be obtained by bioengineers using new engineered proteins and enzymes. One prediction is quite easy and almost certain: When vastly different fields merge, as indicated in Fig. 17, the changes for profound innovations are maximal.

ACKNOWLEDGMENTS

The interaction with H. Scheraga and K. D. Gibson (Cornell University) and F. G. Parak (University of Munster) is most gratefully acknowledged.

REFERENCES

1. E. Clementi, *J. Phys. Chem. 89*:4426 (1985).
2. E. Clementi, S. Chin, and D. Logan, *Israel J. Chem. 27*(2): 127 (1986).
3. E. Clementi and D. Logan, in *Special Purpose Computers* (B. Alder, ed.), Academic Press, New York, 1987.
4. D. Logan, J. Saarinen, and E. Clementi, *Proc. 3rd International Conference on Supercomputing* (L. P. Kartashev and S. I. Kartashev, eds.), International Supercomputing Institute, Inc., St. Petersburg, Fla., 1988, Vol. 1, p. 79.
5. H. L. Nguyen, H. Khanmohammadbaigi, and E. Clementi, *J. Comput. Chem. 6*:634 (1985).
6. E. Clementi, G. Corongiu, and J. Detrich, *Comput. Phys. Commun. 37*:287 (1985).
7. L. Domingo and E. Clementi, *IBM Technical Report KGN-17* (1985).
8. E. Clementi and G. Corongiu, *Advances in Biophysics* (A. Wada, ed.), Elsevier, Limerick, 1985, p. 75.
9. E. Clementi, D. Logan, V. Sonnad, Z. Christidis, and A. Capotondi, *Comput. Mech. Eng.* 5(3) (1986).
10. G. Corongiu, J. Detrich, and E. Clementi, *IBM Technical Report KGN-170* (1987).
11. In addition to ref. 4, there are a number of review papers that deal with both the architectures of LCAP-1, LCAP-2,

LCAP-3, and LCAP-3090 and the system software and applications. The most recent review is in *Biological and Artificial Intelligence Systems* (E. Clementi and S. Chin, eds.), ESCOM Science Publishers, The Netherlands, 1988, p. 319.
12. E. Clementi and H. Popkie, *J. Chem. Phys. 57*:1077 (1972).
13. W. S. Benedict, N. Gailer, and E. K. Plyler, *J. Phys. Chem. 24*:1139 (1956).
14. R. J. Bartlett, I. Shavitt, and G. D. Purvis, *J. Chem. Phys. 71*:281 (1979).
15. K. S. Kim, M. Dupuis, G. C. Lie, and E. Clementi, *Chem. Phys. Lett. 131*:451 (1986).
16. H. Popkie, H. Kistenmacher, and E. Clementi, *J. Chem. Phys. 59*:1325 (1973).
17. O. Matsuoka, E. Clementi, and M. Yoshimine, *J. Chem. Phys. 64*:1351 (1976).
18. G. C. Lie and E. Clementi, *Phys. Rev. 33*(4):2679 (1986).
19. N. Metropolis, A. W. Rosenbluth, M. N. Rosenbluth, A. H. Teller, and E. Teller, *J. Chem. Phys. 21*:1087 (1953).
20. B. J. Alder, *J. Chem. Phys. 40*:2724 (1946).
21. G. C. Lie, E. Clementi, and M. Yoshimine, *J. Chem. Phys. 64*: 2314 (1976).
22. E. Clementi and G. Corongiu, *Int. J. Quant. Chem. Symp. 10*:31 (1983).
23. J. Detrich, G. Corongiu, and E. Clementi, *Chem. Phys. Lett. 112*:426 (1984).
24. M. Wojcik and E. Clementi, *J. Chem. Phys. 85*:3544 (1986).
25. M. Wojcik and E. Clementi, *J. Chem. Phys. 85*:6085 (1986).
26. L. Hannon, G. C. Lie, and E. Clementi, *J. Sci. Comput. 1*: 145 (1986).
27. D. C. Rapaport and E. Clementi, *Phys. Rev. Lett. 57*:695 (1986).
28. L. Hannon, G. C. Lie, and E. Clementi, *Phys. Lett. A 119*: 174 (1986).
29. L. Hannon, G. C. Lie, E. Clementi, and V. Yakhot (to be published); see *IBM Kingston Technical Report KGN-128* (1987).
30. L. Hannon, G. C. Lie, and E. Clementi, *J. Stat. Phys. 51*: 965 (1988).
31. J. Given and E. Clementi, *IBM Technical Report KGN-119* (July 1988); *J. Chem. Phys. 90*:7376 (1989).
32. T. P. Lybrand, I. Ghosh, and J. A. McCammon, *J. Am. Chem. Soc. 107*:7793 (1985); W. L. Jorgensen and E. Ravimohan, *J. Chem. Phys. 83*:3050 (1985); P. A. Bash, U. C. Singh, R. Langrige, and P. A. Kollman, *Science 236*:564 (1987).

33. M. Migliore, G. Corongiu, E. Clementi, and G. C. Lie, *J. Chem. Phys. 88*:7766 (1988).
34. (a) R. Gomer and G. Tryson, *J. Chem. Phys. 66*:4413 (1977); (b) H. L. Friedman and C. V. Krishnan, in *Water: A Comprehensive Treatise*, Vol. 3 (F. Franks, ed.), Plenum Press, New York, 1973.
35. G. Bolis, M. Ragazzi, D. Salvaderi, D. R. Ferro, and E. Clementi, *Int. J. Quant. Chem. 14*:815 (1978).
36. E. Clementi, G. Corongiu, B. Jonsson, and S. Romano, *FEBS 100*:313 (1979).
37. E. Clementi, G. Ranghino, and R. Scordamaglia, *Chem. Phys. Lett. 49*:218 (1978).
38. G. Ranghino and E. Clementi, *Gazz. Chim. Ital. 108*:157 (1978).
39. G. Corongiu and E. Clementi, *Biopolymers 20*:551 (1981).
40. G. Corongiu and E. Clementi, *Biopolymers 20*:2427 (1981).
41. E. Clementi and G. Corongiu, *J. Biol. Phys. 11*:33 (1983).
42. G. G. Price, U. Heinemann, S. Chandresegaran, Lou-Sing Ken, M. L. Kopka, and R. E. Dickerson, *Science 238*:498 (1987).
43. K. N. Swamy and E. Clementi, *Biopolymers 26*:1901 (1987).
44. K. N. Swamy and E. Clementi, *Biopolymers 26*:1901 (1987); ibid. *28*:1497 (1989).
45. S. Lindsay, *Biopolymers 26*:439 (1987).
46. For previous works, see F. Parak, *Methods in Enzymology* Vol. 127 (L. Packer, ed.), Academic Press, New York, (1986), p. 196; H. Hartmann, W. Steigemann, R. Reuscher, and F. Parak, *Eur. Biophys. J.* (1986).
47. K. S. Kim and E. Clementi, *J. Am. Chem. Soc. 107*:227 (1985).
48. K. S. Kim and E. Clementi, *J. Phys. Chem. 89*:3655 (1985).
49. H. S. Shieh, H. M. Berman, M. Dabrow, and N. Neidle, *Nucleic Acids Res. 8*:85 (1980); S. Neidle, H. M. Berman, and H. S. Shieh, *Nature 288*:129 (1981).
50. R. Gomperts and E. Clementi, *IBM Research Report KGN-107* (February 1987).
51. R. Gomperts and E. Clementi, *IBM Research Report KGN-158, KGN-159* (September 1987).
52. K. S. Kim and E. Clementi, *J. Am. Chem. Soc. 107*:5504 (1985).
53. K. S. Kim, H. L. Nguyen, P. K. Swaminathan, and E. Clementi, *J. Phys. Chem. 89*:2870 (1985).
54. See "Unity of Science, Engineering and Supercomputing," a 45-minute computer-animated video with audio commentary in English, French, Italian, Spanish, or Chinese (1986); available from IBM-Kingston, Dept. 48B/Mail Station 428, Kingston, NY 12401.

55. E. Clementi, *Computational Aspects for Large Chemical Systems*, Lecture Notes in Chemistry 19, Springer-Verlag, New New, 1980.
56. E. Clementi, F. Cavallone, and R. Scordamaglia, *J. Am. Chem. Soc. 99*:5531 (1977).
57. M. Ragazzi, D. Ferro, and E. Clementi, *J. Chem. Phys. 70*: 1040 (1979).
58. G. Ranghino, E. Clementi, and S. Romano, *Biopolymers 22*: 1449 (1983).
59. S. Chin, J. J. Perez, M. Re, and E. Clementi, *Proc. 3rd International Conference on Supercomputing*, Boston, Apr. 15–18, 1988, Vol. 1, p. 109.
60. P. Lengyel, *Ann. Rev. Biochem. 51*:251 (1952); S. Pestko and S. Baron, *Meth. Enzymol. 78*:3 (1981).
61. W. P. Levy, M. Rubinstein, J. Shively, U. Del Valle, C.-Y. Lai, J. Moschera, L. Brink, L. Gerber, S. Stein, and S. Pestko, *Proc. Natl. Acad. Sci. U.S.A. 78*:6186 (1981).
62. J. Collins, *Symp. Soc. Gen. Microbiol. 35*:35 (1983).
63. D. L. Miller, H.-F. Kung, T. Staehelin, and S. Pestko, *Meth. Enzymol. 79*:3 (1981).
64. S. Chin, K. D. Gibson, H. A. Scheraga, and E. Clementi: a library of the minima for 400 dipeptides is available from the authors. See also K. D. Gibson, S. Chin, M. R. Pincus, E. Clementi, and H. A. Sheraga, in *Supercomputer Simulations in Chemistry*, Lecture Notes in Chemistry 44 (M. Dupuis, ed.), Springer-Verlag, New York, 1986, p. 198.
65. (a) K. D. GIbson and H. A. Scheraga, *Proc. Natl. Acad. Sci. USA 58*:420 (1967): G. Némethy, Z. I. Hodes, and H. A. Scheraga, *Proc. Natl. Acad. Sci. USA 75*:5760 (1978); Z. I. Hodes, G. Némethy, and H. A. Scheraga, *Biopolymers 18*: 1565 (1979); (b) K. D. Gibson and H. A. Scheraga, *J. Comput. Chem. 8*:826 (1987).
66. D. P. Dolata, A. R. Leach, and K. Prout, *J. Comput.-Aided Mol. Des. 1*:73 (1987); B. Robson, E. Platt, R. V. Fishleigh, A. Marsden, and P. Millard, *J. Mol. Graphics 5*:8 (1987).
67. J. Hermans and D. Ferro, *Biopolymers 10*:1121 (1971).
68. R. DeSantis, E. Giglio, A. M. Liquori, and A. Ripamonti, *Nature 206*:456 (1965).
69. M. Goodman and N. S. Choi, in *Peptides 1968* (E. Bricas, ed.), Wiley, New York, 1968, pp. 1–17.
70. S. Lifson, A. T. Hagler, and P. Dauber, *J. Am. Chem. Soc. 101*:5111 (1979).
71. P. Weiner and P. Kollman, *J. Comput. Chem. 2*:287 (1981).
72. B. R. Brooks, R. E. Bruccoleri, B. D. Olafson, D. J. States, S. Swaminathan, and M. Karplus, *J. Comput. Chem. 4*:187 (1983).

73. R. Scordamaglia, F. Cavallone, and E. Clementi, *J. Am. Chem. Soc.* *99*:5545 (1977).
74. H. Scheraga, in *Biological and Artificial Intelligence Systems* (E. Clementi and S. Chin, eds.), ESCOM Science Publishers, The Netherlands, 1988, p. 1.
75. M. Dupuis, in *Biological and Artificial Intelligence Systems* (E. Clementi and S. Chin, eds.), ESCOM Science Publishers, The Netherlands, 1988, p. 185.

V

Transport

14

Membrane Transport: Insights from Surface Science

MARTIN BLANK Department of Physiology and Cellular Biophysics, Columbia University, New York, New York

I.	Introduction	337
II.	Role of Model Studies	340
	A. Transport through ultrathin layers	340
	B. Interfacial ion transport	343
III.	Surface Compartment Model	347
	A. Equations based on first principles	348
	B. Membrane channels as aggregated protein	350
	C. Ion channel selectivity	350
	D. Oscillating electric fields	353
IV.	Evaluation of the Model	355
	A. Electrodiffusion in membrane systems	356
	B. Protein equilibria	357
	C. Kinetics of membrane processes	361
	D. Concluding remarks	363
	References	363

I. INTRODUCTION

In his monograph *Physics of Bioenergetic Processes*, Blumenfeld asked biologists the pointed question [1]: "What is the reason for the theoretical flimsiness of modern biochemistry when its experimental and practical achievements are so remarkable?" His provocative answer was: "Twentieth century biochemistry uses for theoretical description of biochemical processes nineteenth century physical chemistry, i.e., the concepts developed to describe the behavior of low-molecular compounds in gaseous phase and dilute solutions." This statement is not entirely justified, especially when considering theoretical advances in understanding the properties of biopolymers. However, it does draw attention to the imbalance between theoretical and experimental aspects of biochemistry and the need for greater development of theory.

Focusing on the field of membrane transport, we find essentially the same theories now as shortly after the turn of the century. The theoretical apparatus that was developed for osmotic and ion transport problems in biology was very useful for whole tissues. However, as the focus of biology has sharpened down to the cell membrane and molecular levels, the older theories have proven to be inadequate. There have been important developments in the theory of mitochondrial phosphorylation processes [2] and of ion channel statistics [3], but there is still a need for a general theory of ion transport across membranes that is capable of describing such processes as ion flow during excitation [4]. The recent work reviewed here, on the role of electrical double layers in ion transport, offers promise in this regard. The concepts of surface science have been largely neglected, and by reintroducing them, we have developed a better understanding of membrane transport while establishing a better balance between theory and experiment in biology.

Surface scientists do not have to be told that surface properties are important in biological systems. But it is worth pointing out to them that biological systems, by their very nature, especially the cell- and molecular-level processes that interest researchers today, are in the range of colloidal systems and require an understanding of surface properties.

Consider Table 1. Biological systems are listed there in the order of decreasing size, and the surface-to-volume ratio (S/V) increases markedly when reaching the membrane and biopolymer level. In dissolved and dispersed systems, as S/V increases, the surface energy becomes a larger fraction of the total energy, and surface properties have a greater influence on the properties of the system. Many hydrophobic colloids have typical dimensions of micrometers similar to those of cells and an S/V ratio on the order of 10^4 cm^{-1}. Hydrophilic colloids are like the hemoglobin system and have even higher S/V. It is apparent that the range of cell and biopolymer dimensions is comparable to the colloidal systems stabilized by surface forces. Even at the relatively low S/V of the entire organism, a knowledge of the surface area is needed to account for the energy balance.

Focusing on the energetics of biological processes involving biopolymers, we have found it useful to consider the surface free energy of the molecule as an indicator of the free-energy balance that

TABLE 1 Surface/Volume Ratio in Biological Systems of Different Sizes

System	Surface area, S	Volume, V	S/V
Organism (human)	2 m^2	80 liters	2.5×10^{-1} cm^{-1}
Cell (erythrocyte)	150 μm^2	100 μm^3	1.5×10^{4} cm^{-1}
Membrane (erythrocyte)	300 μm^2	1 μm^3	3×10^{6} cm^{-1}
Biopolymer (hemoglobin)	10^4 Å^2	10^5 Å^3	1×10^{7} cm^{-1}

is reflected by the final stable structure. The electrostatic interactions, hydrophobic interactions, hydrogen bonding, and so on, all contribute to the final configuration of the molecule, but calculating these energies with sufficient accuracy, especially in the absence of detailed information on atomic coordinates and hydration, appears to be impossible. Comparing the surface energy of the different molecular shapes of a particular biopolymer enables us to estimate the energetics of molecular transformations (e.g., conformational changes).

Many years ago, Langmuir [5] found that he could relate molecular interactions to macroscopic thermodynamic quantities by focusing on the contact surface. He showed the relation between surface free energy and the free energy of vaporization in terms of differences in the number of nearest neighbors between surface and interior molecules. The concept of surface free energy, which worked well for small molecules, should apply more readily to the much larger protein molecule.

Aggregation—disaggregation reactions of proteins involve no internal molecular changes, only changes in the interfacial area and in the surface charge density. Therefore, it is possible to estimate the total free-energy change from the change in the surface free energy and to calculate the equilibrium constants of the protein reactions. The opening and closing of the membrane channels involved in ion flow are probably due to aggregation/disaggregation of their protein subunits.

II. ROLE OF MODEL STUDIES

Our approach to the long-standing problem of ion transport and nerve excitation developed as a result of the interplay between experiment and theory. Studies on simple model systems helped determine the physical processes needed in a theory of ion transport across natural membranes. One can ask, and indeed it was frequently asked of me: "Why perform experiments with models when on can study the 'real system'?" The answer was obvious. Research on a biological *mechanism* focuses on the common elements that are present in many systems, not a particular system or technique, and one can pose questions that are more easily answered with simple models. Traditional biochemists were not very receptive to this response, and physiologists considered the experiments irrelevant. After all, what can one possibly learn about axon membranes from measurements on interfacial systems, of nitrobenzene–water and mercury–water, that contain unnatural, long-chain cationic surfactants? The answer to that question, as we shall see below, is that ion concentration changes arise from mobility differences during current flow at all interfaces, and that ion fluxes appear to be directly related to interfacial charge densities, processes that have only recently been demonstrated with natural membranes. These findings suggested that the electrical double layer had a strong influence on interfacial ion transport processes and must be included in the formulation of membrane transport.

In studying the aggregation equilibrium of oligomeric proteins in solution, we are gaining insight into the molecular processes that are probably involved in channel opening and closing. Here, too, the relationship is not immediately evident, but it is possible to obtain important insights about the energetics of channel function from these studies.

A. Transport Through Ultrathin Layers

There is a fundamental difficulty in describing the process of diffusion through natural membranes. Physical laws are based on statistical averages, and they are valid for macroscopic systems because of the many particles or events involved in a process. At long times, the diffusion of molecules *within* a membrane obeys Fick's law and even the Stokes–Einstein equation [6]. However, in diffusion *across* a pauci-molecular membrane, the process is over so quickly that one cannot assume a multicollision process with averaging.

The physical factors that affect membrane permeability have been studied with lipid monolayers that are comparable to natural membranes in size and structure [7]. In these systems, the permeability

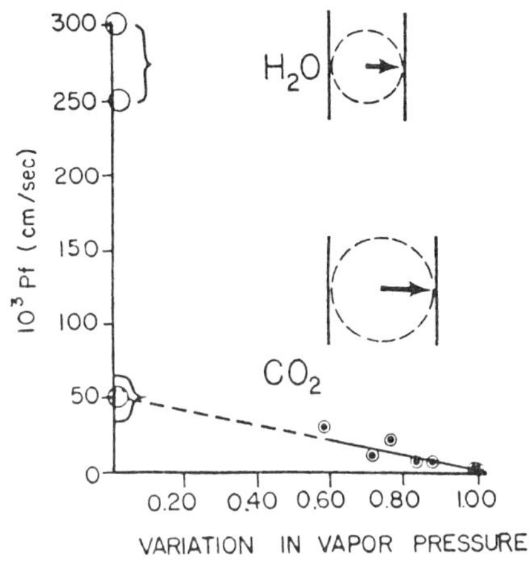

FIG. 1 The permeabilities of an octadecanol monolayer to H_2O and CO_2 are indicated on the ordinate, and the approximate cross-sectional areas of the two molecules are also shown. The smaller H_2O permeates more rapidly. The permeability of an octadecanol monolayer to CO_2 varies with the vapor pressure of the aqueous phase upon which it is spread (due to different concentrations of LiCl), and these data are plotted on the lower part of the figure. [Reprinted with permission from *J. Gen. Physiol.* *52*:1915 (1968).]

varies with the amount of free space in the films. The least permeable monolayers are close-packed lipid films, and the most permeable, such as those of protein films, form relatively open structures. Small molecules permeate more readily than larger ones, and there is also interference between permeants that resembles diffusion in a lattice (e.g., the H_2O traffic across the monolayer affects the CO_2 permeability; see Fig. 1.) The permeability of a lipid film varies more markedly with the length of the hydrocarbon chain than is expected in a classic diffusion process. The large variation of the activation energy with chain length (i.e., monolayer thickness) and the absence of interfacial partition equilibrium effects (i.e., there is no apparent dissolution of the permeant within the monolayer) during permeation, indicate a one-step process as in diffusion from hole to hole in solids.

Monolayer permeation has been formulated in terms of spaces that result from the molecular packing as well as from the equilibrium fluctuations in monolayer density, and the equations account quantitatively for the permeation of a nonadecanoic acid monolayer to water, the one case where it has been examined in detail, as well as for the variation of the permeation rate with temperature and permeant size in general [7]. The equations also predict the self-diffusion coefficient and the frequency of (phospholipid) flip-flop in a bilayer membrane. In fact, the first correct prediction of the self-diffusion coefficient of a lipid molecule in a monolayer was made on the basis of this model [8], several years before the measurements were made.

Natural membranes are thicker and more complex than lipid monolayers, and to study how properties change in the transition from monolayer systems to membrane and bulk systems, we can examine the behavior of multilayer membranes of fatty acids [9] and membrane protein [10]. The permeabilities of lipid multilayers to gases and of protein multilayers to ions all decrease with film thickness, but the calculated diffusion coefficients increase (see Fig. 2). These

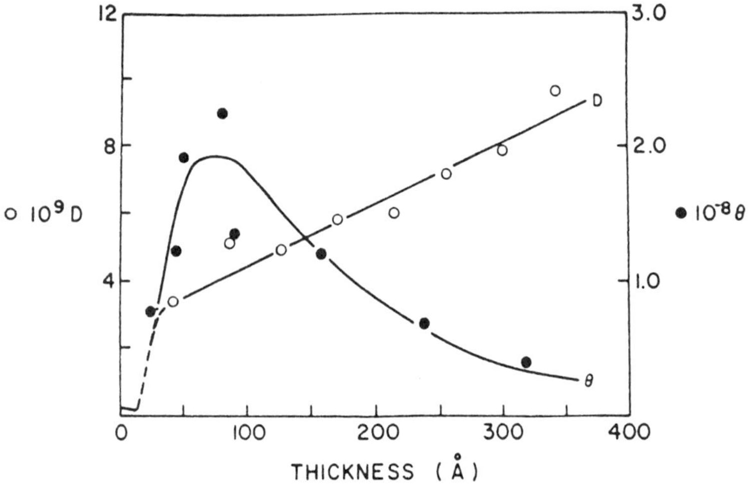

FIG. 2 The bulk properties of multilayers formed from the red cell membrane proteins spectrin and actin, as a function of the thickness of the film (in angstroms). The diffusivity, D, in square centimeters per second (O) and the yield, θ, in dynes per square centimeter (●). Below 20 Å the straight line indicates the range of D values obtained by another technique, and the dashed line joins the two sets of measurements. [Reproduced with permission from *J. Colloid Interface Sci.* 75:433 (1980).]

results are qualitatively different from those obtained by increasing the thickness of a single molecular layer. But monolayer permeation is a one-step process with an absence of partition effects, while multilayer systems are more permeable as a result of less effective packing and greater interactions with the permeants.

(The question of whether a natural membrane should be considered a "phase," an interfacial film, or some intermediate structure with its own properties has been approached as an experimental problem, and readers are referred to a summary of those studies [10] for a discussion of the dependence of membrane properties on thickness. The variation of the mechanical strength, with a maximum at multilayer thickness comparable to that of the natural membrane, also shown in Fig. 2, suggests that this property could have been optimized during evolution.)

The very low diffusion coefficient of a gas through a monolayer as opposed to multilayer also points to the important role of the water layer at the interface in controlling the transport of uncharged molecules. In a single monolayer the major energy barrier is due to the polar group [11]. The variation of the surface potential and the surface energy with film thickness also shows that these properties are due almost entirely to the surface layer [10]. For charged permeants, discussed in the next section, the surface layer is even more important.

From these results it appears that the permeation of small molecules through the membrane bilayer can be thought of as a monolayer process once the permeant is past the interface. The permeability assigned to a particular structure can be viewed as a probability factor that operates in conjunction with particular driving forces (i.e., the chemical potentials) across the structure. What goes on within the membrane bilayer during permeation may very well be similar to the passage of a gas molecule through a lipid monolayer with regard to the speed and the lack of interaction with the structure.

B. Interfacial Ion Transport

Ion transport is driven by differences in the chemical potential (concentration) of the individual ions, but the existence of a charge on the diffusion species leads to sensitivity to imposed electric fields (e.g., electroadsorption, electric migration effects), as well as electrical consequences of the ion movements (e.g., generation of diffusion potentials).

When an electric field is imposed across an interface between two liquids containing ions, there is a change in the interfacial concentration of ions due to an adsorption of cations or anions, depending on the sign of the charge on the surface, and the number adsorbed varies with the potential. When there is negligible transference through the interface and the system is essentially in equilibrium, the variation of the interfacial tension with electrical potential is given by the Lippmann equation of electrocapillary adsorption (see Fig. 3). Current flow through the interface can also cause changes in interfacial ionic concentration when the transport numbers of the ions differ in the two phases. Here, increases as well as decreases in concentration occur at the interface and extend into both phases by diffusion as time passes (see Fig. 4).

Both the equilibrium (electrocapillary) and nonequilibrium (transference) types of adsorption were demonstrated in water–nitrobenzene

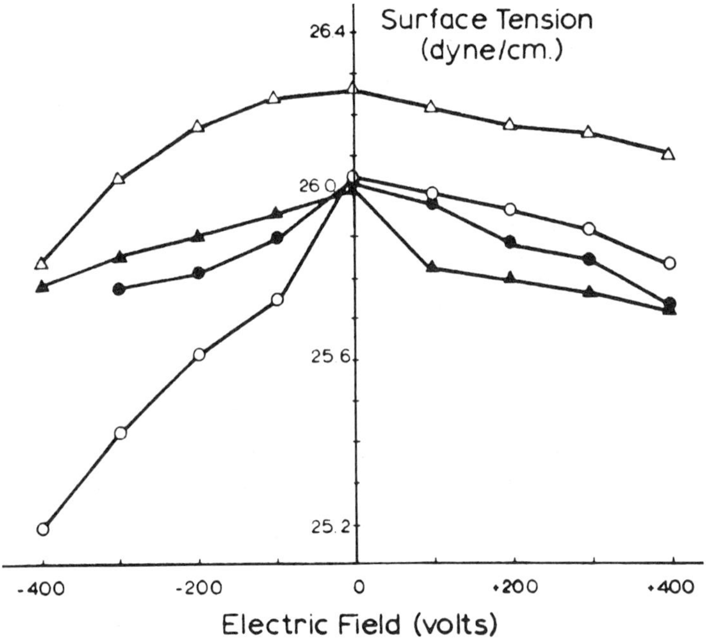

FIG. 3 Interfacial tension as a function of the electric potential imposed across the water–nitrobenzene interface when 10^{-3} M amino acid is dissolved in the water at various pH values. The open symbols are for glycine and for glutamic acid at pH values below their isoelectric points. The shaded symbols are of the same amino acids above their isoelectric points. [Reproduced with permission from *Science 141*:1173 (1963).] O gly·pH 5.71; ● gly·pH 7.96; △ glu·pH 1.06; ▲ glu·pH 6.49.

systems, two immiscible fairly high dielectric materials in which one can dissolve a variety of ionic solutes [12]. If an ion is soluble in the aqueous phase only, it behaves as in electrocapillary adsorption, with cation or anion adsorption, depending on the potential. If the ion is soluble in both phases, there can be increases or decreases in concentration of both cations and anions, depending on the ion transport numbers and the direction of the current. Electrocapillary adsorption is the limiting case of the more general mechanism where ions cross the interface.

The rate of current flow through charged interfaces depends on the charge density. In the transport of cations through oriented charged monolayers at a mercury–water interface [13], the logarithm

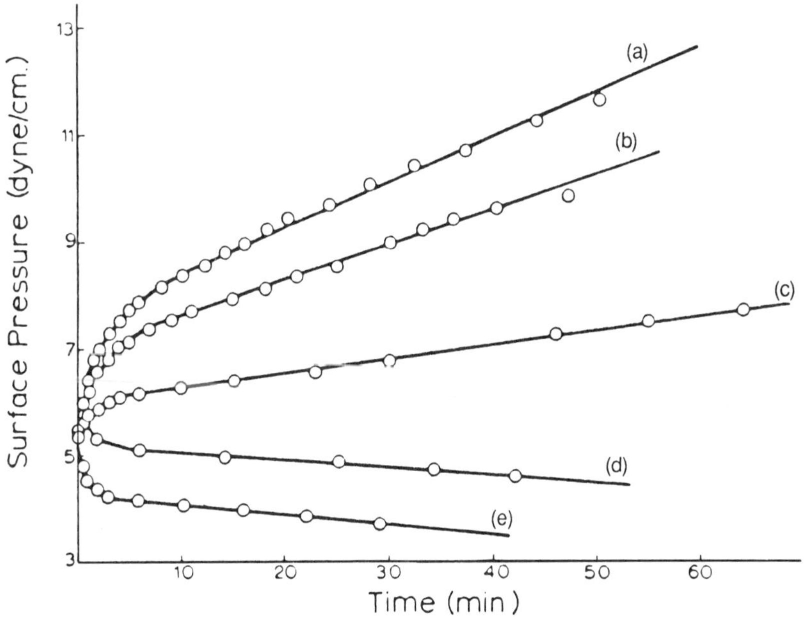

FIG. 4 Variation in CTAB surface pressure with time at a water–nitrobenzene interface when different currents are passed through the interface. The values of the current in microamperes are: a, 150; b, 106; c, 52; d, -93; and e, -208. The CTAB is initially 0.96×10^{-4} M in water and 1.55×10^{-4} M in nitrobenzene, this being the equilibrium partition at 25°C. [Reproduced with permission from *Science 141*:1173 (1963).]

of the rate constant for ion transport was linearly related to the surface charge (see Fig. 5). Similar results were obtained in studies of ion transport through negatively charged erythrocyte membrane protein (spectrin–actin) layers. Our measurements on the membrane proteins indicated that there must be a resistance to anions at the inner face of the membrane and led us to predict that the anion channel present in that membrane must extend beyond the spectrin–actin layer into the cytoplasm of the erythrocyte [14]. The relation between the logarithm of the conductance and the surface charge can be derived from electrical double-layer theory, where the surface charge controls the flow rate through the concentration of ions in the double layer [15].

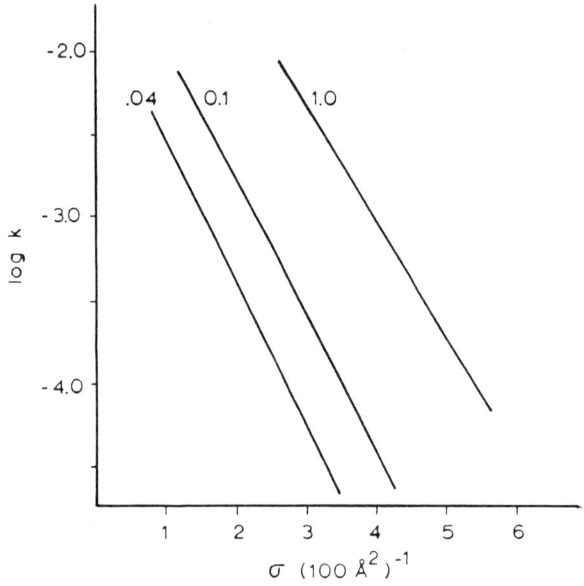

FIG. 5 Logarithm of the ion transport rate constant, k (in cm/s), of Cu^{2+} ions diffusing through a decylammonium monolayer at a mercury–water interface, plotted against σ, the surface charge density (in charges per 100 $Å^2$). The numbers on the lines indicate the ionic strength of the aqueous phase containing 10^{-3} M Cu^{2+} ion. [Reproduced with permission from *Bioelectrochem. Bioenerg.* 4:209 (1977).]

From our studies on interfacial films, it appears that the movement of ions through membranes is bound to be a very slow process unless there are specialized structures that can concentrate ions and provide low-resistance aqueous pathways. Such structures with many different properties (e.g., selectivity, conductance) actually exist in membranes and lie at the basis of characteristic biological properties. We shall focus on the electrically activated sodium and potassium channels of excitable membranes.

III. SURFACE COMPARTMENT MODEL

Theoreticians usually start their analysis of a problem with a statement of first principles (i.e., the fundamental laws of physics and chemistry that we all agree are valid in the macroscopic world) and then apply them to the particular system of interest. In trying to elucidate the mechanisms of ion transport across biological membranes, this approach has thus far not been very successful. Our inability to deal with the complexity of biological structure at the submicroscopic level and the existence of specialized membrane proteins involved in catalysis (i.e., enzymes) and transport (i.e., channels) are often cited as reasons for the lack of progress. I do not agree. Although the properties of membrane proteins are not well understood, there has also been a fundamental problem in the physics and chemistry of our approach to the problem due to an oversimplification of our description of membrane processes (i.e., a neglect of surface processes).

Ion transport in aqueous solution is the result of individual ion flows under electrochemical driving forces. Transport is much more complicated at an interface, a region of discontinuity in physical properties (e.g., ion mobility, dielectric constant, the possible change of current carriers from ions to electrons) as well as the appearance of new properties (e.g., ion partition equilibrium, interfacial capacitance, the inhomogeneities of electrical double layers). Membrane systems have contact with aqueous solutions at two interfaces, and interfacial processes must be involved in ion transport.

Teorell [16] was the first to suggest that the surface potentials at charged membrane surfaces should influence ion transport. However, researchers in the field of passive ion transport have generally modeled the membrane as a simple barrier between two bulk phases, and their descriptions of transport processes have not considered interfacial processes. It has become apparent, after studying ion transport across liquid–liquid interfaces, monolayers, and bilayers, that interfacial processes are critical in membrane transport and must be included in any theoretical description [7,17].

In brief, we have shown that during ion transport there are (1) concentration changes at interfaces due to adsorption and current flow, (2) concentration and electrical potential changes that are controlled by the charge density at a surface, and (3) very large surface capacitances due to the storage of charge at interfaces. In membrane transport, the voltage-dependent permeabilities of ion channels must also be included, and we shall see that surface properties appear to be important there as well.

These and other factors have been incorporated into a description that includes the effects of electrical double layers at membrane surfaces and is known as the surface compartment model (SCM). With this model, the ionic fluxes in excitable membranes are governed by the same equations that apply to ions in solution [18,19]. Insights into the nature of ion selectivity in channels and the effects of ELF (extremely low frequency) electric fields on cells have also resulted from application of the SCM.

A. Equations Based on First Principles

As a first approach to studying the effects of interfacial properties on ion transport, the surface compartment model (SCM) uses the Helmholtz model for the electrical double layers at the surfaces of natural membranes and treats them as compartments (Fig. 6). Each compartment is characterized by a surface potential, surface capacitance (in addition to the dielectric capacitance of the membrane), and ion concentrations that differ from the bulk solution and that change as a result of current flow and cation binding and release processes. The equations for this system are given in Fortran notation in four groups:

1. Conservation of mass for ions:

$$\dot{N}2 = (1/L2) * (JN1 - JN - PN - \dot{N}22)$$
$$\dot{K}2 = (1/L2) * (JK1 - JK - PK - \dot{K}22)$$
$$\dot{A}2 = (1/L2) * (JA1 - JA - PA)$$
$$\dot{N}3 = (1/L3) * (JN + PN - JN3 - \dot{N}33)$$
$$\dot{K}3 = (1/L3) * (JK + PK - JK3 - \dot{K}33)$$
$$\dot{A}3 = (1/L3) * (JA + PA - JA3)$$

2. Conservation of surface charge:

$$\dot{X}2 = -(\dot{N}22 + \dot{K}22) - JX$$
$$\dot{X}3 = -(\dot{N}33 + \dot{K}33) + JX$$

3. Conservation of charge during current flow:

$$I = FA *(JN1 + JK1 - JA1) - CAP1*\dot{E}2$$
$$I = FA *(JN3 + JK3 - JA3) + CAP3*(\dot{E}3-\dot{E}4)$$
$$I = FA *(JN + JK - JA - JX + PN + PK - PA) + CAP2*(\dot{E}2-\dot{E}3)$$

4. Kinetics of ion binding:

$$\dot{N}22 = BK * X2 * N2 - BR * N22$$

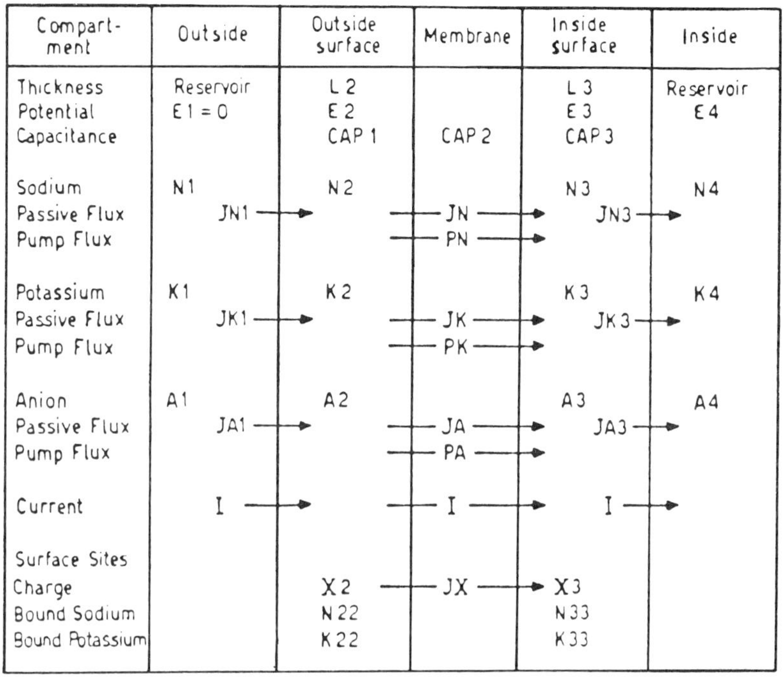

FIG. 6 Symbols used in the SCM equations. [Reproduced with permission from *Bioelectrochem. Bioenerg.* 9:427 (1982).]

$$\dot{K}22 = BF * X2 * K2 - BR * K22$$
$$\dot{N}33 = BF * X3 * N3 - BR * N33$$
$$\dot{K}33 = BF * X3 * K3 - BR * K33$$

In all of the equations the dot above a symbol indicates the time derivative. In the third set of equations FA is the Faraday, which is 96,500 Coulombs per equivalent. In the last set of equations, it should be noted that the ratio BF/BR = BEQ, the binding equilibrium constant.

In the equations, the fluxes (J) are Nernst–Planck expressions where individual ion flows are proportional to the electrochemical driving forces. The numerical values used in the calculations are taken from studies of the squid axon, and except for differences in the permeability of the resting membrane, the two cations are assumed to have identical physical properties: mobility in the double-layer region (M), ion-binding equilibrium (BEQ), and release rate

(BR) constants. The values that must be assumed (BR, M, CAP 1, and CAP 3) are based on measurements with physical models. Tables 2 and 3 list the values of the constants, parameters, and initial values used in the computations.

B. Membrane Channels as Aggregated Protein

Upon depolarization of excitable membranes, initially there are small and rapid shifts of charge (gating currents) that precede the ion permeability changes and that are believed to cause the opening of a channel. We have modeled this process like subunit disaggregation in oligomeric proteins [18]. The relation between the disaggregation constant and charge, together with the number of channels in the axon membrane, lead to a relation between the fraction of channels open and the surface charge density. When combined with the SCM, the gating currents give rise to voltage-dependent changes in ion permeability.

The model for the channel in the resting state is an oligomeric protein that is disaggregated (i.e., open) on the outer surface, where the charge is high, and aggregated (i.e., closed) on the inner surface, where the charge is low (Fig. 7), in accordance with the charge distribution on the squid axon membrane. Upon depolarization, the shift of negative charge (i.e., the measured gating current) from the outer to the inner surface causes the channel to open (Fig. 8). Qualitatively, the properties of this voltage-gated ion channel are compatible with observed membrane properties in terms of the steady-state distribution of charge, the shift of charge during depolarization, the range of surface charge where opening occurs, the cation selectivity of the channel, and the cation binding.

C. Ion Channel Selectivity

The ionic flows during excitation in nerve have been well characterized [4], and they appear not to follow the equations of electrodiffusion. The squid axon, the prototype excitable cell, has two different types of channels with different kinetics that show an apparent specificity for sodium and potassium ions, respectively. The faster-responding channel is selective for sodium ions and the slower one for potassium. The selectivity is difficult to explain because other ions can pass through with ease (Li^+ can be substituted for Na^+ and Tl^+ for K^+), and there is no relation between the selectivity and the conductance of a channel as would be expected with a bottleneck.

The usual analysis of the ion flows in the squid axon is based on the simple barrier model of a membrane, with an empirical voltage

TABLE 2 Constants and Parameters of SCM

Property[a]	Symbol	Magnitude
Outside concentration		
Na	N1	4.4×10^{-4} mol/cm^3
K	K1	1.0×10^{-5} mol/cm^3
Anion	A1	4.5×10^{-4} mol/cm^3
Inside concentration		
Na	N4	5.0×10^{-5} mol/cm^3
K	K4	4.0×10^{-4} mol/cm^3
Anion	A4	4.5×10^{-4} mol/cm^3
Outside (reference) potential	E1	0 mV
Pump flux		
Na	PN	-8.44×10^{-11} mol/(cm^2 s)
K	PK	5.09×10^{-11} mol/(cm^2 s)
Anion	PA	1.11×10^{-11} mol/(cm^2 s)
Membrane conductance to:		
Na	GN	5.00×10^{-10} cm/(s mV)
K	GK	5.00×10^{-9} cm/(s mV)
Anion	GA	1.00×10^{-9} cm/(s mV)
Gating charge conductance		
Na channel	GX fast	5×10^3 (s mV)$^{-1}$
K channel	GX slow	3×10^2 (s mV)$^{-1}$
Na, K, anion mobility in surface solution*	M	1×10^{-6} cm/(s mV)
Na,K binding equilibrium constant*	BEQ	1×10^3 cm^3/mol
Na,K binding release rate constant*	BR	1×10^5 s^{-1}
Surface capacitance*	CAP 1,3	10 μF/cm^2
Dielectric capacitance	CAP 2	0.8 μF/cm^2

[a]For properties marked by an asterisk, values are estimated as described in the text. All other values are taken from the literature.

TABLE 3 Initial Conditions of SCM

Property	Symbol	Magnitude
Surface potential		
Outside	E2	-46 mV
Inside	E3	-78 mV
Measured (outside) potential	E4	-65 mV
Outside surface concentration		
Na	N2	2.73×10^{-3} mol/cm^3
K	K2	6.21×10^{-5} mol/cm^3
Anion	A2	7.25×10^{-5} mol/cm^3
Inside surface concentration		
Na	N3	8.38×10^{-5} mol/cm^3
K	K3	6.70×10^{-4} mol/cm^3
Anion	A3	2.69×10^{-4} mol/cm^3
Outside surface charge density	X2	1.06×10^{-10} mol/cm^2
Outside bound Na	N22	2.89×10^{-10} mol/cm^2
Outside bound K	K22	6.56×10^{-12} mol/cm^2
Inside surface charge density	X3	2.95×10^{-11} mol/cm^2
Inside bound Na	N33	2.47×10^{-12} mol/cm^2
Inside bound K	K33	1.98×10^{-11} mol/cm^2
Outside surface compartment thickness	L2	3.88×10^{-8} cm
Inside surface compartment thickness	L3	6.09×10^{-8} cm
Resting flux		
Na	JN	8.44×10^{-11} mol/(cm^2 s)
K	JK	-5.09×10^{-11} mol/(cm^2 s)
Anion	JA	-1.11×10^{-11} mol/(cm^2 s)

dependence but no surface processes. When the SCM is used to describe the ionic processes in nerve, it is possible to show that the unusual ion flows characteristic of electrical excitation arise from the electrodiffusion equations. A typical result showing the response

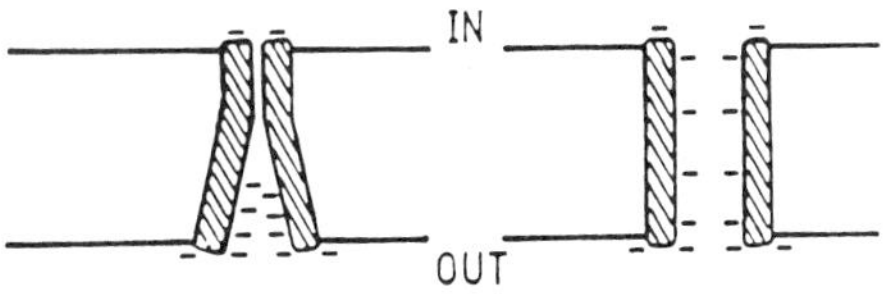

FIG. 7 Model of a voltage-gated oligomeric channel within the membrane matrix: (a) in the resting state, the charge density causes the outer region to be dissociated while the inner region is associated; (b) upon depolarization, the charge shifts cause the inner region to dissociate and the channel to open. [Reproduced with permission from *Bioelectrochem. Bioenerg.* 9:615 (1982).]

of SCM voltage-gated channels to depolarization is given in Fig. 9. The ionic currents in both fast ("sodium") and slow ("potassium") channels can be described by the electrodiffusion equations given above. Both cations pass through each channel, but because the ion gradients change with time, the magnitude and direction of the ion current depends on the gradients at the time of channel opening. Since channels have different opening/closing kinetic constants, the observed fluxes can be very different. A sodium channel is characterized by rapid gating currents and a potassium channel by slower gating currents, as observed in excitable membranes. It is likely that channel kinetics are the basis of selectivity in other systems.

D. Oscillating Electric Fields

The success of the SCM has effectively reoriented our thinking toward the essential validity of the electrodiffusion description of ion transport processes at the membrane level. The ion flows during excitation and the apparent selectivity of different channels can be explained in terms of electrochemistry. We therefore have a useful membrane model for approaching the problem of transduction of externally applied electric fields.

The excitation problem has shown how dc electric fields can cause charges to shift, channels to open and close, and ions to flow. Similar effects have been demonstrated with ac fields. It is possible to show that oscillating electric fields lead to periodic changes in the electrical double layers that are functions of the frequency. All of the ionic concentrations change, but the relative change is greatest

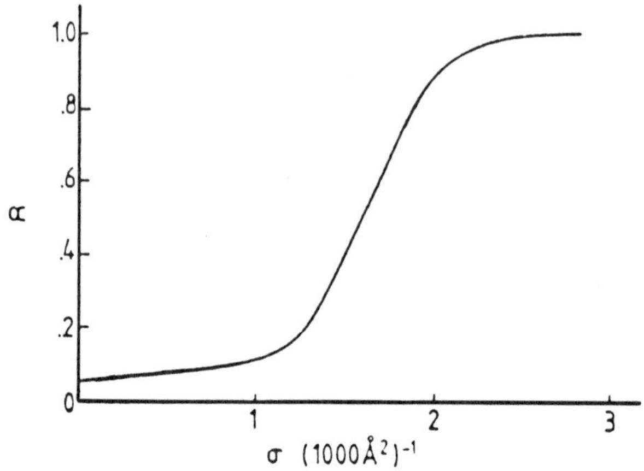

FIG. 8 Fraction of channels that are open, α, as a function of the charge density on the inner surface of the membrane, σ, in charges per 1000 Å^2. The dissociation equilibrium constant for oligomeric channels is assumed to equal 1×10^{-4} pmol/cm^2 at $\sigma = 0$. [Reproduced with permission from *Bioelectrochem. Bioenerg.* 9:615 (1982).]

in those concentrations with the lowest steady-state values. In particular, the sodium on the inner surface of a channel and potassium on the outer surface show maximal effects at 100 to 200 Hz [20] (see Fig. 10). These two ionic concentrations normally control the activity of the ion pump enzyme, the Na,K-ATPase of cell membranes, and increases at particular frequencies should stimulate the enzyme and result in enhanced ion transport. This has been observed in the ouabain-sensitive accumulation of rubidium by erythrocytes [21] in the same frequency range. Calculations show an optimal amplitude for ion accumulation, also in line with observations. It should be noted that the same mechanism that leads to low-frequency optima in terms of ion transport processes between electrical double-layer regions can also account for the additive effects of very low amplitude signals over time.

Since the largest changes occur in those cations with the lowest steady-state concentrations, one would expect relatively large changes in protons and calcium ions. Both are known to affect membrane (and cellular) properties with very small changes in ambient concentrations. This idea opens up possibilities for linking the physicochemical model for ion fluxes with biochemical processes associated with the Na,K-ATPase and various calcium-stimulated protein kinases. It appears likely that the ion concentration changes

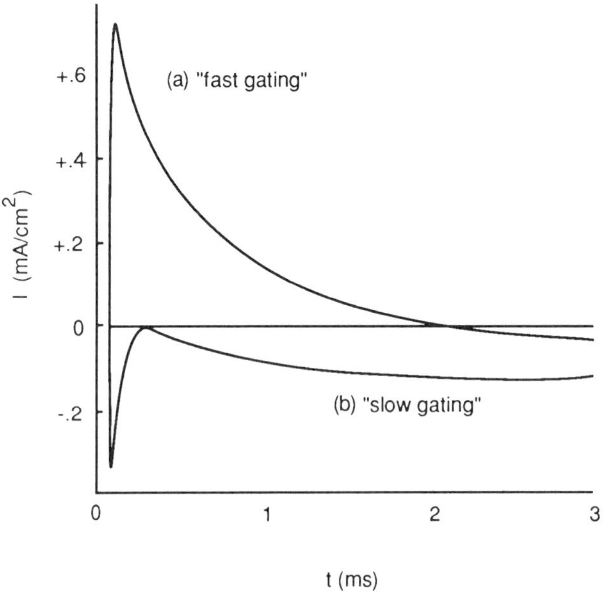

FIG. 9 Ionic current I (in mA/cm^2) versus the time t (in ms) for different gating rates. The parameters of curves (a) and (b) are identical except for the one that controls the gating current. Curve (a) for fast gating is about 15 times faster than curve (b) for slow gating. [Reproduced with permission from *Bioelectrochem. Bioenerg.* *13*:94 (1984).]

simulated by externally imposed electric fields could be acting as second messengers in the stimulation of biochemical reactions.

IV. EVALUATION OF THE MODEL

Our application of the concepts of colloid science to the problems of biological ion transport have led to insights into both the non-specific as well as the specific aspects of the process. The non-specific development is in the role of electrical double layers in membrane transport and the applicability of electrodiffusion to excitation and channel selectivity. The specific aspects of transport, the specialized reactions of the protein channels that span the membrane matrix, have also been elucidated as a result of analyzing their functions in terms of surface chemistry. In the following

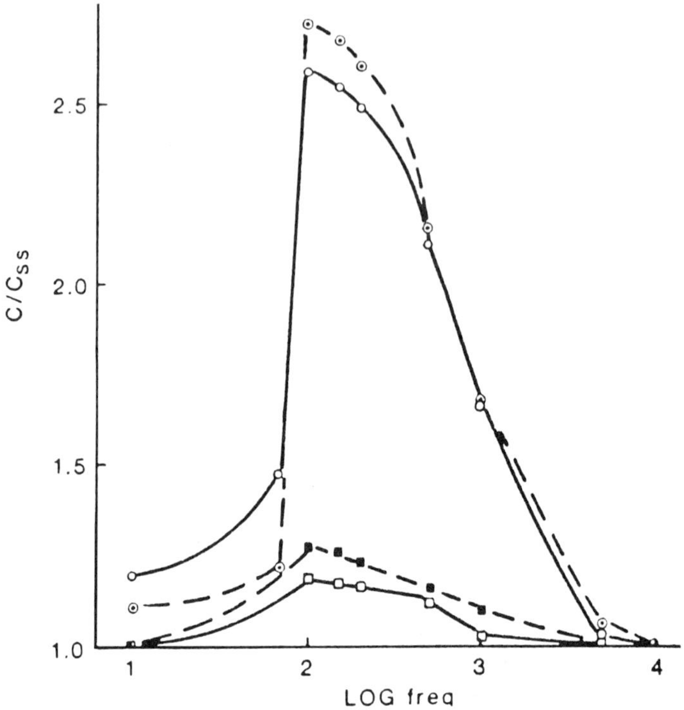

FIG. 10 Plot of C/C_{SS}, the ratio of the maximal to the steady-state concentrations as functions of the frequency for four species: internal free sodium (O) and bound sodium (⊙), and external free potassium (□) and bound potassium (■). The stimulation amplitude is 20 mV. Although all concentrations show optimal values at 100 Hz, the sodium changes more than the potassium and the bound ions more than the free ones. [Reproduced with permission from *J. Electrochem. Soc. 134*:1112 (1987).]

evaluation of the contributions of our approach, we shall focus on the particular problems solved.

A. Electrodiffusion in Membrane Systems

The SCM has proven to be a relatively simple way of applying the ideas of surface chemistry to the problem of ion transport in membranes. A surface compartment is essentially the Helmholtz model of the electrical double layer with its linear gradient of concentration away from the charged surface, and the compartment idea is a good way to approximate the concentration changes that occur during transport. By applying the SCM, three major aspects of biological transport processes have been clarified:

1. Up until now the ion fluxes in excitable membranes have been described empirically [4], because they were not believed to follow the laws of electrodiffusion. We have shown that the same laws govern ion movement across membranes as in solution when one takes into account the ionic processes that occur at charged surfaces.
2. It has always been a problem to explain the selectivity of the sodium and potassium channels in excitable membranes without invoking a Maxwell demon. Surely the smaller ion should fit through either channel when it opens! The SCM channel allows both ions to fit through both channels (with different conductances), and selective currents appear because the electrochemical gradients driving the fluxes change with time. The SCM results suggest that the kinetic contribution to selectivity is at times critical, and acts in addition to pore size and charge density.
3. There are a number of biological effects of oscillating electric fields that appear to have a strong frequency dependence and optimal effects in the ELF (extremely low frequency) range. By applying oscillating electric fields to the SCM sodium channel, we have shown frequency-dependent ion concentration changes with ELF optima that can explain the observations. This is another indication of the strength of the SCM approach and its ability to focus on the relevant physical aspects of biological processes.

These demonstrations of the ability of the SCM to account for unexplained aspects of biological ion transport are indications that the theory is able to describe the major charge movements that occur during membrane transport.

B. Protein Equilibria

1. Aggregation–Disaggregation Reactions

The oligomeric reactions of proteins appear to be governed by nonspecific electrical properties (surface charge, ionic strength). Despite very different compositions and structures, oligomeric proteins tend to disaggregate as the pH is farther away from the isoelectric point, or the ionic strength increases. The tobacco mosaic virus (TMV), a frequent model for the study of protein aggregation reactions, is made up of 2130 copies of an 18,000-kD protein, arranged in a helix. The TMV protein "self-assembles" at the isoelectric point and breaks up into smaller aggregates until it becomes a monomer as it is progressively charged. (There can still be appreciable aggregation at high charge when the ionic strength is increased.)

Since aggregation–disaggregation reactions do not involve rearrangements within the subunits, but rather interactions between the subunit surfaces, it is reasonable to approximate the total free-energy changes in these processes by the surface free energy, which can be calculated simply in terms of the area of the molecule in contact with the aqueous solution and the surface charge density [22]. (Because dissolved proteins have low interfacial energies, the electric charge is an important contributing factor to the change in the total free energy.)

Assuming this model, we have estimated changes in surface free energy in terms of molecular area and charge of hemoglobin, and calculated values of the equilibrium constant of the tetramer–dimer reaction that are quite accurate despite the simplicity of the model [23]. Our measurements and predictions, shown in Fig. 11, indicate that hemoglobin tetramers disaggregate into dimers at all degrees of charge, and that above pH 9.5 there is further disaggregation into monomers. (The process depends on total hemoglobin concentration, and one would expect changes with both hemoglobin and salt concentrations, as well as with pH.)

This quantitative test of the surface free-energy model (SFE) of protein aggregation indicates the usefulness of the model when applied to hemoglobin. For other proteins, where quantitative tests have not yet been performed, the model can account for qualitative aspects of the behavior of the system. For example, the model indicates why disaggregation occurs only in certain ranges of pH and why the ionic strength changes tends to oppose the effect of increasing charge, as seen with TMV.

2. Hemoglobin Reactions

The surface free-energy model (SFE) has been used to describe other hemoglobin reactions that involve changes in the area or surface charge:

(a) We have shown that changes in charge and area cause the equilibrium constant for the reaction of hemoglobin with oxygen to be a function of the pH and ionic strength. For hemoglobin, the pH dependence is known as the Bohr effect, and the SFE model gives both the acid and alkaline Bohr effects with the maximum equilibrium constant below the isoelectric point, as observed [24]. A comparison of the SFE calculations and measurements are given

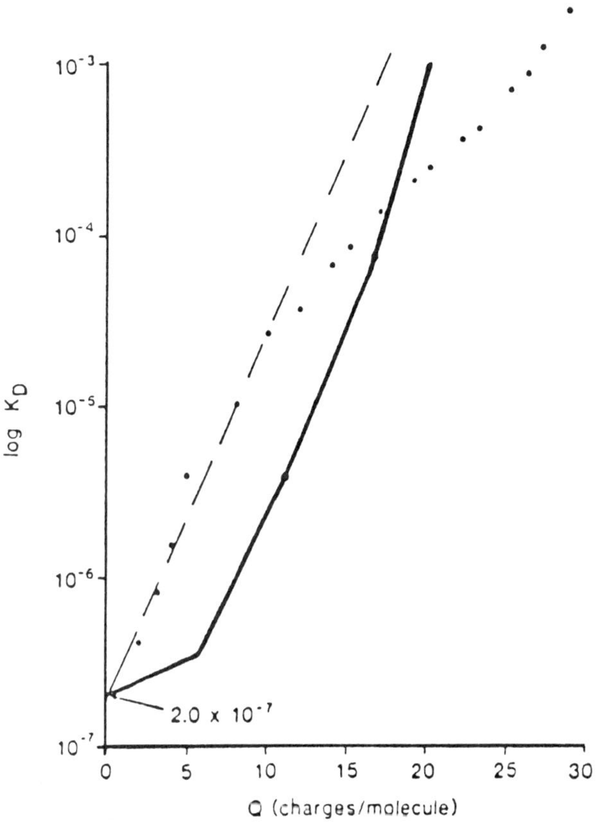

FIG. 11 Logarithm of the disaggregation equilibrium constant, K_D, as a function of the charge, Q, on the hemoglobin molecule. The points and dashed line are experimental, while the solid line is based on the surface free-energy calculation. [Reproduced with permission from *Bioelectrochem. Bioenerg.* *17*:349 (1987).]

in Fig. 12. Here again the calculated points are remarkably close to the observations indicating that the simple model is quite powerful.

(b) We have developed an alternative quantitative description for hemoglobin reactions based on the extra free energy of the conformational change: the change in area and surface charge of the protein when the heme groups combine with oxygen. One can describe

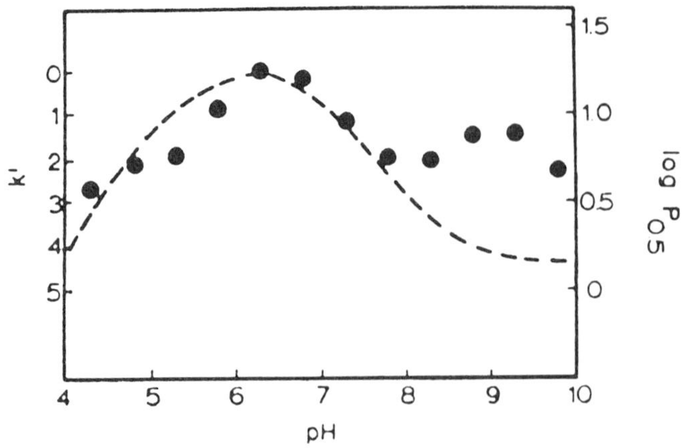

FIG. 12 Calculated values of the oxygen binding equilibrium constant (●) in units of k', which is equal to the surface free energy divided by RT, superimposed on the dashed curve of observed values in units of log $P_{1/2}$, the oxygen pressure for the half-saturation of hemoglobin. [Reproduced with permission from the *J. Theoret. Biol.* *51*:127 (1975).]

the displacement of the equilibrium constant in terms of the surface free energy of the conformational change rather than by introducing the Hill coefficient, which has been used as a measure of cooperativity but is poorly understood. Comparing the SFE approach with the conventional one, we see that the Hill coefficient is directly related to a Gibbs surface excess [25] and therefore not a constant, as it is frequently described.

(c) We have described the viscosity increase when the tetrameric hemoglobin continues to aggregate beyond the tetramer to the n-mer as the concentration of the solution increases [26]. The various aggregated species contribute to an increase in the viscosity of the solution as a result of the increased chain length and also due to the formation of a rigid structure when the chain becomes long enough to fold upon itself. The calculations assume the same loss of area and increase in charge density when a subunit adds at any stage. From the distribution of molecular species, it is possible to estimate the viscosity of the solution and to show that it increases markedly at higher hemoglobin concentrations. This model indicates the possibility of higher aggregates of normal hemoglobin and provides a rationale for the kinds of processes that do occur when the abnormal sickle hemoglobin aggregates.

3. Membrane Channels

The SFE model has described quite a variety of reactions involving hemoglobin with reasonable quantitative agreement between calculations and observations. The ideas of the model appear to apply more generally, as we saw with the properties of the TMV protein, so there is reason to believe that they would also describe the energetics of the channel proteins in contact with aqueous phases.

The integral membrane proteins associated with ion transport in excitable cells (e.g., the Na channel, the acetylcholine receptor, the Na,K-ATPase) appear to be generally composed of associated oligomers or the associated helices of a single chain, organized in a cylindrical geometry. The mechanism of gating in such assemblies is not known, but the opening of a channel involves the formation of a protein–aqueous interface and should depend on the same physical principles that govern aggregation and disaggregation in oligomeric structures.

Our model for a voltage-gated channel [18], based on aggregation phenomena in multi-subunit proteins, appears to account for a number of important properties of channels (see Fig. 7). In the squid axon, the membrane has a much higher negative charge on the outside than on the inside, so we can consider the channel subunits disaggregated (i.e., open) on the outer surface, where the charge is high, and aggregated (i.e., closed) on the inner surface, where the charge is low. Upon depolarization, the shift of some negative charge from the outer to the inner surface causes the channel to open. The properties of this voltage-gated channel have many of the qualitative features of the sodium and potassium channels, as described earlier. More important, there also appears to be quantitative agreement in (1) the relation between the channel gating conductance (that determines the gating current speed) and the channel selectivity (in excitable membranes, a sodium channel is fast gating and a potassium channel is slow gating, as observed; (2) the effectiveness of the electrodiffusion equations in predicting the observed direction and time course of the ionic currents through a population of open channels; and (3) the cooling–heating that accompanies channel opening–closing, expected on the basis of protein reactions, can be due to only 3% of the channel protein surface.

C. Kinetics of Membrane Processes

At this stage it would be desirable to summarize the two major approaches described in this chapter (the SCM for electrical double layers and the SFE for channel proteins) and indicate how they interact. It is obvious that electrical double-layer ion sorption processes are very fast in comparison with the gating and ion transport that are stimulated simultaneously by changes in polarization.

Depolarization not only causes the charge shifts in the channel proteins but also starts the changes of ion concentration gradients between the electrical double layers on the two sides of a membrane. If channel opening is fast, one obtains ion fluxes that reflect the ion gradients present in the steady state or at short time. If the channel opens slowly, the gradients have changed by diffusion (under C) and migration (under E) and one obtains fluxes that reflect the altered cation gradients.

Figure 13 is a diagram of the charge movements that occur in the SCM during membrane channel processes driven by imposed electric fields. The processes are listed in Table 4 along with the characteristic times and the ac frequencies that would be expected to affect those processes. It is apparent that the individual processes of charge movement have very different characteristics, and that the combined effects are different from the component processes. The biological effects that we usually focus on are probably most closely related to the changes in ion concentration, the slowest of these processes.

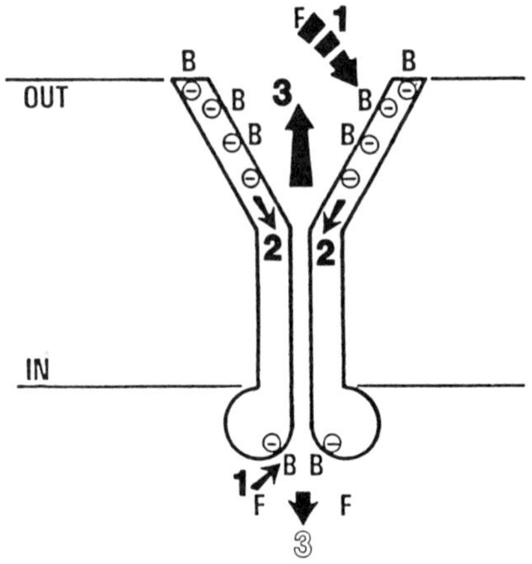

FIG. 13 Diagram of the charge movements that occur as a result of changes in polarization across ion channel proteins. The three processes are: 1, cation binding from free (F) to bound (B) form in the electrical double layer; 2, charge movement within the protein, such as gating currents; and 3, ion movements in the aqueous channel when a pore opens.

TABLE 4 Charge Movements in Channels

Process	Characteristic time (μs)	Resonant frequency (Hz)
EDL adsorption	< 1	$> 10^6$
Gating current	$10–10^2$	$10^4–10^5$
Ion transport	$10^2–10^3$	$10^3–10^4$
Combined effects (Na channel)		10^2

D. Concluding Remarks

Application of the ideas of surface chemistry to the problems of membrane transport have proven useful. Since electrical forces are particularly potent across the membranes and in the ionic solutions of living systems, one would expect electrochemistry to play a central role in the elucidation of biological processes. Surface chemistry and electrochemistry overlap considerably and frequently merge when applied to biological systems. The emerging interdisciplinary field, bioelectrochemistry, which utilizes the techniques and concepts of both disciplines, offers considerable promise for elucidating many of the problems in biology.

REFERENCES

1. L. A. Blumenfeld, *Physics of Bioenergetic Processes*, Springer-Verlag, Heidelberg, 1983, pp. 1–3.
2. P. Mitchell, Keilin's respiratory chain concept and its chemiosmotic consequences, *Science 206*:1148–1159 (1979).
3. B. Hille, *Ionic Channels in Excitable Membranes*, Sinauer Associates, Sunderland, Mass., 1984.
4. A. L. Hodgkin and A. F. Huxley, A quantitative description of membrane current and its application to conduction and excitation in nerve, *J. Physiol. (Lond.) 117*:500–544 (1952).
5. I. Langmuir, Force near the surface of molecules, *Chem. Rev. 6*:451–479 (1929).
6. R. K. Small, M. Blank, R. Ghez, and K. H. Pfenninger, Components of the plasma membrane of growing axons. II. Diffusion of membrane protein complexes, *J. Cell Biol. 98*:1434–1443 (1984).

7. M. Blank, Monolayer permeability, *Prog. Surf. Membr. Sci. 13*: 87–139 (1979).
8. M. Blank and J. S. Britten, Transport properties of condensed monolayers, *J. Colloid Sci. 20*:789–800 (1965).
9. G. D. Rose and J. A. Quinn, Composite membranes: the permeation of gases through deposited monolayers, *Science 159*: 636–637 (1968).
10. M. Blank, The thickness dependence of properties of membrane protein multilayers, *J. Colloid Interface Sci. 75*:435–440 (1980).
11. R. J. Archer and V. K. LaMer, The rate of evaporation of water through fatty acid monolayers, *J. Phys. Chem. 59*:200–208 (1955).
12. M. Blank and S. Feig, Electric fields across water–nitrobenzene interfaces, *Science 141*:1173–1174 (1963).
13. I. R. Miller and M. Blank, Transport of ions across lipid monolayers: reduction of polarographic currents of Cu^{++} by decylammonium monolayers, *J. Colloid Interface Sci. 26*:30–40 (1968).
14. M. Blank, L. Soo, and R. E. Abbott, The ionic permeability of adsorbed membrane protein monolayers, *J. Electrochem. Soc. 126*:1471–1475 (1979).
15. J. S. Britten and M. Blank, The effect of surface charge on interfacial ion transport, *Bioelectrochem. Bioenerg. 4*:209–216 (1977).
16. T. Teorell, An attempt to formulate a quantitative theory of membrane permeability, *Proc. Soc. Exp. Biol. Med. 33*:282–285 (1935).
17. M. Blank and J. S. Britten, The surface compartment model of the steady state excitable membrane, *Bioelectrochem. Bioenerg. 5*:528–540 (1978).
18. M. Blank, The surface compartment model (SCM): role of surface charge in membrane permeability changes, *Bioelectrochem. Bioenerg. 9*:615–624 (1982).
19. M. Blank, Ion channels as short circuits between electrical double layers, *J. Electrochem. Soc. 134*:343–346 (1987).
20. M. Blank, Theory of frequency-dependent ion concentration changes in oscillating electric fields, *J. Electrochem. Soc. 134*: 1112–1117 (1987).
21. E. H. Serpersu and T. Y. Tsong, Stimulation of a ouabain-sensitive Rb^+ uptake in human erythrocytes with an external electric field, *J. Membrane Biol. 74*:191–201 (1983).
22. M. Blank, A surface free energy model for protein structure in solution: hemoglobin equilibria, *Colloids Surf. 1*:139–149 (1980).
23. M. Blank and L. Soo, Surface free energy as the potential in oligomeric equilibria: prediction of hemoglobin disaggregation constant, *Bioelectrochem. Bioenerg. 17*:349–360 (1987).

24. M. Blank, A model for calculating the Bohr effect in hemoglobin equilibria, *J. Theoret. Biol. 51*:127–134 (1975).
25. M. Blank, Hemoglobin reactions as interfacial phenomena, *J. Electrochem. Soc. 123*:1653–1656 (1976).
26. M. Blank, Molecular association and the viscosity of hemoglobin solutions, *J. Theoret. Biol. 108*:55–64 (1984).

15

Transport Mechanisms, Physiological Communication, and Intelligence

MAX BENDER Department of Chemistry, Fairleigh Dickinson University, Teaneck, New Jersey

I.	Introduction	367
II.	Transport and Intelligence	372
III.	Energetics and Transport	374
	A. Physiological energy in contradistinction to intrinsic inanimate energy	374
	B. Physiological energy mechanism manifestations in particle, cell, and flagellar motility	377
IV.	Transport-Associated Electrical Properties: Electrical Diffuse Double Layer, Electrokinetics, and Electrogenics	384
V.	Interactions and Transport	385
	A. Inanimate interactions	385
	B. In vivo interactions	392
VI.	Concluding Remarks	396
	References	397

I. INTRODUCTION

In a physiological system, moieties throughout are in directed translocation, as structure with its complicated array of tremendous interface is maintained. Communication, indispensable for growth and sustenance, is accomplished by signals and messages of physical, chemical, electrical and vibrational nature. The purpose of this chapter is to present concepts resulting from the integration of findings for the complicated transport phenomena, supporting details being available in abundance through the literature.

Transport processes in the living system are of a wide variety. The travel is directed and intricately organized, the energetics leaving relatively little to randomness. Much as bee and ant colonies [34,35,72] may be regarded as superorganisms, cells, vesicles, organelles, microtubules, and protein molecules carry out necessary biotransport functions of maintenance, growth, neutralization of inimical agents, and transmission of intelligence for the biological system. Singular travel is performed, and communication tasks are achieved on the basis of individual and collective intelligence. Actually, the immune system is looked upon by Young and Cohn [65] in the same sense as an insect colony and by Smith [66].

Considering the complex network of interfaces and the wide spectrum of molecular species, ranging from simple ions to particles with very high dalton numbers, the living system as a whole is certainly heterogeneous; but homogeneous aspects must be taken into account. Actually, in view of the many different degrees (and types) of interaction that exist within and between molecules and their moieties, there can be nearly homogeneous sections where phase boundaries are not sharp. Semantically, solubility is considered to be the state when atomic intimacy exists everywhere between the molecules. Under these homogeneous conditions there is no phase boundary. One molecule can move relative to another as in the case of sugar in water. Solubility begins to decrease as interaction develops between atoms or groups of atoms within one species of molecules, so that other species are excluded from its atomic intimacies. Heterogeneity with corresponding phase difference results in the formation of an interface. Understandably, within a (large) molecule there can be (areas of) heterogeneity.

Water is a major biocomponent. It is, necessarily, an important factor in transport processes in different ways. It itself is transported and acts as a transport solvent, dispersing and suspending medium, hydrophilic "lubricant," and fluidizing or plasticizing agent. Water molecules in transported material are held with varying degrees of affinity and influence conformation, as in the case of proteins. Transport can be through a watery medium and along or through hydrated walls.

The nature of the transport differs according to the physiological system; that is, vascular, digestive, lymph, endocrine, renal, glandular, nervous, and immunological, and according to the organs involved, such as the heart, liver, pancreas, lungs, brain, and skin. Also, each system and organ is served by a variety of transport functions.

Transport volume flow ranges from bulk to but a procession of molecules or cells. The pathways, differing in structure and composition, range in width from "pipeline" sizes, through capillary dimensions, to the channels in protein molecules or intermolecular space. Under many circumstances the conduit walls participate in bringing about flow, examples being the endothelial cells on vascular walls, capillary areas, and microtubules. The consistency of bulk material in transport ranges from gaseous to liquid to semisolid.

Moving material may be a solution (homogeneous), a suspension (heterogeneous), or particles on their own. Cells, organelles, microsomes, vesicles, proteins, other polymeric molecules, chemicals, water, simple ions, protons, and electrons are all agents of transfer. In given homogeneous areas, particular molecules and particles are in vectorialized motion unrestricted by boundaries. The actual flux of microtubules is discussed by Burns [26] and referred to by Kamimura and Kamiya [29]. Although they are not part of the actual physiological workings of a system, viruses and bacteria move through it vectorially and impose their influence on it.

Classification of the travel rates in a living system include those in the range of the speed of light, the less rapid but extremely fast protein conformation changes, and the still less rapid biochemical reactions and transport. Some examples of "speed-of-light" travel follow.

When an ovum is fertilized, the entry of one sperm "immediately" transforms it to a state of nonacceptance of other sperm, no matter what the sperm population or their closeness to the ovum surface. Adey [2] obtained results indicative of electromagnetic signaling in the cell interior, similar to that through T cells, where it was observed by Lindsten et al. [36] that stimuli received at the cell surface alter gene expression by induction of specific changes in messenger RNA degradation. Immunological studies by Matsunami et al. [40] and Rathbun et al. [39] point to the enormous diversity of antigen-binding specificities expressed by B and T cells in strictly regulated site-specific genomic recombination, assembling new genes that provide danger-sensing radar to ferret out invading viruses and bacteria. Analogous to tsunamis which are solitary waves of longevity, dissipating very little as they travel long distances, and which are called "solitons," Scott [67] discusses microscopic solitons as a rapid means of biocommunication. In oscillatory traversal these solitons could be the basis of nerve impulses with their energy of passage through the protein of the cell. Possibly they involve and utilize chemical bond frequencies even to the extent of intramolecular reshaping.

Proteins, ubiquitously essential to the very state of being alive, contributing function and structure to muscle, skin, hair, and blood and performing as hormones, antibodies, and enzymes, necessarily are major factors of the transport system. They transport themselves, power the transport of other protein species, and constitute media (i.e., ionic channels and microtubules) along which transport takes place. Given proteins have more than one role. This is illustrated in the findings of Schulte et al. [17], where mitochondrial proteins were seen to be involved in both electron transport and protein building. With respect to proteins moving on their own, the research of Nuchtern et al. [18] is pertinent. It is a study of antigens crossing the membrane from the cell cytosol in which they were processed to the secretory system. Wolf et al. [79] show that the function of the 30-kD movement protein of tobacco mosaic virus is to facilitate cell-to-cell movement of viruses. It alters the size-exclusion limit of the plasmodesmata of the infected plant.

In the light of the wealth of published research on physiological transport, given sample papers are briefly highlighted in the following material, exemplifying intelligence-directed structured (nonrandom) transport. Subject areas include a variety of physiological circumstances, such as off and on (gated) protein-channeled ionic transport, triggering, binding, interaction, transmitters (neuro), targeting, transduction of signals (i.e., G-proteins) [68,71], translocation, signaling pathway, activation, receptors, mediation, rapid intracellular communication, regulation, second messenger in neural and nonneural cells, intracellular signaling, membrane surface receptors, recognition including molecular recognition, protein transport, channel mobility, protein building, intracellular transport, axonal transport, microtubules, chemical messengers, messenger RNA, enzyme cooperativity, motility factors in the regulation of cell movement, and more.

K^+ and other ions being instrumental in the conduction of electrical impulses in nerve cells, Alkon [11] in his discussion of memory storage and neural systems refers to neuronal site receptivity as being greatly enhanced by protein kinase C enzyme (PKC), a Ca-sensitive protein that affects K^+ flow. In the cell cytoplasm it increases K^+ flow as a result of which the excitability of the cell is reduced. As the PKC translocates to the cell membrane and binds there, it regulates the ion channels, reducing K^+ flow. Now, as a result, the cell's excitability is increased, input signals triggering neuronal firing more readily. Neuronal branches (i.e., dendritic trees) can, due to their tremendous surface area and complexity, receive 100,000 to 200,000 signals from the separate input fibers ending there.

The PKC acts by adding phosphate groups to protein substrates. Another calcium-activated enzyme, (CAM) kinase II, also phosphorylates the membrane protein and reduces K^+ flow. It cooperates with PKC in platelet aggregation, insulin secretion, and muscle contraction. PKC translocation prolongs and enhances responses to electrical, chemical, and hormonal signals in different physiological circumstances. As to pathways, the input information from a conditioned stimulus can flow along collateral routes that are not hardwired or genetically laid out but are formed in the course of a learning experience.

With plant systems there is physiological flow up from the roots [i.e., sap (latex)] and downward flow such as that which carries the products of synthesis from the leafy areas. An example of how complicated plant transport processes can be is described by Diamond [12] for the *Arum* lily at blooming time. In the process, heat is generated to a level of 22°C above the ambient temperature, volatilizing compounds to attract pollinators. It takes place when the production and travel of salicylic acid trigger a metabolic explosion by the mitochondrial respiratory transport system.

K^+ ionic channel activation was studied by Kim and Clapham [13] and Ordway et al. [14] in work with cardiac cells of newborn rats and the smooth muscle cells of the toad stomach, respectively. It was observed that the K^+ ionic flow across the cell membranes was enhanced by arachidonic acid and phosphatidylcholine present in the membrane. By lowering the pH from 7.2 to 6.4—6.8, three- to tenfold increases in channel activity were found, indicating fatty acids and fatty acid derivatives as second messengers. By responding to external signals, these compounds regulate the activity of the channel, mediating neurotransmission.

In studies by Stühmer et al. [15,16] on the opening and closing mechanism of Na ionic channel gating in *Xenopus* oocytes, site-directed mutagenesis combined with patch clamp recording led to a clearer picture in terms of channel conformation. With voltage control across the cellular membrane, positive charges in segment S4 of the protein channel were found to account for the voltage-sensing mechanism for channel activation, and the region between III and IV in the protein had to do with inactivation. Voltage-dependent gating is an aspect of cellular activity as in the relay of nervous system information and the activation of skeletal muscle, cardiac muscle, and secretory cells.

Opposed to the viewpoint that protein ionic channels are fixed and embedded, work by Jones et al. [20] indicates that given the circumstances, channels can move and become placed in appropriate positions in cells according to their intelligence mechanism, which requires particular ionic flow intensity for necessary levels of cell function. They furnished evidence for voltage-dependent Ca channels being *mobile* in hippocampal CA 1 neurons. Specific mechanisms act before synaptic contact to recruit channels, segregating them, translocating them, and targeting them to, and immobilizing them on, bodies in the neuron and on specialized dendritic sites as required. The ability of the neuron to orchestrate and regulate Ca signaling is thus expanded, with synaptic plasticity, oscillatory behavior, rhythmic firing, and neuronal integrative properties being appropriately modulated through transduction.

II. TRANSPORT AND INTELLIGENCE

While there is much to the semantics of the term *intelligence*, the definition in *Webster's New Collegiate Dictionary* (1949), ". . . the ability to apprehend the interrelationships of presented facts in such a way as to guide action towards a desired goal," applies to living system operations. These operations are multitudinous and specific in the extreme. There is recognition. Components know where to go and how to perform. Signals are given and acted upon. Cybernetic balance takes place. Organelles are directed along a microtubule network. Intelligence enables the conversion and apportionment of the energy from electromagnetic radiation with regard to vegetation, and from food, the essential sources being vegetation and animals, to the needs throughout the biological system. Illustrating intelligence, Buss and Grosberg [75], in their studies of intraspecific competition in the simple hydroid *Hydractinia symbiolongicarpus*, find a repertoire of aggressive sophisticated discrete behaviors, as colonies in their interaction with each other conduct strategic maneuvers.

Consider the synthesis of protein within a cell. Genetic information is contained in the deoxyribonucleic acid (DNA) in the cell nucleus; its two strands separate by unwinding to act as a template for building the protein, first by being copied onto a molecule of messenger ribonucleic acid (RNA). The messenger RNA then translocates to a ribosome, which is an RNA-containing structure and is the area where the protein will be synthesized. The amino acids of which this protein will be built are brought in by multitudes of transfer RNA molecules, which assemble along the messenger RNA and link according to the template code in the presence of appropriate enzymes and other cell components. Then the protein chain readily translocates through the cell membrane to its destination.

Intermediate state details of the transfer RNA ribosome transport have been researched by Moazed and Noller [38].

In the hormone system, messages from external or internal sources are received by the brain, the brain informs the appropriate gland, and counteracting compounds are secreted that move or are carried to receptors essential for signal transduction. Not least is the wide array of enzymes with their specific powers in the selection of particular bonds to break and to build when and where according to need. With vegetation, the selective ability of the root system to take up the particular elements and minerals in the appropriate quantities for the plants' requirements is another example.

Red blood cells know when and where to pick up oxygen and to give it up. Considering that multitudes of newly born red blood cells are continually emerging from the bone into the bloodstream, the older ones have to become eliminated so that status quo in numbers is maintained. Thus a physiological decision is made as to when a particular cell is to be removed. There is the know-how of removal methodology such that essential ingredients of the cell are preserved.

With respect to the subject of immunity, cells recognize inimical organisms and render them inactive. The immunological considerations and work of Rathbun et al. [39] and Matsunami et al. [40] have already been referred to in connection with the transport speed of messages. This subject is brought up again to demonstrate the intelligence of the operation. The research points to a given gene as the master builder of the immune system. It is called the recombination activity gene (RAG-1). Through its reshuffling of the genetic library in new white blood cells, a limited amount of DNA "scraps" is built into an endless assortment of antibodies according to the bacteria and viruses (including new ones) to be neutralized. Indications are that RAG-1 makes the "recombinase" enzyme, which cuts and splices the DNA to form the new genes.

Work of Scheller and his associates [19] emphasizes particular intelligence in nerve cells. It was shown how controlled release of neuropeptides (2 to 40 amino acid residues) mediates rapid intercellular communication. Their experiments were with the snail *Aplysia*. While the electrical activity in a single cell released a variety of neuropeptides, specific neuropeptides targeted different synaptic vesicles within the cell, each particular reception determining appropriate extracellular signaling via electrical impulses along the nerves.

The intelligence of selection and decision in chemically and physically breaking and making bonds, direction of translocation, recognition, timing, communicating, production, and controlled release that is displayed in physiological systems is at a very high level compared to the "memory" capacity of inanimate systems (e.g., crystal growth). The recent metal ion template synthesis of macrobicyclic lanthanide (amine) complexes by McMurry et al. [1] is illustrative of advanced inanimate molecular recognition, but biological levels are from all appearances in a special class of their own.

Of interest in this connection is consideration of the Maxwell's demon concept as a threshold between animate and inanimate intelligence. This "gatekeeper" defeats entropy. He achieves a bulk temperature difference in a gas through the intelligence of his decisions in separating fast molecules from slow ones. He has to have the "living" energy to make the decisions, and to follow through with the molecular separation. Particular views on Maxwell's demon under biological circumstances are afforded by McClare [82].

Biological intelligence does not necessarily involve the brain. There are the living processes in vegetation. Physiological activity continues in skin and other tissue after excision, an electrical potential continuing to be generated (see Chapter 3).

III. ENERGETICS AND TRANSPORT

A. Physiological Energy in Contradistinction to Intrinsic Inanimate Energy

Intrinsically, matter, whether animate or inanimate, manifests energy in various respects. In addition to the energies inside the atomic nucleus, there are energies of translation, spin, rotation, oscillation, and orbiting. Molecular translation and rotation, and bond oscillation, are referred to as thermal energy, kT, the product of the Boltzmann constant and the absolute temperature, being its classical reference. Whether in the gaseous or the liquid state, particle translation is randomly chaotic.

Classically, for a gas, the average translational energy per molecule is 3/2 kT, 1/2 kT per degree of freedom. Similarly for the gaseous molecule, its rotational energy averages 1/2 kT per degree of freedom. The classical energy equivalent of one of the modes of vibration of a chemical bond is kT. Brownian movement is the microscopically visible manifestation of the translational and rotational energy of particles. These energies are presumably the outward display of the atomic intranuclear dynamics.

Electromagnetic radiation frequencies are associated with discrete quantized jumps in the levels of energy of molecular rotation; electron, proton, and nuclear spin; oscillatory motion; and electron orbitals. While the translational energy of molecules in random motion increases smoothly with temperature, nevertheless, quantized interpretation is applicable, the quantum values being exceedingly small.

Rotational quantum jumps are associated with far-infrared spectral frequencies. Chemical bond vibrational mode quanta are considerably larger than the rotational quanta, the spectra being infrared. Electron energy changes show in the visible frequencies. On a general basis, quantum magnitude sequence increases proceeding from molecular translation to nuclear spin to electron spin to molecular rotation to the interatomic vibration of a chemical bond to electron orbital change.

At room temperature the average energy level of a chemial vibrational mode is well below kT. This is due to the wide quantum spacing, the result being a low probability of occupation at higher levels. Vibrational energy differs depending on the bond. Polymeric molecules have a wide range of intrinsic oscillatory energy. Besides the chemical (strong) bonds, weaker (physical) attractions between atoms and between segments of the molecule involve not only vibrations but also motions such as twisting and bending (of chains). Thus quanta for all these likely modes can vary in magnitude down to those as small as for translational energy.

With respect to protein molecules, the deliberation of the intrinsic energy is a complex proposition. There is much room for study. Pain [83] discusses the application of compressibility (Gekko and Hasegawa) and x-ray crystallography (Frauenfelder et al.) to study "free" space in proteins in terms of their intrinsic thermal energy. In attempting elucidation of the intrinsic energy in protein, Frauenfelder [27] makes a comparison with a much simpler system, glass.

Doster et al. [37] employ inelastic neutron scattering in an examination of rapid oscillations in myoglobin. While elastic scattering gives only the geometry of the protein molecular dynamics, the use of inelastic scattering opens up information on the time-dependence factor. Nienhaus et al. [28] make their examination of the myoglobin crystal with Mössbauer γ-rays. This affords a precise ratio of elastic-to-inelastic diffuse scattering of photons. As a result, information is obtained distinguishing inter- and intramolecular dynamics.

King [69] gives an extensive review of protein molecular conformation changing (i.e., folding) on the basis of restrictions in reconfiguration. He states that arrangement in the molecule cannot take place randomly. There are favored pathways, conformational intermediates forming and directing the folding operation. "Rules" of protein folding involve peptide and hydrogen bonds, ion pairing, and van der Waals interactions, and their geometric orientation. The amino acid sequence determines the folded configuration in the final structure, and also how rapidly this state will be attained.

The living biological system has the capacity to convert the free energy in food to its needs for growth and maintenance. An adequate potential is built up, assuring the energy for intelligence in directing its multitudinous operations. Intrinsic translation and rotation are derandomized. Molecular configuration is changed. There is sufficient energy to break chemical bonds readily and make new ones. In bioluminescence, electrons are elevated in energy out of orbitals they would occupy under inanimate conditions. An intricate signaling system is maintained, and there is appropriate response, in terms of the particular functions of the organism, to outside influences such as sonic and electromagnetic oscillations, thermal changes, and the contact of touch, smell, and taste.

Exemplification of the derandomization of translation is in the observation of live bacteria about 1 μm in size speeding through an aqueous medium in a smooth path with not an erratic flicker [30]. Also apparently devoid of Brownian movement behavior is the propelling of vesicles along microtubules [3–8]. In this light, depending on the structure of the membrane and the solubility of the diffusing material in it, a concentration gradient that is no more effective for transport than kT can be reduced appreciably in its potency for effecting translocation in membranes.

Physiological energy being at a level sufficient for electronic orbital jumping and commensurate with the activation for cutting and splicing molecular chemical bonds, the potential is there to make changes in molecular configuration as required by the system. Molecules evidently stretch out when blood cells squeeze through microscopic bone channels from the marrow to the bloodstream, and also as they travel through the spleen. The translocation of protein through polar, then nonpolar parts of a membrane could be understood in the rearrangement of its polar and nonpolar groups to form a hydrophilic or hydrophobic molecular exterior for compatibility with the surroundings. In this vein, reference is made to Clarke et al. [84]. They observed marked rearrangement in the structure of lactate dehydrogenase at the site of binding to pyruvate anion substrate. This substrate-induced conformational change apparently was in the direction of a regional polar atmosphere in the protein, more suitable to the substrate.

While their deliberations on the conformation of DNA double helix pertained to the crystalline state, Shakked et al. [74] bring out how dependent its molecular structure is on the environment by which it is surrounded. The DNA adopts a range of conformations depending on its hydration, molecular packing, and temperature. In this connection, reference was made to the modes of DNA recognition, by proteins, of a highly variable structure. Stern et al. [76] find evidence through chemical probing methods that the folding and function of 16S ribosomal RNA depends on its interactions with proteins.

In his 1968 book on blood, Vroman [41] pays attention (Fig. 38, p. 101) to this requirement for molecular conformation change to achieve compatibility with the environment. He describes a phospholipid micelle in the blood taking up plasma. As the plasma enters the interior of the micelle, the phospholipid molecules rearrange to form a membrane bilayer around the absorbed plasma. Protein molecules are shown stretched out across this membrane, polar and nonpolar parts of the protein molecule matching, respectively, the polar and nonpolar membrane parts. His studies of the interaction of blood proteins with glass and other surfaces (see Chapter 6) indicate that molecular configuration change is an important aspect of the sorption. Apparently, this is at least one of the factors in the law of mass action hardly being applicable. Anderson et al. [64] show in their studies with proteins of the transferrin family that human apalactoferrin undergoes substantial conformation change in its binding and releasing of iron.

Various research findings portray that more than just molecular configuration change takes place in many cases of transport through physiological media. Enzymes are ubiquitously present. Thus it is conceivable, as with gene recombination [39,40] and antigen protein reprocessing in the cell [18], that molecules can be rearranged by cutting and splicing them to enable their translocation in compatible fashion with the surroundings.

A paper by Verner and Schatz [42] is of interest in this connection. They point to physiologically newly synthesized proteins reaching their destinations as a result of signaling. The signaling is associated with a given part of the molecule. Translocation involves molecular loose conformation, adequate energy, the existence of receptor-like components in the cytosol and on the target membrane, and proteolytic cleavage. In Laine and Esser's studies [33], conclusions are drawn with respect to the transfer of human complement protein C9 from aqueous environment to the hydrophobic interior of a membrane. For the membrane entry, their deliberations include that the protein unfolds in part and that polymerization takes place.

B. Physiological Energy Mechanism Manifestations in Particle, Cell, and Flagellar Motility

Reference has been made above to the power of proteins to migrate and to bring about the transport of other bodies. On a general basis, particle, cell, and flagellar motility are understood to involve a motor protein that converts the chemical energy of adenosine triphosphate (ATP) hydrolysis to mechanical work. The process consists of a cyclic reaction between the motor protein and cytoskeletal polymer and particles when they are transported. Myosin, dynein, and kinesin are known to be motor proteins. Myosin reacts with the actin filaments of the cytoskeleton; dynein and kinesin react with the cell's microtubule filaments.

When binding occurs at a given site on the filament, the motor protein supposedly changes conformation to bring about an increment of motion. Then there is release of the binding at this site, but attachment takes place at another site with corresponding conformation change so that propulsion of particle or filament results. The on and off binding and the timing of conformation change supposedly are controlled in the ATP hydrolysis power source.

1. Particle Motility

For quite some time there has been evidence, including radioactively labeled results, of transport of materials in nerve cells. This is described by Ochs [45]. However, it is only recently and thanks to the utilization of highly developed video-enhanced contrast light microscopy with modern computers that it actually became possible to observe the motion. Vesicles and organelles were seen to move at high speed (i.e., fast axonal transport), being propelled and directed along lengthy slender fibers arranged in a network throughout the cytoplasm. Brownian movement was not evident for the particles unless they were disassociated from the fibers.

The fibers are identified as microtubules of the protein tubulin. Besides their role in the propulsion of matter, microtubules give structure to the cytoskeleton, being ubiquitously associated with cells in general, including blood cells. They make up the spindles of cell mitosis.

Microtubules have the capability to convey materials of diverse nature. They carry mitochondria, surplus membrane, protein, nerve transmitter substances, viruses, and agents for color change in fish. It is conceivable that in plant life, the basic transport of solution or sap and other essentials is along a network of microtubule nature, osmotic and capillary rise processes being secondary. Similarly, when coated with kinesin enzyme, inanimate particles such as glass or carboxylated latex beads are propelled along microtubules.

Allen [43] describes the axonal transport along microtubules that brings about the exchange of substances between the nerve cell body and the synaptic terminal. Vesicles carrying neurotransmitters formed by the Golgi in the cell body go to the synapse, while surplus membrane of the synapse, made up into multivesicular bodies, passes to nerve cell body lysosomes for degrading. Mitochondria move in both directions according to the energy needs of the cell. Mitochondria transport is intermittent, that of the vesicles being continuous.

Particulars of Allen's paper include: Any of the thin-thread microtubules can move particles in opposite directions at the same time, the particles passing each other with no interference. A particle has to be "associated" with the microtubule in order to be transported. In what is considered as fast axonal transport, measurements in chick neurons showed rates of travel of particles in both directions of some 200 mm per day. Larger particles move more slowly than do smaller ones.

Microtubule diameter was observed to be about 250 Å. The tubule itself has the capability of moving, but this is limited to sidewise and lengthwise movement, the lengthwise motion being according to the polarity of the microtubule in the plus-to-minus direction.

The fact that electron microscopy has shown arms and cross-bridges on vesicles and on microtubules leads to the concept of sidearms as the means of propulsion, whether for particles over the microtubules or microtubules per se. These sidearms contain a motor protein enzyme that under the power of the ATP in the mitochondria generates the force of locomotion by way of its undergoing a cyclic change of shape.

Additional information on microtubule action includes the notation by Gilbert et al. [5] that in fast axonal transport, membranous organelles travel at rates from 1 to 5 $\mu m\ s^{-1}$ and for as long a distance as 1 m. Their experimentation, using a reconstituted model of squid axoplasmic vesicles on a microtubule containing axonemes of *Arbacia punctulata* spermatazoa, indicated that this transport process was not mediated by dynein. A molecule similar to (mammalian brain) microtubule-associated protein 2 (MAP 2), a polypeptide of 292 kD molecular weight (H1), was identified by Gilbert and Sloboda [6] as apparently having the properties of cross-bridging organelles to microtubules and acting as mechanochemical transducer through ATP association. They extracted it from squid axoplasm and optic lobe.

In later work Gilbert and Sloboda [7] identify another high molecular-weight polypeptide (H2), also isolated from the squid axoplasm and optic lobe. Since H1 and H2 are MgATPase active, they could be dissociated from the microtubules using MgATP. They promote the motility of salt-extracted vesicles and latex beads along isolated microtubules. The actual molecular weight of the complex of H1 and H2 is about 435 kD. Examination of this complex reveals it to be composed of two globular heads with stems in contact with each other extending to a common base. This complex is a vesicle-associated ATPase, its properties indicating it to be the squid isoform of dynein. According to the data, it brings about membranous organelle motility in connection with fast axoplasmic transport.

Howard et al. [46] conclude from their experiments that a single molecule of kinesin is capable of moving a microtubule for several micrometers. The microphotos also indicate that the kinesin molecule interacts over a distance of 5 μm.

In the paper by Paschal et al. [47], reference is made to two main types of MAP. The fibrous ones, of which MAP 2 and MAP tau are members, organize and regulate the gathering together of microtubules. The other class of MAPs, which includes kinesin, dynein, and dynamin, are mechanochemical ATPases involved with the bidirectional movement of organelles along the microtubules. They have to do with the movement of chromosomes as well. Experimental evidence leads Paschal et al. to consider the MAP 2 class members as moderators of microtubule movement in general. Their work with site-directed antibodies pointed to a specific region of the α- and β-tubulin molecules where they interact with dynein and MAP 2.

2. Cell Motility

Cell motility in skin is described in Chapter 1. It is referred to in Chapter 3 with regard to electrogenics and an understanding of the various bathing solution effects on the outside versus inside skin surfaces. Mitotic cell halves and the cells they become necessarily translocate. A dynamic spindle of microtubules brings this about. A description is given by McIntosh and McDonald [78]. Sawin and Mitchison [80] discuss research on protein motors, such as kinesin, which generate the force bringing about the dynamics of cytoplasmic microtubules to act in spindle formation and chromosome movement with regard to cell division.

Warn and Dowrick [21] identify given proteins produced by the cell as *motility factors*. These proteins regulate and stimulate cell translocation. The binding of the cell to the substrate matrix is mediated by these factors, as was observed in embryogenesis experiments, tissue migration, and the remodeling of wound healing. The formation of pseudopodia is induced. The pseudopodia contain prominent F-actin microfilament bundles and an increased concentration of cell surface receptor molecules for reacting with laminin and fibronectin of the substrate matrix along which the motility and binding take place. The motility depends on the binding, greater interaction slowing it down.

As Sammak and Borisy [73] believe that microtubules act in setting forth the leading lamellum in cell locomotion and maintain the polarity of the cell, they looked in the living fibroblast cell using fluorescence-labeling microscopy. It was observed that the microtubules grew by elongation of their ends and were replaced one by one, turnover being most rapid at the periphery of their network. Lengthening and shortening, which took place at the plus end of the microtubule, and lateral movement all occurred in diverse fashion for the microtubules. Thus there was assembly and disassembly. What seemed to be randomness in the individual microtubule action, nevertheless, as a whole, represented their efficient organization. The entire network does not have to change. The rearrangement effectively reorients the advancing lamellum regarding the motility of the cell.

Haft and Edidin [22], Sheetz et al. [23], and Kucik et al. [24] considered glycoproteins on living mouse peritoneal macrophage cell surfaces and on the dorsal surfaces of lamellipodia of fish epidermal keratocytes, respectively. These proteins were labeled with gold and followed by video imaging. Their transport was observed to be active directed and not the result of membrane lipid flow. They moved relative to the membrane. Thus the authors deduced that the cell cytoskeleton was involved with cell surface organization in a complicated way, protein-actin cytoskeleton forces controlling the mobile glycoprotein and directing its movement. Associated with the transport of the glycoprotein is the control of cell locomotion.

The gold labeling enabled observation of the glycoprotein at the molecular level, and the intermittent reversible interaction of the particles with the cytoskeleton could be seen. When associated with the cytoskeleton, the movement of the particles was directed. But if this association was disrupted, the movement became random Brownian.

As to the mechanism, the authors deemed that the directed action was probably due to myosin I powering movement either through direct linkage to membrane proteins or indirectly by binding to actin filaments that attach intermittently to the glycoproteins. Thus essential proteins are actively transported on the cell surface as needed in cell locomotion, cell adhesion, and assembly of the extracellular matrix.

Membrane transport of the glycoprotein was observed to be both forward and retrograde. The forward progression of glycoprotein along with membrane (although not at the same rate) resulted in concentration of membrane and cytoplasmic proteins at the front of the cell in connection with the cell's motility.

Fukui et al. [44] studied the mechanicochemical aspects of the motility of the eukaryotic cell *Dictyostelium discoideum* amoeba, which moves along a substrate by extending lamellipodia and pseudopodia at the anterior of the cell and retracting its posterior. In using immunofluorescence microscopy technique, they observed that nonfilamentous myosin I protein is present in front at the lamellipodial projections, while filamentous myosin II exists to the rear of the cell. Actin and both myosins have been purified from the amoeba. The myosin I is a single-headed globular protein that does not form filaments, while myosin II self-assembles into bipolar filaments. Both the myosins being mechanochemical enzymes that produce force in complexation with F-actin via the hydrolysis of adenosine triphosphate (ATP), the authors ascribe the cell motility to actomyosin I powering the leading edge of the cell and the contraction of the actomyosin II in the rear delivering a forward push.

Adams and Pollard [48] give evidence, which they consider along with Mooseker [49], that points to myosin I as a prime mover in cell motility, whether of cytoplasmic moieties or of the cell itself. Myosin I was first extracted from the protozoan *Acanthamoeba* and later identified in other cells. The single-headed myosins I do not have the carboxyterminal α-helical tail domain that the double-headed myosins II have. Instead of the tail, they have a non-α-helical domain that contains a second actin-binding site. While the binding site on the head is ATP active, the second one is not. By virtue of the ATP-insensitive site the myosin I can cross-link and generate power.

Five possible ways in which myosin I interacts with membrane are portrayed in Fig. 1. In scheme 1 myosin I bound to the membrane of an organelle moves the organelle along actin filaments in the cytoplasm. With respect to scheme 2, the myosin I is tethered to membrane and brings about contraction on actin filaments, or concentrates the actin into an area of activity. Scheme 3 depicts how

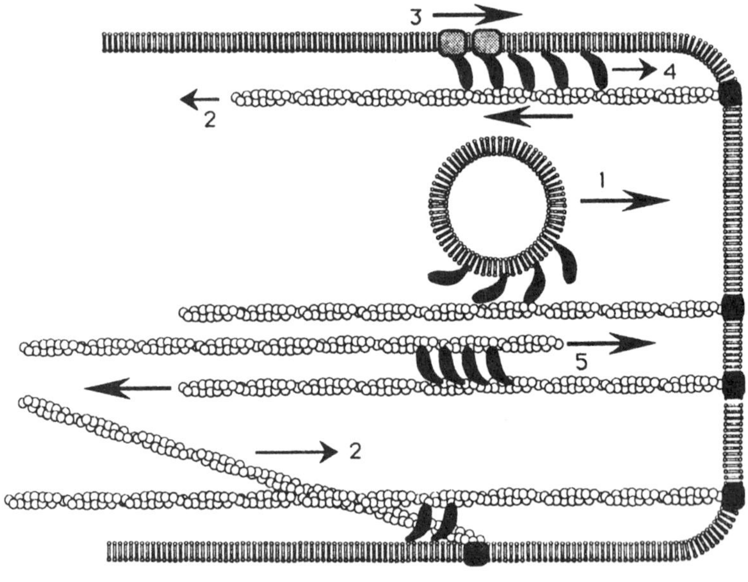

FIG. 1 Myosin I—membrane interaction schematics. (From ref. 48 reprinted with permission of Macmillan Magazines Ltd.)

myosin I brings about the movement of plasma membrane relative to actin in the cytoplasm. This could explain the rotary motor that is understood to drive flagella, microvilli, or filopodia. In scheme 4, the myosin I and membrane molecules to which it is attached would be free to diffuse in the plane of the membrane. The simultaneous association with tethered cortical actin filaments could bring about (partial) membrane flow and thus cell translocation. Scheme 5 indicates how a network of actin filaments might be contracted in cytoplasm.

Scholey [50] has presented a comprehensive review of molecular motors in cell motility, and his summary of actin and microtubule-based activity is given in Table 1. Further in connection with cell

TABLE 1 Classification of Molecular Motors from Eukaryotic Cells

	Description	Function
	Actin-based	
Myosin-1	Monomeric ~110–140k heavy chain	Moves membranes on actin filaments Drives active sliding between actin filaments
Myosin-2	Filaments ~170–240k heavy chain	Muscle contraction, cortical tension, capping surface molecules, cell polarity, cytokinesis
	Microtubule-based	
Axonemal dynein	~400–500k heavy chain	Drives active sliding between flagellar microtubules; moves glass surfaces toward "minus" end of microtubule
Cytoplasmic dynein	~400–500k heavy chain	Moves glass surfaces, anionic beads and membranes towards "minus" end of microtubule
Kinesin	110–140k heavy chain	Moves glass surfaces, anionic beads and membranes towards "plus" end of microtubule
Dynamin	100k heavy chain + cofactor	Drives active sliding between microtubules

Source: Ref. 50, reprinted with permission of Macmillan Magazines Ltd.

motility, the assembly of myosin filaments based on dynamic molecular-exchange mechanisms between polymeric and monomeric pools of the molecules, polymerization and folding, is discussed by Small [81].

3. Flagellar Motility

Bacteria propel themselves through liquid media by means of their flagella. Ideas on the mechanism of centrifugal rotary motors, which are understood to drive the flagella, have been given in Sec. III.B. 2. The general discussion as to the development of motion by ATP-driven enzyme molecular motors interacting with actin and microtubule filaments applies. For some extra dimension on flagellar and cilia propulsion, brief highlights of the research of Namba et al. [51] and Kamimura and Kamiya [52] are presented.

Namba et al. carried out an x-ray fiber diffraction analysis of the structure of the central channel at the core of *Salmonella typhimurium* bacteria flagella. This work is interesting from the viewpoint of transport in that not only does it touch on the rotary motor process by means of which flagella propel the bacteria, but also the transport of the flagellin protein, of which the bulk of the flagellum is composed. This protein moves up through the core of the flagellum to the distal end for filament growth. Kamimura and Kamiya examined the high-frequency subnanometer scale of amplitude vibration in flagella axonemes of sea urchin sperm for its possibilities as an elementary aspect of the movement of the individual dynein arms propelling cilia and flagella, as they are ATPase activated in the presence of microtubules.

IV. TRANSPORT-ASSOCIATED ELECTRICAL PROPERTIES: ELECTRICAL DIFFUSE DOUBLE LAYER, ELECTROKINETICS, AND ELECTROGENICS

Electric currents and potentials, and the corresponding magnetic and electrostatic effects being intrinsic to biofunctions, the features of bioelectrogenics are a very important consideration. Although thermodynamic gradients are thought of as driving forces for electrical passage, it has been proposed, originally in ref. 53 and subsequently in refs. 54 and 55, that electrogenics is associated with active living intelligent electrically charged biologically energized propagation. Conceptually, studies with frog skin [56] led to this conclusion; a relatively complete description is provided in Chapter 3. The studies of Leach [85] seem to be in agreement. He found electric currents in plants which he associated with the physiological movement of organic materials through them. Electrical potentials of up to 120 V were measured between the ground and the surfaces of bean, cucumber, and cherry leaves. The electrical activity was correlated diurnally with photosynthesis and transpiration.

Propagated electricity-transporting particles include electrons, protons, simple ions, and macroions such as proteins, organelles, and vesicles. Surfaces, including pathway walls, are electrically charged, in vivo isoelectric conditions not being very likely. They may be in a state of ionization; at least preferential ionic sorption takes place. The resulting electrical diffuse double layers would account for electrokinetic potentials (i.e., fall and streaming) developing due to propelled components speeding through the medium and liquid moving past charged surfaces. These effects are greater the narrower the pathways. Reference is made in an early paper [32] to postulations that electrokinetic phenomena occur in connection with body functions such as those in the brain.

Electrokinetic effects should apply as well to physiological transport when the conditions are homogeneous. Double-layer properties were apparent for cellulose and cellulose acetate in which the electrolyte solution was essentially homogeneously absorbed [9,10]. It was concluded that the electric layers radiate out from the molecular chains of which the synthetic cellulose and cellulose acetate membranes are composed.

V. INTERACTIONS AND TRANSPORT

In this section we emphasize how transport is affected by interaction. A distinction is made between animate and inanimate conditions. The forces recognized in connection with inanimate systems, ranging in strength from physical to chemical bonding, are basic to the living system as well. But there are unique interaction effects under the circumstances of being alive. Molecules, particles, filamentous segments, and interfaces are involved, which require intelligence data regarding whether to react, when to react, how to react, and to what degree.

A. Inanimate Interactions

Fundamentally, the rate of translocation of a particle is slowed as attraction interaction between it and its surroundings is increased, restricting its freedom. The Brownian movement displacement of a particle ceases when it becomes attached to a substrate [30]. On the other hand, particle translocation proceeds more readily when the interaction is that of repulsion, the greater force of collision recoil increasing the acceleration. For instance, in quantitative studies of the Brownian movement of ZnS particles in an aqueous medium, negatively charged particles in a given suspension were observed to show some 60% greater displacement than when the suspension was changed to isoelectric point conditions [30]. Although positively charged particles were followed only qualitatively, it was evident that their activity was considerably greater than those at the isoelectric point.

Various observations indicate that there is considerably more to attraction interaction than the "usual recognized" physical chemical forces of induced dipole, dipole–dipole, and hydrogen bonding [54]. For water taken up in cellulose and in other materials, molecules released upon drying leave progressively less easily the less the proportion of water remaining, the last traces being essentially chemically held. The spectrum of degree of interaction of the water and substrate as a function of water concentration is continuous and smooth and not readily inferred from "a set" of different distinct modes of attraction interaction.

Smooth, continuously increasing attraction interaction with decreasing concentration is deduced from steady-state diffusion measurements of the complex anionic compound disodium 16,17-dimethoxyviolanthrone leuco solute in cellulose [25]. This is indicated by the curve on the right in Fig. 2, the conditions being high aqueous ionic strength maintained with a constant level of alkaline $Na_2S_2O_4$ concentration. Fick's first law interpretation of the slope becoming increasingly negative with lower leuco concentration shows strong diffusion concentration dependence, the slower diffusion indicating stronger interaction between the leuco and the cellulose membrane. Actually, in desorption studies, the last traces of the solute could not be leached out of the cellulose by clear alkaline $Na_2S_2O_4$. Chemical bonding at the lowest solute concentrations is thereby indicated.

Residual amounts of simple electrolyte were also found to be nonleachable from cellulose [9] and cellulose acetate [10]. This is a further illustration of attraction interaction increasing with decreased solute concentration. It was observed for these aqueous systems that the nonleachable electrolyte could be replaced by different electrolyte through ion exchange. As application of the "new" electrolyte solution increased *total* electrolyte concentration, membrane interaction, including that for the "old" electrolyte, was thereby lessened. With subsequent leaching, residual amounts of essential "only new" electrolyte were left behind, the "old" concentration being very low in proportion to the "new."

The diffusion and sorption observations of the simple electrolyte in these membranes [9,10] indicated the existence of a diffuse electric double-layer network made up of the ionic content throughout the membrane. Evidently, while the double-layer ions vary in their interaction such that their mobility is of differing degree, including

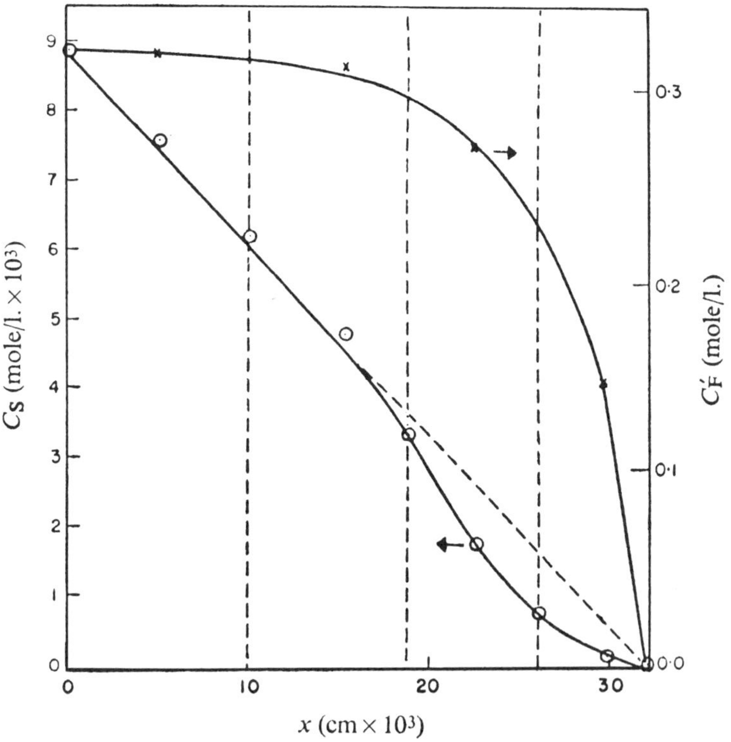

FIG. 2 Steady-state diffusion concentration gradients for disodium 16,17-dimethoxyviolanthrone leuco solute in four-layer cellulose membrane [25]. C_F^-, marked (X) is the solute concentration in the membrane. C_S, marked (O), is the equivalent aqueous solute concentration taken from equilibrium solute absorption data in cellulose. x is the distance into the membrane from the high-solute-concentration side, the vertical dashed lines showing the swollen thickness of each membrane layer. The diagonal (solid and dashed) represents concentration-independent diffusion. (From ref. 25 reprinted with permission of the Royal Society of Chemistry.)

no translocation, the overall level of attraction interaction is increased with their lower concentration.

Further, it would appear that double-layer interaction is the reason for the extreme concentration dependence of diffusion found for these systems of simple electrolyte in cellulose and in cellulose acetate.

Actually, localized barriers are indicated in the steady-state membrane concentration gradients by the precipitous concentration drops. The solid delineation in Fig. 3 shows this for KCl in cellulose. In Fig. 4, barrier periodicity is seen for NaCl in cellulose acetate. As fewer ions are present at the lower concentrations to counter the net membrane charge, their mobility is decreased, and a drastic change in charge over a short distance through the membrane interfering with electrolyte diffusion signifies a barrier.

Electric charge periodicity with changing concentration such as that implied by the multibarriers of Fig. 4 has been observed in previous experiments by the author [30]. This is shown in Fig. 5, which portrays the cataphoretic velocity of ZnS particles in media of changing pH. Two other examples of cataphoretic velocity periodicity are given in ref. 57, $AlCl_3$ concentration in the suspensions having been varied.

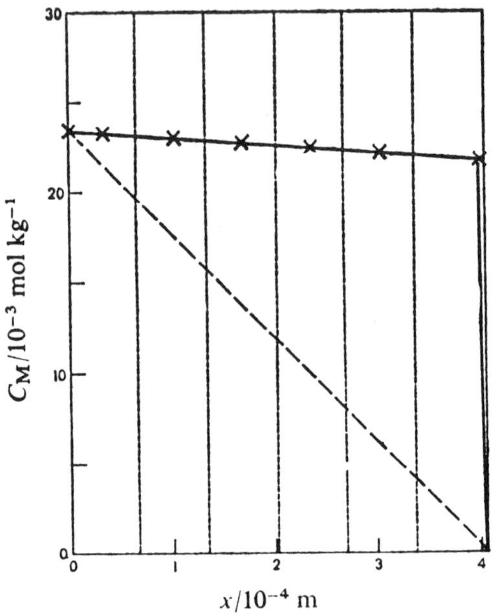

FIG. 3 Steady-state diffusion concentration gradient for KCl in six-layer cellulose membrane. C_M is the KCl concentration in the membrane. x is the distance into the membrane from the high-solute-concentration side, the vertical lines denoting membrane layer width. The diagonal represents concentration-independent diffusion. (From ref. 9 reprinted with permission of the Royal Society of Chemistry.)

Diffuse double-layer electrostatic interactions are long range in nature, the effects reaching well beyond atomic size distances. For example, Bedzyk et al. [70] using long-period x-ray standing waves, found the Debye length of the diffuse layer extending from polymerized phospholipid membrane into dilute aqueous $ZnCl_2$ to be between 3 and 58 Å. It is likely that double-layer mechanics account for a good proportion of the numerous examples of long-range interaction observed. Presumably, the structure of gels, especially those

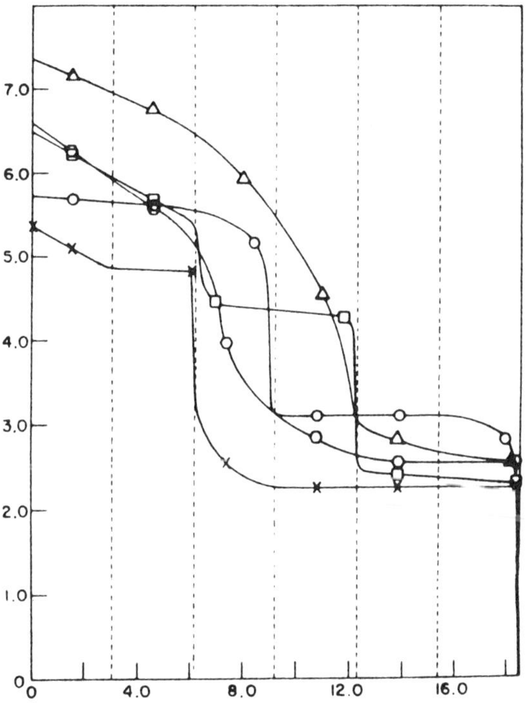

FIG. 4 Steady-state diffusion concentration gradients for NaCl in six-layer cellulose acetate membranes. Ordinate is membrane concentration in moles $kg^{-1} \times 10^3$. Abscissa is the distance into the membrane from the high-solute-concentration side in $m \times 10^5$, the vertical lines denoting membrane layer width. High-side aqueous solution concentrations were 0.005 M (X), 0.010 M (O), 0.030 M (□), 0.050 M (⬡), and 0.10 M (Δ), respectively. (From ref. 10 reprinted with permission of Academic Press.)

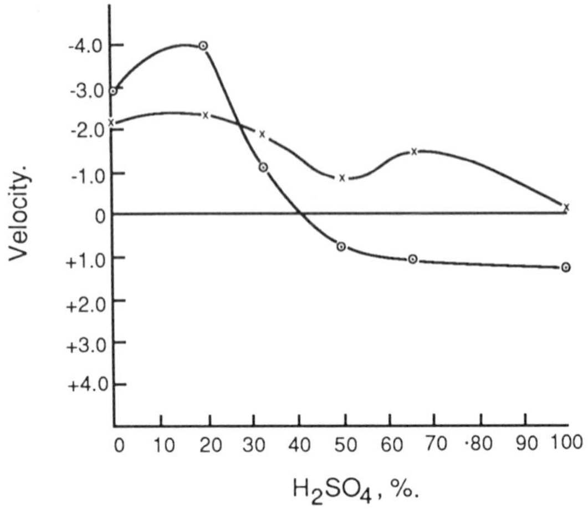

FIG. 5 ZnS particle cataphoretic velocity and liquid velocity at glass surface in 10^{-6} m/s/V/cm for different mixtures of 0.001 M Na_2CO_3 and 0.002 M H_2SO_4. x C.V. (true) of particles; ʘ velocity of liquid at glass surface (sign corresponds to charge of glass). (From ref. 30 reprinted with permission.)

formed with very small proportions of gelling agent, is based on long-range forces. Bubbles on the surface of water gather together from macroscopic distances. A chapter in Adamson's book [58] is devoted to long-range forces.

For improved understanding of the interactions between vesicles and microtubules under living conditions, a summarization of given observations made by the author [30] on ZnS particles in aqueous suspension exposed to glass surface, is made. Given features of the explanations offered regarding interactions between the particles themselves and with the glass could well be applicable to the microtubule systems and the particles associated with them. ZnS - ZnS and ZnS-glass interactions were apparently long range in nature, cataphoretic and electroosmotic measurements pointing to an involvement of electrical diffuse double layers at the surface as the basis of the interactions. Under both weakly alkaline and weakly acidic conditions the effects that took place required a balance of attractive and repulsive forces for explanation.

When the medium was alkaline, the particles and glass surface had an appreciably negative charge. There was no tendency toward flocculation or for adherence to the glass. But long-range attraction was evident in given smaller particles acting as if they were tethered by a string. Albeit in active Browning motion, they never left the locale of a particle, which although too large to be in suspension, was "Brownian rocking" on the glass. Another manifestation of attractive forces for this system was the comparatively hard cake formed upon sedimentation over a period of several days. The fact that the measured deep-cell Brownian movement intensity of this alkaline suspension was less than classical is also evidence of overall attraction.

In acidic medium there was a tendency to flocculate. These flocs were easily broken up with shaking, as was the cake of flocs formed by settling. Particle charge was still negative, but appreciably less than that in the alkaline medium. The charge on the glass was now positive. Despite the like charges on the particles, the attractive forces bringing about flocculation necessarily had to be long range, the observation of floccules with the microscope indicating elementary particles of a floc to be at a distance from each other. Particles in a floc showed no Brownian movement, as was the case for particles now found to be in a state of adherence to the glass. With due consideration of the attraction electrostatics between the positively charged glass and the negatively charged particles, the interaction was, nevertheless, long range. This was indicated when water was allowed to evaporate out from under a microscope coverglass that sandwiched the ZnS in the acid solution. Particles remained in the liquid phase, moving with the receding air—liquid interface line, but under the attraction for the glass which attraction was sufficiently great to quell visible Brownian movement.

Although a factor in the adherence of these particles to the glass is due to opposite charge electrostatics, this is not the case for tethered particles in dilute alkali or those flocculated together in an acid medium, where the particles have the same sign of charge. It being concluded by the ion exchange and diffusion—absorption experiments referred to above that ions and gegenions are held with varying degrees of tenacity in the makeup of a double-layer network, there is a suggestion of interaction between the ions of the diffuse double layers belonging to the two adjacent surfaces. "Common ions" now share the electrostatic emanations from both surfaces, thereby serving in their attractions to bind the surfaces together. This interaction can be strong, depending on the closeness of the surfaces. For example, gravity in conjunction with these attraction forces brought about a compact cake with settling in the dilute alkali solution, there being no flocculation to prevent particles from packing closely.

Gilbert's observation of vesicles exhibiting Brownian motion while tethered to an end of a microtubule [8] is in parallel with the description above of Brownian active ZnS particles in dilute alkaline solution never leaving the locale of a larger one on glass surface. Her living system was reconstituted, isolated vesicles being added to mammalian brain microtubule-associated protein-free (MAP-free) microtubules.

But the visible Brownian movement of this vesicle ceases when it is in close association with the microtubules such that according to the live conditions, it undergoes directed motion. The adherence is not so great as to interfere with the vesicle mobility requirements of the physiological system. This is parallel with the ZnS particles described above adhering to glass in acidic medium and moving inside the aqueous medium with the water—air interface line over the glass surface. They did not show Brownian movement. Long-range interaction between the ZnS and the glass was inferred. The vesicle is at an appreciable distance (i.e., some 14 to 20 nm) from the microtubule to which it adheres, as Fig. 6 indicates [5]. According to the figure, there is filamental contact between the vesicle and the microtubule.

B. In Vivo Interactions

Definitive statements on the process of being alive are offered to help differentiate the interactions described in this section from the inanimate physics discussed above. The author owes a debt to L. Vroman for an interchange in this connection. The biological system as a whole, or down to its smallest living component, displays physiological intelligence. There is directed translocation. Chemical and physical reactions are guided, and electrical potential is generated. There is growth and/or maintenance against harmful influences of the environment. Evolution is a feature of the living processes.

Factors in interaction are component size and shape. Translocation is influenced by the reaction. Also, the component translocates in order to react. Pertinent to interaction is the question of what the smallest dimensions of a living moiety might be. Essential atoms, at least in the case of protein, apparently are nitrogen, carbon, oxygen, and hydrogen. The microtubule shown in Fig. 6 is noted to have a diameter of 240 Å. It is likely that the cross-bridging filaments displayed in this micrograph function in the alive state. Their width appears to be some 60 Å, judging their size from that of the microtubule.

Bacterial virus F2 is one of the smallest. It is reported to have a diameter of 200 Å [31,60]. Its molecular weight is 3.6×10^6. Viroids, known to be the smallest agents of infectious disease, are described by Diener [60]. He identifies potato spindle-tumor viroid as a single-strand ribonucleic acid (RNA) molecule of 500 Å length and 0.13×10^6 molecular weight. As the viroid diameter is of single-strand RNA molecular width, it follows understandably that biological filaments, per se, should be alive.

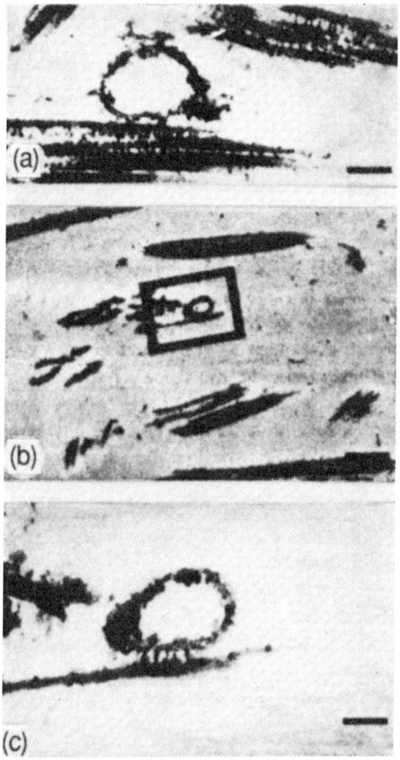

FIG. 6 Transmission electron micrograph illustrating vesicle association with axonemal microtubules; (a) the vesicle (230 × 346 nm) illustrated is in contact with an axoneme; (b) a typical field of axonemes with one vesicle in association with an axonemal microtubule; (c) higher magnification of the area enclosed in the rectangle in (b). The vesicle (180 × 309 nm) is in association with an axonemal microtubule (24 nm) and seems to be attached by cross-bridging filaments; the spacing between the filaments is 14 to 20 nm. (From ref. 5 reprinted with permission of Macmillan Magazines Ltd.)

Biologically essential inanimate particles (i.e., H_2O, simple ions, CO_2, O_2, CH_4, simple amino acids, carbohydrates, fats, etc.) translocate under alive directed conditions. At given sites there may be spontaneous diffusion and capillary action, but the gradient is of physiological energy origin. As mentioned in Sec. III, B, part 1, inanimate material such as glass beads and latex, are physiologically transported along microtubules utilizing kinesin. Sanders and Lord [77] demonstrate that the secretory matrix of the stylar-transmitting tract of angiosperm flowering plants guides pollen tubes to the ovary. Since latex beads and pollen tubes were seen to travel at comparable rates with no evident effect due to the force of gravity, it was concluded that the energy and mechanics of propulsion were purely a function of the matrix. Water and dilute solutions such as sap in live vegetation could well be whisked along by flagellar or fibrillar pumping (dynein) on a filamentous network. In the same way that flagella drive a bacterium by action—reaction in aqueous medium, the same whiplike process by appendages associated with "anchored" tissue should propel the fluid.

Given ions can be held with different degrees of binding in protein, and can be released as well, depending on phýsiologcal circumstances. For instance, according to Alkon [11], neural firing brings on protein kinase C enzyme translocation, and this results in K^+ flow. Alkon also refers to the phosphorylation of protein by calcium-activated kinase (CAM kinase II), which is promoted at post-synaptic sites. Regehr et al. [61] and Cornell-Bell et al. [62] find Ca accumulations taking place. In Regehr's work the buildup results from synaptic activation, while with respect to Cornell-Bell, glutamate-sensitive ion channel signaling occurs.

Large molecules such as the proteins evidently reconfigurate in their role as vehicles for simple ions and small molecules, including O_2, CO_2, and H_2O. For example, Anderson et al. [64] demonstrate how lactoferrin, like all members of the transferrin family of proteins, regulates the free iron inside animals. The molecular conformation includes jawlike "iron traps" that hold onto and release the Fe, commensurate with biological needs. Proteins involved in the transport of small inanimate particles do not necessarily have to move with them. Undulating configuration could produce on and off "pocket" formation with appropriate geometry and polarity [84] holding inanimates and releasing them, thrusting them into directed transport.

The directed transport of water and solutions, whether by flagellar action or by protein molecular undulation, is not necessarily on the basis of molecule by molecule of the fluid. Operation in (microscopic) bulk is conceivable considering the evidence for attraction at long range. In this connection the observations of Howard et al. [46], which include the fact that a kinesin molecule can maintain its grip on a microtubule 5 μm away, are elucidating.

Under these conditions of ample space, the evidence of the way in which vesicles undergoing two-way travel (i.e., orthograde and retrograde) pass each other on a single microtubule is understandable. Motion studies utilizing computer-enhancement techniques for the light microscope [43] show no visible evidence of alleviation of speed or deviation from the path of either particle where an encounter would have been expected [63]. Adherence to the microtubule is necessarily strong enough for accomplishment of the directed motion of the vesicles while they have adequate space to move freely past each other.

These observations of orthograde and retrograde moving vesicles passing each other on the same microtubule, with what appears to be no interference, suggest analogous dynamics for living vegetation. The essential material in transport up from the roots can move along the same filamentous network as that used by substances transported downward from the leaves. Components are transport-directed to their particular function.

The filamentous state is basic to the living system. For example, DNA and RNA are fibrillar; their action involves their strands. Protein chains translocate through membrane. They unfold to a chain upon adsorption to a surface (see Chapter 6). Muscles and nerves are fibrous. Motility is brought about by filamentous activity. Organelles move along a network of microtubules. The cytoskeleton is fibrous, and a spindle of microtubules is the agent for mitosis. Growth takes place along filaments, mobile proteins, enzymes, carbohydrates, and so on, interacting chemically and traveling the fibrillar length. A rough analogy is the formation of icicles. Vegetation is filamentous. In Chapter 4, saliva is described as having a filament structure.

Thus, in this light, an important mode of biological interaction and binding must involve filaments. In being alive there is a decision as to whether to attack or not to attack. The attack can, as directed, entangle an object with the filaments. This could involve sticking to the surfaces of the object. Furthermore, these filaments would be the avenue for injecting toxins and lysing agents. A specific example is the T-2 bacteriophage virus, which has tentacles that bring it close to its host bacterium for virus injection of nucleic acid [31].

Lindberg et al. [59] show in Fig. 7 the pili surrounding a bacterium of a strain of *Escherichia coli*. These filaments consist of Pap proteins. Given Pap proteins (referred to as "adhesin") located at the tips of the pili are receptor binding. They are used to stick to the digalactoside-containing glycolipids present in the membranes of the host epithelial cells. This actual atom-to-atom adhesion evidently should be the result of basic physicochemical attractions, the filaments directing and presenting appropriate degrees of polarity [84] for binding the glycolipids. Figure 8

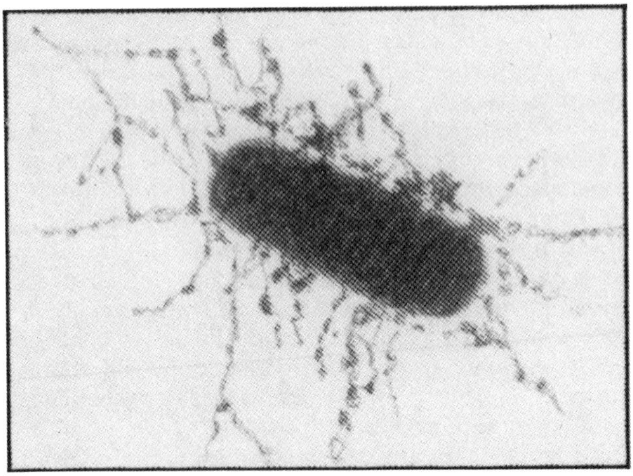

FIG. 7 Filaments surrounding a bacterium of a strain of *Escherichia coli*. (From ref. 59 reprinted with permission of Macmillan Magazines Ltd.)

illustrates how an immune system cytotoxic T cell (top) binds to a target tumor cell by filaments attaching themselves. Perforin protein is injected to break through the membrane, resulting in the demise of the cell.

VI. CONCLUDING REMARKS

Key processes of living system communication and transport have been given attention in terms of the relevance of interfacial and surface aspects. Deliberations in Chapter 3 are in conjunction. Supplying the driving force is physiological energy, which, it seems apparent, is in an energy class of its own. Intelligence is an essential feature of this energy. There is inherent guidance at appropriate energy levels for carrying out all the organism's necessary operations of communication, organized growth for achieving purposeful goals, maintaining the system, neutralizing inimical onslaughts, and determining when given components are not needed further. While an inanimate particle has internal vibrations, and bounces, wriggles, and rotates, its displacement being vectorless, live conditions produce oriented translocation. Rapid molecular

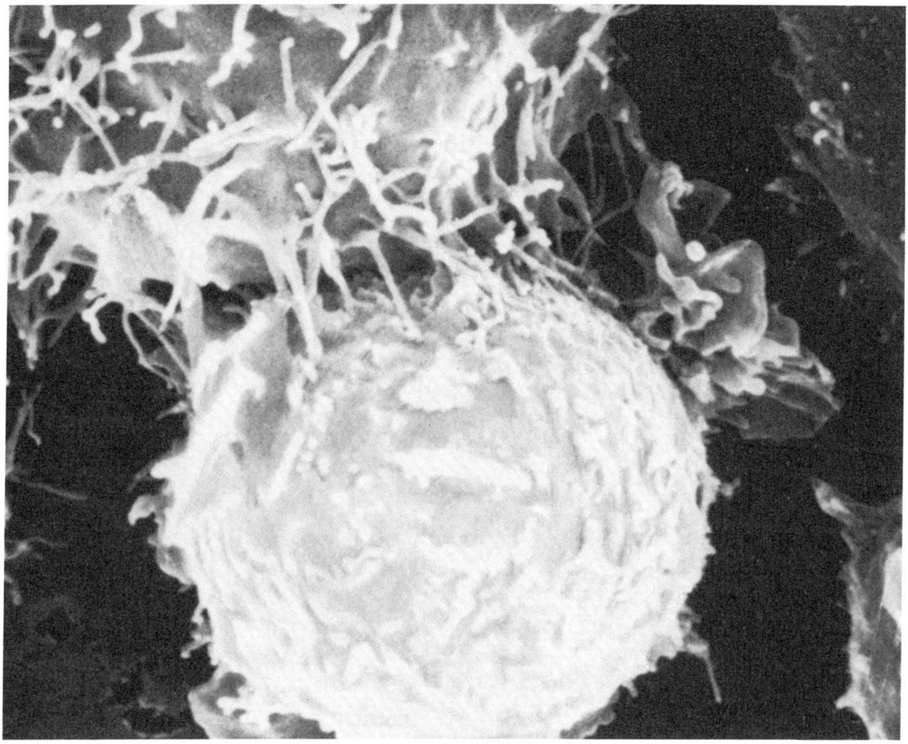

FIG. 8 "Killer" cytotoxic T cell (top) makes contact with target cell. Scanning electron micrograph at a 5700-diameter enlargement. (From ref. 65 reprinted with permission from Scientific American.)

configuration changes enable locomotion. Molecules are readily built and rebuilt to suit translocation and other purposes as required, "old" chains being severed, the segments with additions and deletions being rearranged.

REFERENCES

1. T. J. McMurry, K. N. Raymond, and P. H. Smith, *Science* *224*:938 (1989).
2. W. R. Adey, Paper 153, abstract, Division of Colloid and Surface Chemistry, *ACS Spring Meeting*, Denver, Colo., Apr. 9, 1987.

3. R. D. Allen, J. Metuzals, I. Tasaki, S. T. Brady, and S. P. Gilbert, *Science 218*:1127 (1982).
4. S. P. Gilbert and R. D. Sloboda, *J. Cell Biol. 99*:445 (1984).
5. S. P. Gilbert, R. D. Allen, and R. D. Sloboda, *Nature 315*: 245 (1985).
6. S. P. Gilbert and R. D. Sloboda, *J. Cell Biol. 103*:947 (1986).
7. S. P. Gilbert and R. D. Sloboda, *J. Cell Biol. 109*:2379 (1989).
8. S. P. Gilbert, personal communication of 6/19/89.
9. M. Bender, J. K. Moon, J. Stine, A. Fried, R. Klein, and R. Bonjouklian, *J. Chem. Soc. Faraday Trans. 1 71*:491 (1975).
10. M. Bender, B. Khazai, and T. E. Dougherty, *J. Colloid Interface Sci. 63*(2):346 (1978).
11. D. L. Alkon, *Sci. Am. 261*:42 (1989).
12. J. M. Diamond, *Nature 339*:258 (1989).
13. D. Kim and D. E. Clapham, *Science 244*:1174 (1989).
14. R. W. Ordway, J. V. Walsh, Jr., and J. J. Singer, *Science 244*:1176 (1989).
15. W. Stühmer, F. Conti, H. Suzuki, X. Wang, M. Noda, N. Yahagi, H. Kubo, and S. Numa, *Nature 339*:597 (1989).
16. R. W. Aldrich, *Nature 339*:578 (1989).
17. U. Schulte, M. Arretz, H. Schneider, M. Tropschug, E. Wachter, W. Neupert, and H. Weiss, *Nature 339*:147 (1989).
18. J. G. Nuchtern, J. S. Bonifacino, W. E. Biddison, and R. D. Klausner, *Nature 339*:223 (1989).
19. G. Byrne, *AAAS Observer*, July 7, 1989, p. 16; *No. 6 Supplement to Science*, July 7, 1989.
20. O. T. Jones, D. L. Kunze, and K. J. Angelides, *Science 244*: 1189 (1989).
21. R. M. Warn and P. Dowrick, *Nature 340*:186 (1989).
22. D. Haft and M. Edidin, *Nature 340*:262 (1989).
23. M. P. Sheetz, S. Turney, H. Qian, and E. L. Elson, *Nature 340*:284 (1989).
24. D. F. Kucik, E. L. Elson, and M. P. Sheetz, *Nature 340*:315 (1989).
25. M. Bender and W. H. Foster, Jr., *Trans. Faraday Soc. 61*:159 (1965).
26. R. Burns, *Nature 340*:511 (1989).
27. H. Frauenfelder, *Nature 338*:623 (1989).
28. G. U. Nienhaus, J. Heinzl, E. Huenges, and F. Parak, *Nature 338*:665 (1989).
29. S. Kamimura and R. Kamiya, *Nature 340*:476 (1989).
30. M. Bender, *J. Phys. Chem. 56*:272 (1952).
31. E. Katchalski, *Man and His World*, the Noranda Lectures, Expo 67, University of Toronto Press, Toronto, 1968, p. 231.
32. M. Bender, *Am. Sci. 46*:368(373) (1958).

33. R. O. Laine and A. F. Esser, *Nature 341*:63 (1989).
34. B. K. Hölldobler and E. O. Wilson, *The Sciences*, Nov–Dec. 1989, p. 18.
35. T. D. Seeley, *Am. Sci. 77*:546 (1989).
36. T. Lindsten, C. H. June, J. A. Ledbetter, G. Stella, and C. B. Thompson, *Science 244*:339 (1989).
37. W. Doster, S. Cusack, and W. Petry, *Nature 337*:754 (1989).
38. D. Moazed and H. F. Noller, *Nature 342*:142 (1989).
39. G. Rathbun, F. W. Alt, and G. D. Yancopoulos, *Nature 342*: 863 (1989).
40. N. Matsunami, Y. Hamaguchi, Y. Yamamoto, K. Kuze, K. Kangawa, H. Matsuo, M. Kawaichi, and T. Honjo, *Nature 342*: 934 (1989).
41. L. Vroman, *Blood*, American Museum Science Books ed., The Natural History Press, Garden City, N.Y., 1968.
42. K. Verner and G. Schatz, *Science 241*:1307 (1988).
43. R. D. Allen, *Sci. Am. 256*(2):42 (1987).
44. Y. Fukui, T. J. Lynch, H. Brzeska, and E. D. Korn, *Nature 341*:328 (1989).
45. S. Ochs, *Science 176*:252 (1972).
46. J. Howard, A. J. Hudspeth, and R. D. Vale, *Nature 342*:154 (1989).
47. B. M. Paschal, R. A. Obar, and R. B. Vallee, *Nature 342*:569 (1989).
48. R. J. Adams and T. D. Pollard, *Nature 340*:565 (1989).
49. M. S. Mooseker, *Nature 340*:505 (1989).
50. J. M. Scholey, *Nature 343*:118 (1990).
51. K. Namba, I. Yamashita, and F. Vonderviszt, *Nature 342*:648 (1989).
52. S. Kamimura and R. Kamiya, *Nature 340*:476 (1989).
53. M. Bender, The application of the concepts of electrochemistry and surface and colloid chemistry to biological systems, Paper 16, Division of Colloid and Surface Chemistry, *ACS Spring Meeting*, Denver, Colo., Apr. 8, 1987, copyright obtained.
54. M. Bender, Water and ions in biological systems, *Proc. 4th International Conference*, Bucharest, Romania, May 24–28, 1987, Birkhäuser, Verlag, Basel, p. 415.
55. M. Bender, Water and ions in biological systems, *Interfacial Phenomena in Biotechnology and Materials Processing* (Y. A. Attia, B. M. Moudgil, and S. Chandes, eds.), Elsevier, Amsterdam, 1988, p. 105.
56. M. Bender, K. Hillman, L. Tirri, M. Graben, D. Fenster, and M. Rosensaft, The influence of the composition of the bathing solutions on frog skin membrane potential, in *Bioelectrochemistry: Ions, Surfaces, Membranes*, Advances in

Chemistry Series 188 (M. Blank, ed.), American Chemical Society, Washington, D.C., 1980, pp. 411—443.
57. M. Bender, *J. Phys. Chem.* *57*:466 (1953).
58. A. W. Adamson, *Physical Chemistry of Surfaces*, 4th ed., Wiley-Interscience, New York, 1982.
59. F. Lindberg, B. Lund, L. Johansson, and S. Normark, *Nature* *328*:33, 84 (1987).
60. T. O. Diener, Chap. 12, *Life at the Edge* (J. L. Gould and C. G. Gould, eds.), W. H. Freeman, New York, 1989.
61. W. A. Regehr, J. A. Connor, and D. W. Tank, *Nature* *341*:533 (1989).
62. A. H. Cornell-Bell, S. M. Finkbeiner, M. S. Cooper, and S. J. Smith, *Science* *247*:470 (1990).
63. S. Gilbert, Seminar, Dept. of Biology, Princeton University, 5/9/89 (one of the occasions of observations by M. Bender).
64. B. F. Anderson, H. M. Baker, G. E. Norris, S. V. Rumball, and E. N. Baker, *Nature* *344*:784 (710) (1990).
65. J. D. Young and Z. A. Cohn, *Sci. Am.*, Jan. 1988, p. 38.
66. K. A. Smith, *Sci. Am.* *262*, Mar. 1990, p. 50.
67. A. C. Scott, *The Sciences* *30*(2):28 (1990).
68. R. L. Rawis, *Chem. Eng. News*, Dec. 21, 1987, p. 26.
69. J. King, *Chem. Eng. News*, Apr. 10, 1989, p. 32.
70. M. J. Bedzyk, G. M. Bommarito, M. Caffrey, and T. L. Penner, *Science* *248*:52 (1990).
71. E. J. Neer and D. E. Clapham, *Nature* *333*:129 (1988).
72. N. R. Franks, *Am. Sci.* *77*:139 (1989).
73. P. J. Sammak and G. G. Borisy, *Nature* *332*:724 (1988).
74. Z. Shakked, G. Guerstein-Guzikevich, M. Eisenstein, F. Frolow, and D. Rabinovich, *Nature* *342*:456 (1989).
75. L. W. Buss and R. K. Grosberg, *Nature* *343*:63 (1990).
76. S. Stern, T. Powers, L. Changchien, and H. F. Noller, *Science* *244*:783 (1989).
77. L. C. Sanders and E. M. Lord, *Science* *243*:1606 (1989).
78. J. R. McIntosh and K. L. McDonald, *Sci. Am.* Oct. 1989, p. 48.
79. S. Wolf, C. M. Deom, R. N. Beachy, and W. J. Lucas, *Science* *246*:377 (1989).
80. K. Sawin and T. Mitchison, *Nature* *345*:22 (1990).
81. J. V. Small, *Nature* *331*:568 (1988).
82. C. W. F. McClare, *J. Theor. Biol.* *30*:1 (1971).
83. R. H. Pain, *Nature* *326*:247 (1987).
84. A. R. Clarke, D. B. Wigley, W. N. Chia, D. Barstow, T. Atkinson, and J. J. Holbrook, *Nature* *324*:699 (1986).
85. S. Weisburd, *Sci. News* *132*:53 (1987).

Index

Absorption (*see also* Adsorption, Desorption, Sorption)
  surfactants by skin, 46–47
Acids, dental adhesion, 78–79
Acids, dental surfaces
  hydrophilicity-hydrophobicity, 78–79
  self repair, 79
Activation
  clotting by surfaces, 140, 144
  complement system, blood, 145–146
  surfaces, clotting, 140, 144
Activation, chain
  blood, immunological, 145–146
Activity, biological, proteins adsorption, 289–290
Activity, enzyme at interface, 295–296
Adherence (*see also* Adsorption)
  DLVO theory, 109–110
  vesicle-microtubule, directed vs. Brownian motion, 391–392
Adherence distance, filament directed transport, 395
Adhesin, 396
Adhesion (*see also* Adsorption)
  acids, dental, 78–79
  blood platelets, 140–143
  dental plaque, 77, 80, 83
  dentistry, restorative, 79
  DLVO theory, 109–110
  fibrillar, 111
  filamentous, 111
  vs. fluoridation of tooth surfaces, 80
  re, fouling, oral, 85–91
  integration of tissue, 117–118
  mechanisms, 109–117
  microorganisms on dental surfaces, 80–82
  microorganisms on oral surfaces, 79–80
  oral, 77–91, 96
  pellicle, 80–84
  plaque, dental, 77, 80, 83
  restorative dentistry, 79
  tissue integration, 117–118
Adhesion, tooth surfaces, organic films, 78
Adhesive proteins, in tissue integration, 117–118

Adsorption (*see also* Adherence, Adhesion, Desorption, Sorption)
albumin on red cells, 195
alkyl sulfates on callus, 39–40
biological activity of proteins, 289–290
on biomaterials, 118
and blood complement system, 146
bovine serum albumin, 291
conformation change, proteins, 146
dextran on erythrocyte, 195
DLVO theory, 109–110
endothelium, 138–139
enzymes, 295–296
fibrinogen on erythrocytes, 195
immunoglobulins, 145–146
kininogen blood protein, 140–145
lysozyme, 291–292
mechanism, proteins, 290–292
ovalbumin, 291–292
protein, unfolding, 293
proteins, 161, 289–295
proteins on artificial surface, 138–143
proteins and their biological activity, 289–290
proteins and mass action law, 141
receptor sites, 139–143
solid surfaces, proteins on, 294–295
surfactants on proteins, 37–40
surfactants on skin, 35–40
Adsorption on erythrocytes
aggregation mechanisms, 239–244
mathematical description, 198–204, 220–221
pellet analyses, 220–244
proteins, 237–241
[Adsorption on erythrocytes]
results, 204–216, 221–235
solution analyses, 198–220
solution vs. pellet analyses, 235–237
survey, 216–220, 235–244
Adsorption, ionic
at bio-layers, 343–345
electrocapillarity in bio-layers, 344–345
Adsorption isotherms, proteins, 294–295
Adsorption measurement
pellet analysis, 196
solution analysis, 196
trapped volume, 203–204
Adsorption, non-specific, by erythrocytes, 193–250
Adsorption parameters, protein, 294–295
Adsorption, proteins
barrier, energy, 291–292
ellipsometer measurements, 141–142
endothelium, 138–139
interfacial pressure, 290–292
and isoelectric conditions, 293–294
kinetics, 290–292
radiolabeling, 143
surface viscosity, 292
Aggregation, cells, mechanism, nonadsorption, 194
Aggregation-Disaggregation
hemoglobin, 358–360
proteins, 357–359
proteins, and surface energy 339
tobacco mosaic virus, 357–358
Aggregation, erythrocytes
macromolecular cross-bridging, 193–194
mechanism, 239–244
proteins in, 237–241
Aggregation, hemoglobin viscosity, 360

Aggregation, reversible, erythrocytes, 193–196
Albumin adsorption, in aggregation of erythrocytes, 239–241
Allografts, 105–136
Alloy, titanium, biomaterial, 117–120
Aluminum biomaterial, 119
Amino acid-amino acid potentials, 324
Amphipathicity (*see* Hydrophilicity, Hydrophobicity)
Anaerobic corrosion, bacterial, 114
Analyses comparison, solution vs. pellet, 235–237
Anaphylatoxins, 146
Anions
  concentration gradient in frog skin effect, 67
  influence on frog skin potential, 67
Artificial heart, 121–122
  clinical aspects, 121–122
  infections, 122
  materials of construction, 121
Artificial intelligence, 304
Artificial surfaces (*see* Biomaterials)
ATP hydrolysis re, protein motors, 377–384
Attachment (*see* Adherence, Adhesion, Adsorption)
Attachment of bacteria, mechanism to surface, 110–112
Autografts, 105–136
Axonal transport on microtubules, 378–380
Axonal transport rates, 379

Bacteria (*see also* Microorganisms)
  adhesion, conditioning film, 115–117
[Bacteria]
  anaerobic corrosion, 114
  attachment to surfaces, 110–112
  biomaterial sepsis, 107–108
  and biomaterials, 105–136
  conditioning film adhesion, 115–117
  corrosion, 114
Bacterial growth, iron in, 114–115
Bacterial microcolonies, 113–115
  and biomaterials, 114–115
Bacterial slime, 111–113
  composition, 111–113
  polysaccharides, 111–113
  properties, 111–113
  purposes, 111–113
  re, surface attachment, 111–113
Bacteriorhodopsin
  and electric field effects, 174
  water in, 161
Barrier, energy, adsorption, proteins, 291–292
Barrier periodicity, 388–389
Barriers to diffusion, 388–389
Basale layer, 4–7
Bathing solution composition
  influence on frog skin electric potential, 51–76
Bathing solutions
  compositions for frog skin, 51–65
  frog skin ionic gradients, 53–67, 69–70
Bilayers
  d-spacing, 16–26
  hydrocarbon chain length, 19–20
  lipid structural model, 11–16
Bilayers, hydrated lipid, neutron diffraction, 163
Bilayers, lipid, 161–163
  partial volume evaluation, 162–163
  Raman spectra, 162–163

Binding vs. motility, 380
Bioactivity of proteins, water in, 159
Bioadhesiveness, pellicle (*see* Adhesion)
Bioalteration and electric fields, 172–173
Biochemical equilibria and water, 156–157
Biochemical feedback, 284
Biocommunication speeds, 369–370
Biocompound properties and electromagnetic fields, 175
Bioconformation change and electromagnetic fields, 172
Bioelectrogenics, 384–385
Bioelectrogenics and electrokinetics, 384–385
Bioelectrogenics and fall potential, 384–385
Bioelectrogenics and streaming potential, 384–385
Bioenergetics, surface, 338–339
Bioenergized directed propagation electrogenics, 384–385
Biofilaments, 395–396
Biofunctions, specialized, saliva, 98–99
Biohydration and thermal energy, 320–321
Biolayers
  ionic adsorption, 343–345
  ionic transport, 343–347
Biological (*see* Physiological)
Biological activity, protein adsorption, 289–290
Biological communication ramifications, 369–372
Biological electrogenics, 69–74
Biological feedback, 144
Biological films on microorganisms, dental fouling, 85–91
Biological interactions via filaments, 396
Biological molecules, water interaction, 156–168
Biological particulates in saliva, 92–101
Biological processes, electromagnetic effects on, 179–180
Biological systems
  magnetosensitivity, 173
  surface to volume ratios, 338–339
Biological transport, intelligence directed, 372–374
Biomaterials, 105–136
  adsorption, 118
  aluminum, 119
  artificial heart, 121–122
  and bacteria, 105–136
  and bacteria microcolonies, 114–115
  bacterial sepsis, 107–108
  biological substrata on, 121
  characteristics, 117–121
  clinical aspects, 121–125
  heart, artificial, 121–122
  and infections, 122–124
  infections, ocular, 122
  interaction with bacteria, 105–136
  interaction with tissue, 105–136
  metal surface characteristics, 117–119
  methyl methacrylate, 120–121
  ocular implants, 122
  polymer surface characteristics, 118
  polymers, 119–121
  protein adsorption, 138–143
  sepsis, 107–108
  substrata, biological on, 121
  substrata on implants, 121
  surface energy in adsorption, 117
  and tissue, 105–136
Biomembranes (*see* Membranes)
Biomolecular hydration, 319–322

Biomolecules, interaction simulation, 301–333
Bioreactivity and electric field coupling, 181–182
Biosystems
closed circuits by magnetic field, 187
and electrical parameters, 173
and parameters electromagnetic, 173
and parameters magnetic, 173, 175–176
window effects, 177–178
Biothermal effects and electromagnetic radiation, 187
Bio-transport
ramifications, 369–372
and water, 343
Blood
activation immunology chain, 145–146
coagulation, 140–145
and endothelium, 137–139
extrinsic clotting, 143
feedback in clotting, 144
immune system, 145–146
intrinsic clotting, 143–145
irreversibility of change, 138
proteins, 137–150
serial interactions, 139–146
thrombin in clotting, 144
Blood clotting
and kininogen protein, 140–145
rates, 140–145
Blood complement, 139–143 145–146
Blood complement system and adsorption, 146
Blood plasma
clotting, 139–143
fibrin-fibrinogen clotting, 139–143
Blood plasma proteins at interfaces, 137–150
Blood proteins
conformation change, 140–141 144, 146
Blood vessel wall (*see* Endothelium)
Bonding, resin cements, dental, 79
Bovine serum albumin, adsorption, 291
Brownian movement, 374
Brownian movement vs. interaction, 385–386, 390–392

Calcium ion
and electrical potential, 52–56
and frog skin potential, 52–56
Callus, adsorption of alkyl sulfates, 39–40
Catastrophes, microtubule, 282
Cations, polyvalent stimulation of electrical potential, 58, 60–67, 74
Cavity, oral, 77–103
Cell additions and electroporation, 259–260
Cell aggregation mechanism
nonadsorption, 194
Cell attachment
protein adsorption, in, 296
Cell electroporation, chronological, 265–267
Cell growth medium
interfacial attachment, 296–297
Cell membrane and signal coupling, 186
Cell mitosis, microtubules in, 380
Cell motility, 380–384
Cell motility mechanisms, 380–384
Cell motility molecular motors, 383
Cell observations, electroporation, 265–269
Cell recovery from electroporation, 257–258

Cell separation, protein adsorption in, 296
Cell stress via electroporation, 257
Cell water, freezing, 168
Cells
attachment to surfaces, 296–297
electric pulse effects, 176
electroporation, 253–277
interaction, water, 155–158
surface attachment, 296–297
water in, 155–158
Cellular function control by electroporation, 260
Cellular oscillations, 178, 181–182
Cellular oscillations, and ionic diffuse double layers, 182
Cements, dental, 79
Chain, activation, immunological in blood, 145–146
Channel formation, and water, 165–168
Channel gating, protein aggregation model, 350, 353–354
Channel, molecular, gramicidin A, 165–166
Channels, ionic, selectivity, 350, 352–353, 355
Channels, kinetics, 361–363
Characteristics of biomaterials, 117–121
Chemistry, surface, oral, 77–103
Chemotaxins, 146
Cholesterol in lipid interlayers, 22–26
Circuitry, electrical, in frog skin, 72
Circuits, electrical, physiological, 51–76
Clinical
artificial heart, 121–122
ocular implants, 122
Clinical experience of biomaterials, 121–125
Closed circuits, electrical, magnetic field on biosystem, 187
Clotting (*see also* Coagulation)
activation surfaces, 140, 144
blood plasma, 139–143
feedback, 144
intrinsic interactions, 143–145
and phospholipid micelles, 144
rates, 140–145
receptor sites, proteins, 139–143
Clotting, blood
extrinsic interactions, 143–145
fibrin-fibrinogen, 139–143
kininogen, 140–145
serial interactions, 139–145
thrombin in, 144
Coagulation (*see also* Clotting)
blood, 140–145
protein at interface, 292
Coagulum (*see* Structure)
Collagen, in conditioning films, 116
Colloidal properties, saliva, 91–101
Colloidal structure, saliva, 91–101
Communication, biological ramifications, 369–372
Communication, physiological, 367–400
Complement activation, immune system, blood, 145–146
Complement system, blood, 139–143, 145–146
Composite restoration, 77
Computation, gramicidin, K^+ channel, 306, 322–323
Computer pair-potentials matrix solutions, 322, 324–325
Computer simulations of biomolecule interaction, 301–333

Computer simulations, protein structure, 325–327
Concentration dependence
diffusion, 386–389
electric double layer interaction, 386–388
of interaction, 386–387
Concentration gradients (*see also* Thermodynamic gradients)
ionic, across frog skin, 53–67, 69–70
Conditioning films (*see also* Slime)
on biomaterials, 115–117
composition, 115–117
on implants, 115–117
Conductance, membrane, re, electroporation, 268–269
Configuration (*see* Conformation, Structure, Unfolding)
Conformation (*see also* Folding, Structure, Unfolding)
biomolecular, and surface energy, 339
Conformation change
biological, in electromagnetic ficld, 172
blood proteins, 140–141, 144, 146
hemoglobin, 359–360
proteins, 140–141, 144, 146, 375
proteins and adsorption, 146
proteins in electric field, 185
Conformation, enzymes at interface, 295
Conformation, proteins
on adsorption, 290, 292
and surface viscosity, 292
Contact angle, oral surfaces wettability, 78
Corneal keratitis, 108
Corneocytes, 11–15
Corneum layer, 4
Corpuscular (*see* Macromolecule, Molecule, Particle, Polymer)
Coupling, biochemical reactivity and electric fields, 181–182
Critical surface tension and dental plaque adhesion, 80–82
Cross-bridging, macromolecular, aggregation of erythrocytes, 193–194
Crystalline/liquid crystalline transition, 26
water influence, 26
Cytometry, flow re, electroporation, 268
Cytoskeleton
dynamic forces from wave motion, 283
as electric dipoles, 280–281
electromagnetic field and filamentation, 281–284
filamentation and electromagnetic field, 281–284
mechanical forces from wave motion, 283
nonequilibrium thermodynamics application, 279–285
polymer treadmill by radiation, 282–283
pondermotive forces from wave motion, 283
quantum field theory application, 279–285
structure, 279–285
superconducting pathways in, 283
water in, 282

d'Alembert wave propagation, 283
Definition, living process, 392
d-spacing of bilayers, 16–26
Denaturation, proteins
electroporation, 262
at surface, 292
Dental (*see* Oral)

Dental adhesion, acids, 78–79
Dental, bonding, mechanism, 79
Dental enamel
  surface chemistry, 78
  wettability, 78
Dental film (*see* Pellicle)
Dental fouling, 85–91
Dental plaque, 77, 80
Dental surfaces, acids, self-repair, 79
Dental wettability, acids, 78–79
Dentine
  surface free energy, 78
  wettability, contact angle, 78
Dentine, surface chemistry, 78
Denture, 77
Depression, frog skin electrical potential, 52–70
Dermis, 4
Desorption (*see also* Absorption, Sorption)
  re, fouling, dental, 85
Dextran adsorption
  erythrocyte aggregation, 241–244
  erythrocytes, 193–250
Dielectric breakdown (*see* Reversible electric breakdown)
Diffraction, neutron, lipid bilayers, 163
Diffuse double layer (*see also* Electric double layer)
  and frog skin electric potential, 67–69, 73–74
  potential in bilayer lipids, 161
Diffusion
  on biomaterial surfaces, 118
  concentration dependence, 386–389
  re, protein adsorption, 290–291
  surface of biomaterials, 118
  and ultrathin layer permeability, 340–343
  water in skin, 26–30
Diffusion barriers, 388–389
Dimensions, least, living moieties, 392
Dipole membrane potential, bilayers, lipids, 161
Dipoles, electric, the cytoskeleton, 280–281
Directed bioenergized propagation electrogenics, 384–385
Directed particle motility, 378–380
Directed transport, filament adherence distance, 395
Distance of adherence in filament directed transport, 395
Distribution ratio, erythrocyte adsorption, 197–204, 220–221
DLVO theory, 109–110
DNA
  amino acids interaction computations, 324
  computed hydration, 319–320
  in electric field, 180
  electroporation take-up, 267
  water bridging, 164
Domains, biomaterial surface, 105–136
Dynamic forces and cytoskeletal waves, 283
Dynamic polymerization by radiation pressure, 282–283
Dynamic mechanism structure, cytoskeleton, 279–285

Electric activity, membranes, water in, 155–156
Electric charge
  periodicity, 388, 390
  and surface activity, proteins, 293–294
Electric charged intelligent propagation electrogenics, 384–385

Electric circuits
in frog skin, 72
physiological, 51–76
Electric dipoles
the cytoskeleton, 280–281
Electric double layer (*see also* Diffuse double layer)
bioelectrogenics, 384–385
and cellular oscillations, 182
and frog skin potential, 67–69, 73–74
interaction, concentration dependence, 386–388
length, 389
in long range interactions, 389–392
and membrane transport, 340
membranes, 386–388
Electric effects
on protein, 181
thresholds in biochanges, 188
Electric field
and bioalteration, 172–173
coupling with bioreactivity, 181–182
effect on bacteriorhodopsin, 174
effect on DNA, 180
effect on ions in biosystems, 184
mechanisms of effect on membranes, 184–188
effect on polyriboadenylate, K^+ salt, 174–175
and protein conformation change, 185
Electric formulas in electroporation, 260–263
Electric parameters
electroporation, 260–263
and biosystems, 173
Electric potential
cation, polyvalent stimulation, 58, 60–67, 74
depression, 52–71
[Electric potential]
diffuse double layer in frog skin, 67–69, 73–74
effect of anions, 67
frog skin, 51–76
frog skin, recovery, 52–69
influence of bathing solution composition, 51–76
influence of electrolyte, 51–76
and ionic strength, 52–56
lipid bilayers, 161
and pH, 56–57
polyvalent cation stimulation, 58, 60–67, 74
profiles in frog skin, 70–71
reversal, 53, 55, 58–59, 67–70, 72–73
stimulation by polyvalent cations, 58, 60–67, 74
and surface energy, 344–345
toxic elements effect, frog skin, 63, 65–67
vs. Ca, 52–56
vs. K, 52–56
vs. Na, 52–56
Electric properties, biological
and electrocapillarity, 69
Hofmeister relative ionic influence, 68
Electric pulse effects, on cells and vesicles, 176
Electric resistance vs. potential in frog skin, 69
Electrocapillarity
re, biological electrical properties, 69
ionic adsorption, 344–345
Electroconformation, 260, 264–265
Electrodiffusion in membranes, 356–357
Electrogenics, 69–74, 384–385
biological, 70–74
living, 70–74
mechanisms, 73–74
thermodynamic, 69–71

Electrokinetics
bioelectrogenics, 384–385
and frog skin potential, 67–69, 73–74
Electrolyte, influence on frog skin potential, 51–76
Electrolyte in saliva, 95
Electromagnetic effect
on biological processes, 179–180
on kinase, 181
on protein, 180–181
Electromagnetic field
and biocompound properties, 175
and bio-conformation change, 172
cytoskeletal filamentation, 281–284
and membranes, 171–191
Electromagnetic parameters and biosystems, 173
Electromagnetic radiation
and bioheat effects, 187
Electromagnetic signalling, 369
Electromanipulation, 176
Electron microscope, saliva studies, 94–101
Electronic components simulation, 327–329
Electrophoresis of microorganisms, re, fouling, oral, 85
Electroporation, 253–277
cell additions, 259–260
cell and membrane recovery from, 257–258
and cell stress, 257
cells observations, 265–269
and cellular function control, 260
DNA take-up, 267
electrical description, 254
experimental, 260–269
flip-flop molecular, membranes, 267
flow cytometry, 268
[Electroporation]
historical, 265–267
gene transfer, 254, 258–259, 267
importance, 258–260
kinetics, 267
mechanisms, 254, 264–265
membrane conductance and permeability, 268–269
methodology, 254, 260–263
observations, 263–269
and permeability, 257, 259–260
protein denaturation, 262
pulsed electric field, 260
studies, chronological, 265–267
temperature, 256–257, 262–263, 265, 269
of tissue, 259–260
uptake-release, 268
Electroporative transport, 268
Ellipsometer, protein adsorption, 141–142
Emulsification, proteins in, 297
Enamel, dental
free energy, 78
surface chemistry, 78
wettability, 78
Endothelium, 137–139
adsorption, 138–139
blood, 137–139
surface, 138–139
Energetics
biological surfaces, 338–339
of electrogenics, 69–74
and transport, 374–384
Energy barrier, protein adsorption, 291–292
Energy, biological, 375–384
Energy for intelligence, 375–376
Energy, surface, pellicle, 80–84
Entropy of water-hydrophobic biostructure, 157–161
Enzymes
activity at interface, 295–296
adsorption, 295–296
interfacial conformation, 295

[Enzymes]
modification, 296
in saliva, 93, 99
Epidermis, 4
Equilibrium
biochemicals and water, 156–157
hemoglobin-oxygen, 358–360
Erythema by surfactants, 40–44
Erythrocytes
adsorption of dextrans, 193–250
adsorption of proteins, 193–250
aggregation, 193–196
albumin desorption, 195
dextran adsorption, 195
fibrinogen adsorption, 195
nonspecific adsorption, 195
rouleaux formation, 193–250
Erythrocytes, adsorption
mathematical description, 198–204, 220–221
pellet vs. solution analyses, 235–237
proteins, 237–241
results, 204–216, 221–235
survey, 216–220, 235–244
trapped volume, 203–204
Erythrocytes aggregation
cross-bridging, macromolecular, 193–194
proteins in, 237–241
Erythrocytes, dextran adsorption aggregation mechanism, 241–244
Erythrocytes, protein adsorption aggregation mechanism, 239–241
Evolution, membrane in water, 154
Exocytosis, 7
Exopolysaccharide (*see* Polysaccharide)
Extracellular signalling, 373
Extrinsic interactions in clotting, 143

Fall potential, 74
Fall potential from bioelectrogenics, 384–385
Fast axonal transport, 378–380
Fatty acids, lipid structure model, 11–16
Feedback, biological, 144, 284
Fibrillar (*see* Filamentous)
Fibrin, blood, 139–143
Fibrinogen, blood, 139–143
Fibrinogen adsorption in red cell aggregation, 239–241
Fibronectin in conditioning films, 115
Filament directed transport adherence distance, 395
Filamentation, cytoskeletal and electromagnetic field, 281–284
Filamentous adhesion, 111
Filamentous biological interactions, 396
Filamentous biostate, 395–396
Filamentous network, saliva, 92–101
Filamentous protein growth, 384
Filaments, cytoskeletal, superconductance, 283
Film, organic dental (*see* Pellicle)
Films
biological on microorganisms, dental fouling, 85–91
conditioning on biomaterials, 115–117
salivary in oral fouling, 85–91
Flagellar motility, 384
Flagellar rotary motor, 383–384
Flexibility (*see* Conformation)
Flip-flop, molecules in membranes, electroporation, 267

Flow cytometry, electroporation, 268
Fluoridation of dental surfaces and adhesiveness, 80
Folding (*see* Conformation, Structure, Unfolding)
Folding pathways, proteins, 327
Formation
 oral film, 80–84
 pore, 253–277
 rate of, pellicle, 80–84
Formulas, electric in electroporation, 260–263
Fouling, oral, 85–91
 adhesion, 85–91
 desorption, 85
 electrophoresis of bacteria, 85
 microorganisms, 85–91
 pellicle, 85–91
 plaque, 85–91
 polysaccharides, 89–91
 salivary films, 85–91
Free energy
 ionic hydration, 318–320
 tooth surfaces, 78
Freezing, water in membranes and cells, 168
Frog skin, 51–76
 potential determining ions, 68
 recovery, electric potential, 52–69
 surface, 51–76
 under thermodynamic gradients, 53–67, 69–70
Frog skin electric resistance vs. potential, 69
Frog skin potential, 51–76
 anions effect, 67
 vs. calcium, 52–56
 cation, polyvalent stimulation, 58, 60–67, 74
 depression, 52–70
 and electrokinetics, 67–69, 73–74
[Frog skin potential]
 influence of bathing solution composition, 51–76
 influence of electrolyte, 51–76
 and ionic gradients, 53–67, 69–70
 and ionic strength, 52–56
 isoelectric conditions, 67
 and pH, 56–57
 polyvalent cation stimulation, 58, 60–67, 74
 vs. potassium, 52–56
 reversal, 53, 55, 58–59, 67–70, 72–73
 vs. sodium, 52–56
 stimulation by polyvalent cations, 58, 60–67, 74
 toxic elements effect, 63, 65–67
Functions, biological, specialized, saliva, 98–99
Functions of saliva, 91–99

Gastric cytoprotective functions in saliva, 99
Gating, channel, aggregation of protein as model, 350, 353–354
Gating mechanism, 361–363
Genetic material and electroporation, 254, 258–259, 267
Global simulation
 description, 302–310
 water, 310–317
Glycocalyx, 186
Glycoproteins
 adhesive in tissue integration, 117–118
 in conditioning films, 115–117
 in saliva, 96
Golgi, 7
Gradients (*see also* Thermodynamic gradients)
Gradients, thermodynamic
 anions in frog skin, 67

[Gradients, thermodynamic]
and frog skin electric potential, 53–67, 69–70
ionic across frog skin, 53–67, 69–70
Gramicidin channel
computation, 306, 322–323
ionic transport, 165–166
Granulosum layer, 4–7
Growth factors in saliva, 93, 99

Heart (*see also* Artificial heart)
Heart, artificial, 121–122
Heat wave propagation, cytoskeletal, 283
Hofmeister ionic series re, biological electric properties, 68
Hemoglobin
aggregation-disaggregation, 358–360
conformation change, 359–360
viscosity change and aggregation, 360
Hemoglobin-oxygen equilibrium constant, 358–360
Hydration (*see also* Water)
biomolecular, computed, 319–322
free energy computed, of ions, 318–320
Hydrocarbon chain length in bilayers, 19–20
Hydrogen bonding sphingolipids, 156
Hydrophilicity-hydrophobicity dental surfaces, 78–79
Hydrophobic effect, 157
Hydrophobic grouping-water biostructure, 157–161
Hydrophobicity scale, 157
Hydrophobicity, re, surface activity in proteins, 292–293
Hygiene, oral, 79–80
Hypodermis, 4

Icelike water structure in proteins, 158–160
Immune system, blood, 145–146
Immunity
biomaterials disruption, 122–124
disruption, 122–124
tissue transplant disruption, 122–124
Immunity mechanisms
and biomaterials, 122–124
and tissue transplants, 122–124
Immunoglobulins adsorption, 145–146
Immunoproteins, adsorption and conformation change, 146
Implants (*see also* Biomaterials)
biomaterial, 105–136
Inanimate interactions, 385–392
Inanimates, living transport of, 393–394
Inanimates, transport by reconfiguration of proteins, 395
Infections
artificial heart, 122
and biomaterials, 122–124
ocular implants, 122
and tissue transplants, 122–124
Integration
biomaterials and tissue, 107, 117–118
tissue and tissue transplant, 117–118
Intelligence
artificial, 304
directed transport, 372–374
energy for, 375–376
physiological, 367–400
threshold, Maxwell's demon, 374
Intelligent electric charged bioenergized particle electrogenics, 384–385

Interactions
adhesive, 77
bacteria and biomaterials, 105–136
biological via filaments, 396
biological molecules and water, 156–168
bio-surface and ions, 70
biosystems and electric fields, 172–173
biosystems and magnetic fields, 173
blood clotting, intrinsic, 143–145
and Brownian displacement, 385–386, 390–392
cells and water, 155–158
clotting, extrinsic, 144
concentration dependence, 386–387
dental surfaces and microorganisms, 80–82
electric double layer, concentration dependence, 386–388
electric double layer, long-range, 389–392
electric fields and biosystems, 172–173
energies, amino acids-DNA, 324
gramicidin A channel, water, 165–166
hydrophobic grouping-water, 157–161
inanimate, 385–392
in vivo, 392–396
living, 392–396
long range, electric double layer, 389–392
magnetic fields and biosystems, 173
membranes and water, 154–156
nucleic bases and water, 163–164
proteins and water, 157–161
[Interactions]
sequences, blood proteins, 139–143
simulation of biomolecules, 301–333
skin and surfactants, 33–49
strength and vibrational energy, 375
vs. transport, 385–396
water and biological molecules, 156–168
water and cells, 155–158
water in gramicidin A channel, 165–166
water-hydrophobic grouping, 157–161
water and ions, 154–156
water and membranes, 154–156
water and nucleic bases, 163–164
and water structure, 153–154
Interface (*see also* Surface)
and blood plasma proteins, 137–150
cell attachment, 296–297
definition, 368
enzyme activity, 295–296
and proteins, 137–150
Interfacial pressure, protein adsorption, 290–292
Interfacial properties, re, the surface compartment model, 347–348
Interfacial protein
coagulation, 292
conformation, 292
Interferon, structure, 325–327
Interior structure, saliva, 94–101
Interlayer spacing, lipids, 16–26
Intrinsic energy, 374–375
Intrinsic interactions in clotting, 143–145
In vivo interactions, 392–396
In vivo transport, inanimates, 393–394

Ion channel selectivity, 350, 352–353, 355
Ionic adsorption
at bio-layers, 343–345
Lippmann electrocapillarity, 344–345
Ionic double layers and oscillations, cellular, 182
Ionic exchange in membranes, 386
Ionic gradients and frog skin electric potential, 53–67, 69–70
Ionic hydration free energy, 318–320
Ionic sorption bilayers, lipids, 161–162
Ionic strength and frog skin potential, 52–56
Ionic transport
actin-spectrin layers, 346
and bio-layers, 343–347
in biosystems under electric fields, 184
gramicidin A channel, 165–166
and oscillating electric fields, 353–355
surface compartment model, 347–363
water in membrane pores, 166–168
Ions
bio-surface interactions, 70
protein binding variation, 394–395
release by signalling in frog skin, 72–74
Iron, re, bacterial growth, 114–115
Irreversibility of change, in blood, 138
Isoelectric conditions, frog skin electrical potential, 67
Isoelectric point, proteins and adsorption, 293–294
Isotherms, adsorption, proteins, 294–295

Keratinization, 4–7
Keratinocytes, epidermal, 4
Keratohyalin synthesis, 4–7
Kinase, electromagnetic effects on, 181
Kinetics
adsorption of proteins, 290–292
electroporation, 267
protein channels, 361–363
Kininogen protein in blood, 140–145

Lamellar lipid structure model, 11–16
Lipid interlayers, role of cholesterol, 22–26
Lipid layers, water in, 16–26
Lipid monolayers, permeability, 341–342
Lipid structure model
bilayer, 11–16
fatty acids, 11–16
pH, 11–16
phase combination, 11–16
trypsin test, 11
water in, 11–16
X-ray diffraction, 11–16
Lipids
bilayers, 161–163
bilayers, electric potentials, 161
bilayers, ionic sorption, 161–162
bilayers neutron diffraction, 163
bilayers, partial specific volume evaluation, 162–163
bilayers Raman spectra, 162–163
bilayers spacing vs. component penetration, 18–20
crystalline/liquid crystalline states, 36
interlayer spacing, 16–26

[Lipids]
lamellar, 11–16
penetration into bilayers, 17–19
in skin, functions, 8–10
skin structure location, 16–26
in stratum corneum, 3–32
structure, 3–32
surfactant skin removal, 35–37
and water, 11–16, 161–163
Lipids in skin composition, 7, 9–10
Lipoid droplets in saliva, 93–94, 98
Lipoproteins pseudoemulsions, 297
Lippmann electrocapillarity, 344–345
Liquid crystals
lipid structure model, 11–16
transition to crystalline, 26
Liquid crystalline/crystalline states lipids, 26
Living directed electric charged propagation electrogenics, 384–385
Living electrogenics, 70–74
Living filaments interaction, 396
Living interactions, 392–396
Living moieties, least dimensions, 392
Living process, definitive, 392
Living transport of inanimates, 393–394
Long range interactions, 389–392
Lyotropic ionic series re, biological electric properties, 68
Lysis, osmotic and electroporation, 258
Lysozyme adsorption, 291–292

Macromolecule (*see* Molecule, Particle, Polymer)
Macromolecular cross-bridging erythrocyte aggregation, 193–194
Magnetic field
closed biocircuits, 187
in sickle cell orientation, 179–180
Magnetic parameters and biosystems, 173, 175–176
Magnetosensitivity
biological systems, 173
Magnetotaxis, 284
Manifestations of biological energy, 375–384
Man-made biomaterials, 105–136
Manufactured materials as tissue substitutes, 106
Mass action law and protein adsorption, 141
Materials of construction, artificial heart, 121
Materials, restorative, 77
Mathematical description, adsorption, erythrocytes, 198–204, 220–221
Maxillary denture, 77
Maxwell's demon, intelligence threshold, 374
Mechanical forces and cytoskeletal waves, 283
Mechanical retention restoratives, dental, 78
Mechanisms
adhesion, 109–117
adsorption-aggregation, albumin on erythrocytes, 239–241
adsorption-aggregation, dextran on erythrocytes, 241–244
adsorption-aggregation, fibrinogen on erythrocytes, 231–241
adsorption proteins, 290–292
of bonding, re, dental cements, 79
cell aggregation nonadsorption, 194
cell motility, 380–384

[Mechanisms]
dynamic cytoskeleton structure, 279–285
electric field on membranes, 184–188
electrogenics, 73–74
in electroporation, 264–265
gating, 361–363
immunological, and biomaterials, 122–124
microtubule propulsion, 379–380
transport, 367–400
Membrane, cell and signal coupling, 186
Membrane conductance, re, electroporation, 268–269
Membrane electric double layer, 386–388
Membrane evolution, water in, 154
Membrane ion exchange, 386
Membrane-myosin interactions, 382–383
Membrane permeability re, electroporation, 268–269
Membrane pores
by electroporation, 256
and transport, 166–168
and water, re, ion transport, 166–168
Membrane proteins in electroporation, 254
Membrane rearrangement by electroporation, 263–264
Membrane recovery from electroporation, 257–258
Membrane transport, 337–365
electrical double layer in, 340
Membrane water, freezing, 168
Membranes
biological, water in, 153–170
electrodiffusion in, 356–357
in electromagnetic fields, 171–191
[Membranes]
electroporation, 253–277
mechanisms of electric field effects, 184–188
pore formation, 253–277
reactions to electric fields, 181–184
reactions to electromagnetic fields, 181–184
Singer and Nicolson model, 153
water in structure, 154–156
Metal implants, biomaterials, 117–120
Methodology of electroporation, 260–263
Methyl methacrylate, as biomaterial, 120–121
Micelles, phospholipid in clotting, 144
Microcolonies, bacteria, 113–115
Microelectrophoresis (*see* Electrophoresis)
Microorganisms (*see also* Bacteria)
adhesion on dental surfaces, 79–80
adhesion vs. energy of dental surfaces, 79–82
control in saliva, 97
and dental plaque, 77
electrophoresis, re, oral fouling, 85
films on, in dental fouling, 85–91
Microscope, electron in saliva studies, 94–101
Microtubule catastrophes, 282
Microtubule-vesicle adherence, directed vs. Brownian motion, 391–392
Microtubules, 378–380
and axonal transport, 378–380
in cell mitosis, 380
propulsion mechanisms, 379–380
Mitochondria, 7

Model
cytoskeleton dipoles, 280–281
membranes Singer and Nicolson, 153
lipid structure, 11–16
Modification of enzymes, 296
Moisturizing
role of lipids in skin, 26–30
skin, 26
Molecule (*see* Macromolecule, Particle, Polymer)
Molecular channel, gramicidin A, 165–166
Molecular dynamics and protein structure, 327
Molecular flip-flop, via electroporation, 267
Molecular simulations, electronic devices, 328–329
Molecular weight and protein surface activity, 292
Monolayers, lipid, permeability, 341–342
Motility of cells, 380–384
Motility factors, 380
Motility, flagellar, 384
Motility vs. binding, 380
Motor proteins, 377–384
Motors molecular in cell motility, 383
Mucosa, 77
Multilayers, permeability, 342–343
Myoglobin hydration, 320
Myosin-membrane interactions, 382–383

Natural parallelism in supercomputer simulations, 309
Nerve cell particle transport, 378–380
Neutron diffraction hydration in lipid bilayers, 163
Nonadsorption mechanism, cell aggregation, 194
Nonequilibrium thermodynamics and cytoskeleton structure, 279–285
Nonspecific adsorption on erythrocytes, 193–250
Nucleic bases and water, 163–164

Observations, cells, electroporation, 265–269
Ocular implants, 122
Opsonization in immunological activation, 145–146
Oral (*see* Dental)
Oral film (*see also* Pellicle)
formation, 80–84
Oral fouling, 85–91
Oral hygiene, 79–80
Oral hygiene vs. surface energy, 79–80
Oral surfaces, 77–103
adhesion, 77–91, 96
chemistry, 77–103
wettability, contact angle, 78
Organic film dental (*see* Pellicle)
Organic molecular simulations, 328–329
Organic surface films on teeth, 78
Orientation of sickle cells in magnetic field, 179–180
Oscillating electric fields and ionic transport, 353–355
Oscilations, cellular, 178, 181–182
Oscillatory energy (*see also* Vibrational energy)
and interaction strength, 375
Osmotic lysis and electroporation, 258
Outer structure, saliva, 94–101

Ovalbumin
adsorption, 291–292
as surfactant, 293

P. aeruginosa, corneal keratitis, 108
Pair-potentials matrix, computer solutions, 322, 324–325
Pap proteins, 396
Parameters
electric in electroporation, 260–263
electric, magnetic and biosystems, 173, 175–176
protein adsorption, 294–295
Parallelization in computer simulations, 306–310
Partial specific volume and lipid-bilayer water, 162–163
Particle (*see also* Macromolecule, Molecule, Polymer)
Particle interactions and Brownian displacement, 385–386
Particle motility directed, 378–380
Particulates in saliva, 92–101
Pellet analyses, adsorption on erythrocytes, 196, 220–244
Pellicle
adhesiveness, 80–84
energy, surface, 80–84
measurements, results, 80–84
in oral fouling, 85–91
pH, 84
rate of formation, 80–84
results of measurements, 80–84
saliva, 84
structure, 80–84
surface energy, 80–84
[Pellicle]
thickness, 80–84
wetting, 80–84
Penetration
and lipid bilayer spacing, 18–20
lipids into bilayers, 17–19
Perforin, 306
Periodicity
barriers, 388–389
electric charge, 388, 390
Periodontium, 77
Permeability
vs. diffusion, ultrathin layers, 340–343
by electroporation, 257, 259–260
lipid monolayers, 341–342
membrane, re, electroporation, 268–260
multilayers, 342–343
pH
and frog skin potential, 56–57
lipid structure model, 11–16
pellicle, 84
skin, 11–16
Phospholipid micelles in clotting, 145
Phospholipids, 10
Physiological (*see also* Biological)
communication, 367–400
electric circuits, 51–76
energy manifestations, 375–384
intelligence, 367–400
Plaque, dental, 77, 80, 83
adhesion, 77, 80, 83
and critical surface tension, 80–82
fouling, 85–91
Plasma (*see also* Blood)
blood, 137–150
proteins adsorption on erythrocytes, 193–250
Platelet deposit time window, 140

Platelets, blood, adhesion, 140–143
Poisons (*see* Toxic)
Polymer (*see also* Macromolecule, Molecule, Particle)
Polymer biomaterials, 119–121
Polymer treadmilling, cytoskeletal, 282–283
Polymerization, dynamic and radiation pressure, 282–283
Polyriboadenylate, K^+ salt and electric field, 174–175
Polysaccharides
  bacterial slime, 111–113
  in fouling, dental, 89–91
Polyvalent cations, frog skin potential stimulation, 58, 60–67, 74
Ponderomotive forces and cytoskeletal waves, 283
Poration (*see* Pore formation)
Pores
  formation, 253–257
  formation and water, 165–168
  membrane by electroporation, 256
  membrane and ionic transport, 166–168
Potassium ion and frog skin potential, 52–56
Potential
  in computer simulations, 322, 324–325
  diffuse double-layer, lipid bilayers, 161
  dipole, lipid bilayers, 161
Potential determining ions, re frog skin potential, 68
Potential, electric
  vs. calcium, 52–56
  cation, polyvalent stimulation, 58, 60–67, 74
  depression, 52–70
[Potential, electric]
  diffuse double layer in frog skin, 67–69, 73–74
  frog skin, 51–76
  frog skin, effect of toxic elements, 63, 65–67
  influence of anions, 67
  influence of bathing solution composition, 51–76
  influence of electrolyte on, 51–76
  and ionic strength, 52–56
  and pH, 56–57
  polyvalent cation stimulation, 58, 60–63, 74
  vs. potassium, 52–56
  profiles in frog skin, 70–71
  reversal, 53, 55, 58–59, 67–70, 72–73
  vs. sodium, 52–56
  stimulation by polyvalent cations, 58, 60–67, 74
Potential, fall, 74
Potential, frog skin
  and anions, 67
  vs. calcium, 52–56
  cation, polyvalent stimulation, 58, 60–67, 74
  depression, 52–70
  effect on, toxic elements, 63, 65–67
  and electric resistance, 69
  and electrokinetics, 67–69, 73–74
  influence of bathing solution composition, 51–76
  influence of electrolyte, 51–76
  and ionic strength, 52–56
  and pH, 56–57
  polyvalent cation stimulation, 58, 60–67, 74
  vs. potassium, 52–56
  recovery, 52–69
  reversal, 53, 55, 58–59, 67–70, 72–73

[Potential, frog skin]
vs. sodium, 52–56
stimulation by polyvalent cations, 58, 60–67, 74
Potential reversal, frog skin, 53, 55, 58–59, 67–70, 72–73
Potential, streaming, 74
Processes, biological, electromagnetic effects on, 179–180
Profiles, electric potential in frog skin, 70–71
Proflavine deoxydinucleotide phosphate and water, 322
Propagation electric charged living particles electrogenics, 384–385
Properties of biosubstances and electromagnetic fields, 175
Protein adsorption, 161, 289–295
in cell attachment, 296
in cell separation, 296
diffusion, 290–291
ellipsometer measurements, 141–142
endothelium, 138–139
energy barrier, 291–292
erythrocyte aggregation, 237–241
and mass action law, 141
parameters, 294–295
radiolabeling, 143
viscosity, surface, 292
Protein aggregation as model for channel gating, 350, 353–354
Protein channels, kinetics, 361–363
Protein configuration states, 375
Protein conformation
and surfactants, 42
water in, 160–161
Protein denaturation by surfactants in skin roughness, 37–40
Protein isoelectric point and adsorption, 293–294
Protein motors and ATP hydrolysis, 377–384
Protein reconfiguration in transport of inanimates, 395
Protein structure
via molecular dynamics, 327
from simulations, computer, 325–327
Protein surface activity
and electric charge, 293–294
vs. hydrophobicity degree, 292–293
vs. molecular weight, 292
and unfolding, 293
Protein synthesis cellular, 372–373
Protein thermal energy, 375
Proteins
adhesive in tissue integration, 117–118
aggregation-disaggregation reactions, 357–359
aggregation of erythrocytes, 237–241
binding variation, ions, 394–395
biological activity, adsorption, 289–290
in blood plasma, 137–150
in conditioning films, 115–117
conformation, 160–161
conformation on adsorption, 290, 292
conformation change, 140–141, 144, 146
conformation change in electric field, 185
denaturation and electroporation, 262
denaturation in skin, 37–40
electric effect on, 181

[Proteins]
electromagnetic effects on, 180–181
in emulsification, 297
in erythrocyte aggregation, 237–241
folding pathways, 327
icelike water structure in, 158–160
interaction with water, 157–161
at interfaces, 137–150
ion binding variation, 394–395
kinetics, adsorption, 290–292
kininogen in blood, 140–145
mechanism of adsorption, 290–292
motor, 377–384
rate of adsorption, 138–143
in saliva, 97
serial interactions, 139–143
structure with water, 157–161
surface activity, 289–299
surface energy in aggregation-disaggregation, 339
water in bioactivity, 159
and water, interaction, 157–161
water and structure, 157–161
Proteins, plasma, adsorption on erythrocytes, 193–250
Proteins in skin, 4–7
Protein transport
general, 370
self-directed, 370
Pseudoemulsions, lipoproteins, 297
Pseudopodia, 380
Pulsed electric field in electroporation, 260

Quantum field theory and cytoskeleton structure, 279–285

Radar biocommunication, 369
Radiation pressure and dynamic polymerization, 282–283
Radiolabeling, protein adsorption, 143
Raman spectra, re water-lipid bilayers, 162–163
Rate
of adsorption, proteins, 138–143
of blood clotting, 140–145
of formation, pellicle, 80–84
Reactions
electromagnetic fields on membranes, 181–184
membranes and electric fields, 181–184
Rearrangement, membrane by electroporation, 263–264
Receptor sites
adsorption, protein, 139–143
proteins in clotting, 139–143
Receptors for adhesive proteins, 117–118
Recombination activity gene, 373
Reconfiguration, protein transport of inanimates, 395
Recovery, frog skin potential, 52–69
Red cells (*see* Erythrocytes)
Repair, self, dental surfaces, from acids, 79
Resin cements, dental, 79
Resistance, electric vs. potential in frog skin, 69
Restoration
composite, 77
dental, 77
Restorative dentistry, adhesion, 79
Restorative materials, 77
Restoratives, dental, mechanical retention, 78
Retention, mechanical, dental restoratives, 78

Reversal of potential, frog skin, 53, 55, 58–59, 67–70, 72–73
Reversible aggregation, erythrocytes, 193–196
Reversible electric breakdown, 254, 256, 263, 266
Rheology (*see* Structure)
Rheological properties, saliva, 91–101
Rotary motor flagellar, 383–384
Rouleaux, erythrocytes, 193–250

Saliva, 77
  biofunctions, specialized, 98–99
  biological particulates in, 92–101
  colloid chemistry, 91–101
  electrolyte in, 95
  electron microscope studies of, 94–101
  enzymes in, 93, 99
  filamentous network, 92–101
  functions, 91–99
  gastric cytoprotective functions, 99
  glycoproteins in, 96
  growth factors in, 93, 99
  interior structure, 94–101
  interior transport system, 98
  lipoid droplets in, 93–94, 98
  microorganism control, 97
  outer structure, 94–101
  particulates in, 92–101
  pellicle, 84
  properties, 91–101
  proteins in, 97
  research results, 92–101
  rheology, 91–101
  and secretory products, general, 99
[Saliva]
  specialized biofunctions, 98–99
  surface chemistry, 91–101
  as tissue, 92
  transport, interior system, 98
  ultrastructure, 94–101
  viscoelasticity, 97
  water balance in, 95
Second sound heat wave propagation, 283
Secretory products general and saliva, 99
Selectivity, ionic channels, 350, 352–353, 355
Self-directed transport, proteins, 370
Sepsis and biomaterials, 107–108
Sequences of interaction blood proteins, 139–146
Serial interactions
  blood, 139–146
  clotting, 139–145
Sickle cells orientation in magnetic field, 179–180
Signal coupling and the cell membrane, 186
Signal transduction, 373
Signaling
  in frog skin, 72–74
  ion release in frog skin, 72–74
Simulations
  global, description, 302–310
  global of water, 310–317
  supercomputer of biomolecule interaction, 301–333
  water from electrons and nucleii to hydrodynamics, 310–314
  water hydrodynamics, 314–317
Singer and Nicolson membrane model, 153
Skin (*see* Stratum corneum)
Skin appearance and crystalline/liquid crystalline state, 26
Skin cell transition, 4

Skin proteins denaturation
by surfactants, 37–40
Skin roughness
and protein denaturation
by surfactants, 37–40
by surfactants, 37–40
Skin swelling
and surfactant mixtures, 44–45
vs. surfactant characteristics, 42–44
Skin tightness by surfactants, 35–37
Slime (*see also* Conditioning films)
bacterial, 111–113
by bacterial production, 106
production, 111–113
Sodium gradient across frog skin, 53–56
Sodium ion, influence on frog skin potential, 52–56
Solid surfaces, protein adsorption on, 294–295
Solitons in biocommunication, 369–370
Solution analyses, erythrocyte adsorption, 196, 198–220
Sorption (*see also* Adsorption, Desorption)
ions, lipid bilayers, 161–162
Specialized biofunctions, saliva, 98–99
Spectra, Raman, re, bilayers, lipid-water, 162–163
Spectrin-actin layers, ion transport, 346
Speed, supercomputer, 304
Sphingolipids, 10
Sphingolipids and hydrogen bonding, 156
Spinosum layer, 4–7
Stabilization, membranes by water, 153–170
Stainless steel biomaterial, 117–118
Stimulation of potential, polyvalent cations on frog skin, 58, 60–67, 74
Stratum corneum
filaments in, 4
importance of, 3–4
interaction with surfactants, 33–48
lipids, 3–32
lipids role in water evaporation, 26
lipids role in water transport, 26–30
penetration by surfactants, 34–35
peripheral envelope, 4–7
proteins in, 4–7
structures, 3–32
surfactant interactions with, 33–49
swelling and surfactants, 40–44
water transport, 26–30
X-ray diffraction of, 11–16
Stratum corneum, frog, 51–76
anions vs. potential, 67
vs. calcium, 52–56
depression of potential, 52–70
effect of toxic elements, 63, 65–67
electric potential profiles, 70–71
electric potential reversal, 53, 55, 58–59, 67–70, 72–73
electrical circuitry, 72
electrokinetics, 67–69, 73–74
influence of bathing solution composition, 51–76
influence of electrolyte, 51–76
and ionic strength, 52–56
and pH, 56–57
vs. potassium, 52–56

[Stratum corneum, frog]
potential, depression of, 52–70
reversal of potential, 53, 55, 58–59, 67–70, 72–73
signaling and ion release, 72–74
vs. sodium, 52–56
stimulation of potential, 58, 60–67, 74
Streaming potential, 74
Streaming potential from bioelectrogenics, 384–385
Structure (*see also* Conformation, Folding, Unfolding)
channel, molecular gramicidin A, 165–166
cytoskeleton, 279–285
interferon, 325–327
layered, 11–16
lipid bilayers, neutron diffraction, 163
lipids, 3–32
membrane pores, 166–168
membranes, water in, 154–156
non-polar molecules on water, 156
nucleic bases and water, 163–164
partial specific volume, re, water-lipid bilayers, 162–163
pellicle, 80–84
pores, membrane, 166–168
proteins via molecular dynamics, 327
proteins from computer simulations, 325–327
proteins and water, 157–161
Raman spectra of lipid bilayers, 162–163
saliva, methodology of investigation, 92–101
saliva research results, 92–101
[Structure]
stratum corneum, 3–32
of water, 153–154, 157–158
water adjacent to non-polar, 156
water in membranes, 154–156
water and nucleic bases, 163–164
water and proteins, 157–161
Structure, colloidal saliva, 91–101
Substitutes for tissue, 105–136
Substrata biological on implants, 721
Supercomputer simulations of interaction of biomolecules, 301–333
Supercomputer speed, 304
Superconductive pathways, cytoskeletal, 283
Surface
activation clotting, 140
attachment of bacteria, 110–112
biological, ionic interactions, 70
biomaterials, 105–136
cell attachment, 296–297
clotting activation, 140
diffusion, biomaterials, 118
endothelium, 138–139
films, teeth, 78
frog skin, 51–76
metal implants characteristics, 117–119
oral, 77–103
oral, wettability, 78
polymer implants characteristics, 118
teeth, 77
teeth wettability, 78
Surface activity, proteins, 289–299
vs. conformation, 293
vs. electric charge, 293–294
vs. hydrophobicity, 292–293
vs. molecular weight, 292

Surface chemistry
oral, 77–103
of saliva, 91–101
Surface compartment model
description, 348–342
interfacial properties, 347–348
ionic transport, 347–363
Surface denaturation, proteins, 292
Surface energy
and adsorption on biomaterials, 117
biological systems, 338–339
of biomaterials vs. adsorption, 117
conformation, biomolecular, 339
and ionic adsorption, 344–345
and oral hygiene, 79–80
pellicle, 80–84
protein aggregation-disaggregation, 339
Surface science membrane transport, 337–365
Surface tension, critical microorganism-dental surface interaction, 80–82
Surface viscosity
protein adsorption, 292
and protein conformation, 292
Surface to volume ratios, biological systems, 338–339
Surfactant binding and skin roughness, 37–40
Surfactant characteristics and skin swelling, 42–44
Surfactant mixtures and skin swelling, 44–45
Surfactant, ovalbumin, 293
Surfactants
absorption by skin, 46–47
adsorption to skin, 35–37
effects on skin, 34–35
and erythema, 40
interactions with skin, 33–49
[Surfactants]
skin lipids removal, 35–37
skin penetration, 34–35
and skin protein conformation, 42
skin protein denaturation, 37–40
and skin roughness, 37–40
and skin swelling, 40–44
and skin tightness, 35–37
Swelling of skin by surfactants, 40–44
Swelling by surfactants and erythema, 40–44

Teeth surfaces, 77
Temperature and electroporation, 256–257, 262–263, 265, 269
Thermal energy, 374–375
and biohydration, 320–321
proteins, 375
Thermodynamic gradients (*see also* Concentration gradients)
electrogenics, 69–71
across frog skin, 53–67, 69–70
Thermodynamics, nonequilibrium and cytoskeleton structure, 279–285
Thickness, pellicle, 80–84
Threshold effects, electrical in biosystems, 188
Thrombin, in blood clotting, 144
Tissue
adhesion, conditioning film, 115–117
and biomaterials, 105–136
conditioning film adhesion, 115–117
electroporation, 259–260
integration, 117–118
saliva structure, 92
substitutes, 105–136

[Tissue]
transplants and immunity mechanisms, 122–124
transplants and infections, 122–124
Tissue factor extrinsic clotting, 143
Titanium alloy biomaterial, 117–120
Tobacco mosaic virus, disaggregation-aggregation, 357–358
Tooth contact angles vs. testing liquids, 78
Total artificial heart (*see* Artificial heart)
Toxic elements, effect on frog skin potential, 63, 65–67
Transient pore theory, 254, 264
Transition, crystalline/liquid crystalline, 26
Transmembrane channels, gramicidin A, 165–166
Transmembrane movement by electroporation, 254
Transmembrane potential in electroporation, 254–256
Transplants, tissue (*see* Tissue transplants)
Transport
associated electrical properties, 384–385
bio-ramifications, 369–372
electroporative, 268
and energetics, 374–384
filament directed, adherence distance, 395
inanimates by protein reconfiguration, 395
intelligence directed, 372–374
and interactions, 385–396
interior saliva system, 98
mechanisms, 367–400
in membranes surface science, 337–365
[Transport]
particles in nerve cells, 378–380
pores, membrane, 166–168
proteins, general, 370
rate vs. interaction, 385
rates, 369–370 (*see also* Biocommunication speeds)
saliva interior system, 98
in skin, 26–30
through ultrathin layers, 340–343
water in skin, 26–30
water in vivo, 395
Transport, ionic
and biolayers, 343–347
biosystems by electric fields, 184
gramicidin A channel, 165–166
and oscillating electric fields, 353–355
spectrin-actin layers, 346
Trapped volume, erythrocyte adsorption, 203–204
Trapping
re, adsorption measure, inter/intracellular, 197
erythrocyte adsorption studies, 239
Trypsin test, lipid structure model, 11

Ultrastructure, saliva, 94–101
Ultrahin layers
diffusion and permeability, 340–343
transport through, 340–343
Unfolding (*see also* Conformation, Folding, Structure)
of proteins and adsorption, 293
and protein surface activity, 293

Vesicle-microtubule adherence, Brownian vs. directed motion, 391–392
Vesicles, electric pulse effects, 176
Vibrational energy (*see* Oscillatory energy)
Video-enhanced contrast light microscopy, 378
Viroid, 392
Viscoelasticity, saliva, 97
Viscosity, hemoglobin aggregation, 360
Viscosity, surface, adsorption, proteins, 292
Vitallium biomaterial, 117–118
Volume, partial specific and water-lipid bilayer, 162–163

Water (*see also* Hydration)
- in bacteriorhodopsin, 161
- in biochemical equilibrium, 156–157
- as biocomponent, 368–369
- biological molecules, interaction, 156–168
- in biomolecules vs. their thermal energy, 320–321
- and biotransport, 343
- in cells, 155–158
- cells, interaction, 155–158
- channel formation, 165–168
- in cytoskeleton, 282
- vs. dental bond strength, 79
- diffusion in skin, 26–30
- in DNA, 319–320
- effect on crystalline/liquid crystalline transition, 26
- effect on skin, 26
- electrical activity, membranes, 155–156
- evaporation from skin, 26–30
- freezing in cells and membranes, 168

[Water]
- global simulation, 310–317
- re, gramicidin A channel, 165–166
- icelike structure in proteins, 158–160
- influence of non-polar on structure, 156
- interaction with biological molecules, 156–168
- interaction in cells, 155–158
- interaction with ions, 154–156
- interaction in membranes, 154–156
- interaction, molecular, 153–154
- interaction with proteins, 157–161
- in vivo transport, 395
- and ion interactions, 154–156
- ionic transport, pores, 166–168
- and lipids, 11–26, 161–163
- in membrane evolution, 154
- membrane pores, 166–168
- membrane stabilization, 153–170
- in membrane structure, 154–156
- membranes, electrical, 155–156
- membranes, interaction, 154–156
- molecular interactions, 153–154
- and myoglobin, 320
- and nucleic bases, 163–164
- pore formation, 165–168
- pores, membrane, 166–168
- and proflavine deoxydinucleotide phosphate, 322
- protein bio-activity, 159
- in protein conformation, 160–161
- and proteins, interaction, 157–161
- proteins, structure, 157–161
- simulation, electrons, nuclei to hydrodynamics, 310–314
- simulation, global, 310–317
- simulation, hydrodynamic, 314–317
- re, Singer and Nicolson membrane model, 153

[Water]
structure, 153–154, 157–158
structure of membranes, 154–156
structure due to non-polar molecules, 156
structure, proteins, 157–161
transport in vivo, 395
Water balance in saliva, 95
Water-bridging in DNA, 164
Water-hydrophobic biostructure, 157–161
Water in lipid bilayers
diffraction, neutron, 163
partial specific volumes, 162–163
Raman spectra, 162–163
Wave propagation, heat, cytoskeletal, 283
Wettability
vs. acids on dental surfaces, 78–79
dentine, 78
teeth surfaces, 78
Wetting pellicle, 80–84
Window
effects in biosystems, 177–178
platelet deposit time, 140
time, platelet deposit, 140

X-ray diffraction
lipids location, 16–26
lipid structure model, 11–16
stratum corneum, 11–16